Water Engineering in Ancient Societies

Water Engineering in Ancient Societies

Editor

Charles R. Ortloff

MDPI • Basel • Beijing • Wuhan • Barcelona • Belgrade • Manchester • Tokyo • Cluj • Tianjin

Editor
Charles R. Ortloff
University of Chicago
USA

Editorial Office
MDPI
St. Alban-Anlage 66
4052 Basel, Switzerland

This is a reprint of articles from the Special Issue published online in the open access journal *Water* (ISSN 2073-4441) (available at: https://www.mdpi.com/journal/water/special_issues/Water_Ancient).

For citation purposes, cite each article independently as indicated on the article page online and as indicated below:

LastName, A.A.; LastName, B.B.; LastName, C.C. Article Title. *Journal Name* **Year**, *Volume Number*, Page Range.

ISBN 978-3-0365-4163-1 (Hbk)
ISBN 978-3-0365-4164-8 (PDF)

© 2022 by the authors. Articles in this book are Open Access and distributed under the Creative Commons Attribution (CC BY) license, which allows users to download, copy and build upon published articles, as long as the author and publisher are properly credited, which ensures maximum dissemination and a wider impact of our publications.

The book as a whole is distributed by MDPI under the terms and conditions of the Creative Commons license CC BY-NC-ND.

Contents

About the Editor vii

Preface to "Water Engineering in Ancient Societies" ix

Charles R. Ortloff
Water Engineering at Precolumbian AD 600–1100 Tiwanaku's Urban Center (Bolivia)
Reprinted from: *Water* **2020**, *12*, 3562, doi:10.3390/w12123562 1

Charles R. Ortloff
Caral, South America's Oldest City (2600–1600 BC): ENSO Environmental Changes Influencing the Late Archaic Period Site on the North Central Coast of Peru
Reprinted from: *Water* **2022**, *14*, 1403, doi:10.3390/w14091403 31

Charles R. Ortloff
Inka Hydraulic Engineering at the Tipon Royal Compound (Peru)
Reprinted from: *Water* **2022**, *14*, 102, doi:10.3390/w14010102 69

Kenneth R. Wright
The Masterful Water Engineers of Machu Picchu
Reprinted from: *Water* **2021**, *13*, 3049, doi:10.3390/w13213049 97

Kevin Lane
Engineering Resilience to Water Stress in the Late Prehispanic North-Central Andean Highlands (~600–1200 BP)
Reprinted from: *Water* **2021**, *13*, 3544, doi:10.3390/w13243544 109

Charles R. Ortloff
Hydraulic Engineering at 100 BC-AD 300 Nabataean Petra (Jordan)
Reprinted from: *Water* **2020**, *12*, 3498, doi:10.3390/w12123498 137

Charles R. Ortloff
Roman Hydraulic Engineering: The Pont du Gard Aqueduct and Nemausus (Nîmes) Castellum
Reprinted from: *Water* **2021**, *13*, 54, doi:10.3390/w13010054 171

Paul M. Kessener
Roman Water Transport: Pressure Lines
Reprinted from: *Water* **2022**, *14*, 28, doi:10.3390/w14010028 195

About the Editor

Charles R. Ortloff

Charles R. Ortloff, In his 50 years working for major US and foreign corporations on research topics related to industrial and military defense applications, Charles developed an early interest in the work of ancient engineers. From there, he followed a parallel career path in archaeology to discover ancient water engineering practices. Now a Research Associate in Anthropology at the University of Chicago and Director of CFD Consultants International, Charles continues his research in continues South American, Middle Eastern and Asian countries, exploring the engineering of ancient archaeological sites' urban and agricultural water systems. Charles has published over 100 major journal papers on fluid mechanics problems, participated in several NOVA television documentaries on ancient water systems, published three books on ancient water engineering with major publishing houses, and conducted many conference presentations. His investigative work into ancient water engineering practice continues, with an interest in bringing information on the water systems of World Heritage sites to the public and adding new information to the history of water science, building on the contributions of ancient water engineers.

Preface to "Water Engineering in Ancient Societies"

The Special Issue 'Water Engineering in Ancient Societies' addresses the hydrological engineering underlying the design, construction and operation of many of the ancient World Heritage archaeological sites in South America and the Middle East. Manuscripts published in the literature on these sites mainly focus on describing architectural sites, socio–political–economic issues, anthropological and historical studies and social organizational structure, but infrequently delve into the water-engineering practices ancient engineers used to design, construct and manage the urban and agricultural water systems of these sites.

To address this issue and highlight aspects of ancient water engineering, modern fluid mechanics analysis methods are applied to existing descriptions of ancient canals, aqueducts and channels used in urban and agricultural settings of ancient archaeological sites to determine a modern description of ancient engineering practice. Using this procedure, ancient hydraulic engineering principles used by different ancient civilizations are known in terms of modern hydraulic-engineering solutions and notations. As few ancient technical literature manuscripts describing their water supply and distribution systems have survived to provide an understanding of ancient engineering notations and design principles, it is likely that water-engineering principles do exist in antiquity, used for the design of water-supply and -distribution systems similar to those used in modern-day engineering practice. Ancient water-engineering principles must therefore exist, albeit in manuscript notations presently unknown to present-day scholars.

Of the scarce literature available from Roman sources, there are no direct Latin language counterparts to velocity, flow rate, pressure, velocity head, detailed time-measurements and other modern hydraulic terms. We now can assume that some equivalent terminology for the technical terms vital for the design of water conveyances, together with an understanding of basic hydraulic principles, was available in some elementary form through nature and experimental and observation. As most ancient hydraulic structures functioned at a high level of efficiency, this could only occur as a result of knowledge of basic water engineering principles.

The purpose of this book is to show which ancient design principles, expressed in modern hydraulic-language terms, were used by water engineers to construct their hydraulic works. The following chapters continue my earlier research work on ancient water engineering, as published in the 2010 Oxford Press book 'Water Engineering in the Ancient World' and the later 2020 Routledge Press book 'The Hydraulic State: Science and Society in the Ancient World', and present new findings and interpretations for readers of Water with an interest in historical aspects of water engineering. As there was no known written language in ancient Peru and Bolivia to describe their version of water engineering principles, the chapters related to hydraulic engineering at several of the key ancient South American sites reveal use of modern hydraulic principles well in advance of their 'official' discovery in western science some 2000 years later. To a large degree, for Roman, Nabataean and several South American sites, the modern use of critical flow dynamics is apparent, as several chapters illustrate. Further, in the chapter describing the ancient (600–1100 CE) site of Tiwanaku in Bolivia, the aquifer water table height was maintained throughout seasonal rainfall changes using an elaborate systems of spring-fed water channels, together with an elaborate drainage system for urban and agricultural use. This degree of water-engineering sophistication is indicative of the advancements of water engineering of an ancient society.

This book discusses many new discoveries at other New- and Old-World archaeological sites, illustrating that a deep hydraulic engineering knowledge base existed in ancient times. It therefore helps to fill the reportage gap in the current archaeological literature regarding the technical achievements of ancient societies. Though hydraulic engineers are familiar with modern solution methodologies, the present book contains chapters on the history of water engineering as developed by their water-engineering ancestors.

The chapters in this book arose from participation in several major archaeological projects, as well as many individual trips to remote archaeological sites in South America, the Middle East and Asia. Of special note are associations with Drs. Alan Kolata, Michael Moseley, Ruth Shady Solis, Tom Dillehay, John Janusek, Dora Crouch and Phillip Hammond, who shared their knowledge of world ancient archeological sites with me throughout many years of my work investigating ancient water technology.

Charles R. Ortloff
Editor

Case Report

Water Engineering at Precolumbian AD 600–1100 Tiwanaku's Urban Center (Bolivia)

Charles R. Ortloff [1,2]

[1] CFD Consultants International, 18310 Southview Avenue, Los Gatos, CA 95033, USA; ortloff5@aol.com
[2] Research Associate in Anthropology, University of Chicago, Chicago, IL 60637, USA

Received: 23 October 2020; Accepted: 9 December 2020; Published: 18 December 2020

Abstract: The pre-Columbian World Heritage site of Tiwanaku (AD 600–1100) located in highland altiplano Bolivia is shown to have a unique urban water supply system with many advanced hydraulic and hydrological features. By use of Computational Fluid Dynamics (CFD) modeling of the city water system, new revelations as to the complexity of the water system are brought forward. The water system consists of a perimeter drainage channel surrounding the ceremonial center of the city. A network of surface canals and subterranean channels connected to the perimeter drainage channel are supplied by multiple canals from a rainfall collection reservoir. The perimeter drainage channel provides rapid draining of rainy season rainfall runoff together with aquifer drainage of intercepted rainfall; water collected in the perimeter drainage channel is then directed to the Tiwanaku River then on to Lake Titicaca. During the dry season aquifer drainage continues into the perimeter drainage channel; additional water is directed into the drainage channel from a recently discovered, reservoir connected M channel. Two subterranean channels beneath the ceremonial center were supplied by M channel water delivered into the perimeter drainage channel that served to remove waste from the ceremonial center structures conveyed to the nearby Tiwanaku River. From control of the water supply to/from the perimeter drainage channel during wet and dry seasonal changes, stabilization of the deep groundwater level was achieved—this resulted in the stabilization of monumental ceremonial structure's foundations, a continuous water supply to inner city agricultural zones, water pools for urban use and health benefits for the city population through moisture level reduction in city ceremonial and secular urban housing structures.

Keywords: pre-Columbian; urban Tiwanaku; Bolivia; hydraulic/hydrological analysis; surface canals; CFD; perimeter drainage channel; moat; subterranean channels; societal structure

1. Introduction

Archaeological studies of ancient pre-Columbian Peru and Bolivia have not thus far brought forward the technical engineering achievements at major archaeological sites in those countries.

To address this missing element of Andean archaeology, the new field of PaleoHydrology is intended to bring forward new perspectives on what ancient New and Old-World water engineers accomplished together with the scientific base they used for their hydraulic engineering works. While many urban and agricultural water delivery and transport structures of the ancient world are well known from the archaeological literature, the engineering methodologies and theoretical basis used by ancient water engineers in their design and operation of complex water systems await discovery. Surviving literature from ancient authors on water engineering methodologies reveals the absence of hydraulic engineering principles and parameters vital to any water conveyance design—yet recent analysis using modern hydraulic engineering methodologies to analyze ancient water conveyance structures reveals use of versions of modern water engineering principles—albeit in indigenous formats yet to be discovered. By use of Computational Fluid Dynamics (CFD) methodologies and modern hydraulic engineering principles, ancient water structures of South America, the ancient Mediterranean, and dynastic Asian societies can now be modeled and analyzed to extract ancient versions of the water technologies used in their design. From investigations of this nature, the civil engineering base used by these societies can be brought forward together with the knowledge base used to support their water system designs. Use of CFD methodology involves the numerical solution of the Navier-Stokes mass, momentum and energy conservation equations governing water flow. A CFD model (shown as Figure 1a) of the Tiwanaku urban center showing the network of surface and subterranean canals and channels, the sub-ground surface aquifer and surface drainage channels shows the components of the complete water supply and distribution system of the Tiwanaku urban center. Since these networked water transfer features interact with each other to transfer surface and groundwater in different ways during the rainy and dry seasons, the water flow transfers through these interacting features is computed using CFD methodology to show the water engineering put into practice by Tiwanaku water engineers' design and construction of their urban water system. The investigation to follow then provides insight into the engineering technology that underlies the water control system at the Tiwanaku urban center.

The present paper is intended to bring forward in detail the water engineering used at the AD 600–1100 pre-Columbian World Heritage site of Tiwanaku located on the high (~4000 m) altiplano of Bolivia; this site demonstrated an advanced use of hydrologic and hydraulic science for urban and agricultural applications that is unique in the Andean world. From recently discovered aerial photos taken of the site in the 1930s prior to excavations that began in the early 1980s, new perspectives of the water system of the city that extend previous interpretations of the dividing moat between ceremonial and secular parts of the city is possible based upon a network of water channels not previously known but now displayed from the early aerial photographs surrounding the ceremonial structures of urban Tiwanaku. The perimeter drainage channel served as the linchpin of an intricate network of reservoir and spring supplied surface canals and subterranean water channels that served many hydraulic and hydrological functions. These functions include: (1) collect and drain rainy season rainfall runoff and aquifer seepage from infiltrated rainwater into the nearby Tiwanaku River to limit flood damage; (2) accelerate post-rainy season ground drying by collecting aquifer seepage from infiltrated rainwater into the perimeter drainage channel and transfer collected water to the nearby Tiwanaku River to lessen ground moisture to promote health benefits for the city's population; (3) provide water from a newly discovered M channel to two subterranean channels under ceremonial core structures to flush human waste to the nearby Tiwanaku River; (4) maintain the groundwater level constant throughout rainy and dry seasons to stabilize foundation soils underneath massive pyramid structures to limit structural deformation; (5) facilitate rainy season water accumulation drainage from the floor of the Semisubterranean Temple into the groundwater layer to rapidly dry the temple floor; and (6) provide drainage water to inner city agricultural zones. The water control network in urban Tiwanaku is analyzed by CFD modeling of transient surface and groundwater aquifer flows to illustrate the function of the perimeter drainage channel in both rainy and dry seasons as well as its role in the (1) to (6) functions.

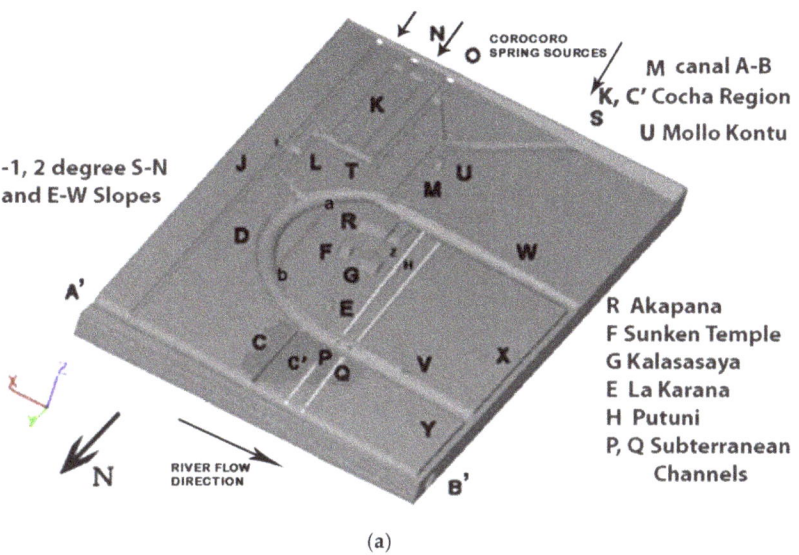

(a)

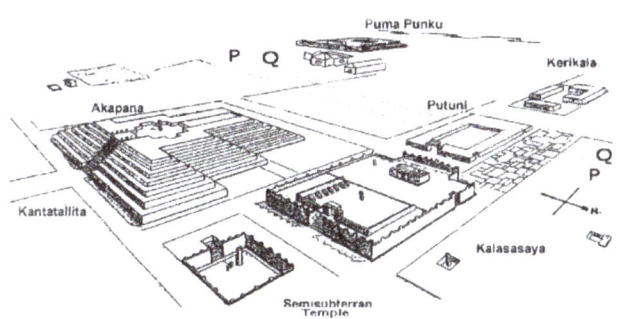

(b)

(c)

Figure 1. *Cont.*

(d)

(e)

Figure 1. (**a**) Representative FLOW-3D Computational Fluid Dynamics (CFD) model of hydrological and architectural features of the Tiwanaku ceremonial center. The CFD model is a best-estimate representation of the site geometry scaled from aerial photos, historical sources and ground survey. Line a–b represents a later drainage path interpretation compared to earlier curved versions by Poznansky 1945 and Bandelier 1911 used for the model. The A'-B' channel represents the Tiwanaku River. (**b**) Schematic of several main ceremonial sites within the perimeter drainage channel; P and Q represent subterranean drainage channels under the Putuni's palace floor. The perimeter drainage channel surrounds this area (Figure 1a). (**c**) Excavated P subterranean channel section located below the Putini floor- note scale from Figure1. (**d**) Two top-slab excavated canals below floor level at La Karaña leading water to subterranean channels P and Q. (**e**) Slab-covered channel originating from the top platform of the Akapana and running at high slope down the side of the Akapana pyramid to drain the room complex located on the top platform.

2. Overview of the Hydrological Regime of Tiwanaku

The pre-Columbian AD 300–1100 city of Tiwanaku located in the high altiplano (~4000 m.a.s.l) region of Bolivia demonstrated use of advanced hydraulic/hydrologic principles to maintain city drainage during the long rainy season through a complex network of surface and subterranean channels coupled into a main perimeter drainage channel. The urban water control system was designed to regulate seasonal deep groundwater levels constantly by regulating rainfall runoff and aquifer drainage into the perimeter drainage channel during the rainy season and by providing supplemental water supplied from a reservoir to surface canals connected to the perimeter drainage channel to maintain constant groundwater level during the dry season. This was accomplished by the perimeter drainage channel bottom designed to intersect the seasonally maintained deep water table top surface so that excess drainage water arriving into the perimeter drainage channel bottom during the rainy season was not infiltrated into the saturated perimeter drainage channel bottom surface but rather drained away to the Tiwanaku River. In the dry season, additional water supplied from a reservoir supplied M channel plus continued aquifer seepage to the perimeter drainage channel maintained a constant deep groundwater height. Excess water entering the saturated bottom perimeter drainage channel was delivered to the Tiwanaku River by a separate channel. Constant groundwater height maintained by the presence of the perimeter drainage channel through seasonal rainfall changes promoted many hydraulic/hydrological engineering functions. These include year-round surface dryness in urban Tiwanaku through continuous aquifer drainage into the perimeter drainage channel that promoted health benefits to city inhabitants together with maintaining constant monument foundation soil strength properties to limit monument settling distortion. The spring and reservoir supplied canal network coupled to the perimeter drainage channel together with dual subterranean channels under the ceremonial center bounded by the encircling perimeter drainage channel provided comprehensive hydraulic system design and demonstrated complex hydrologic engineering not seen at any other pre-Columbian South American site. The city's ceremonial center, composed of monumental architecture and elite residential compounds circumscribed by the perimeter drainage channel, is shown in Figures 1b, 2 and 3 with W-D-V-X describing the perimeter drainage channel. Figures 2 and 3 are derived from aerial photographs taken in the 1930s prior to excavations started in the early 1980s; Figure 4 is derived from Google photographs. Figure 1a is a CFD model representing the Tiwanaku urban water system based upon early explorer diagrams of the site features, CFD model dimensions approximate the scale of the site features as determined from aerial photographs and site exploration measurements.

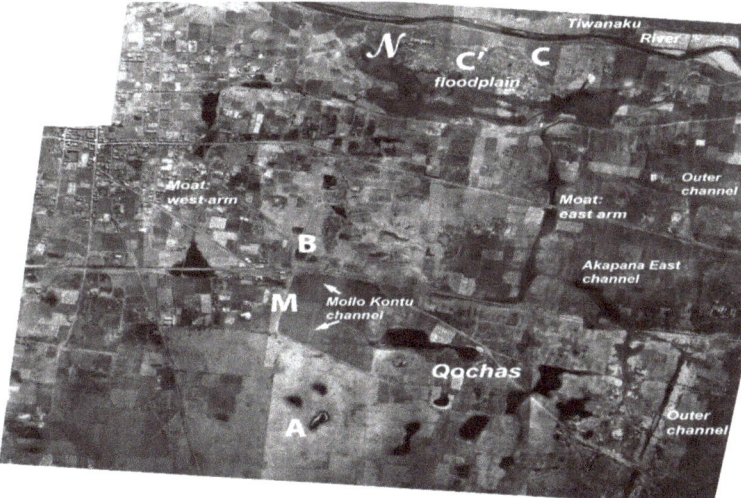

Figure 2. Aerial photograph view of the inland perimeter drainage channel (denoted above as the Moat) surrounding the ceremonial core of Tiwanaku indicating the intersecting Mollo Kontu M canal, qocha regions and the Tiwanaku River to the north.

Previous researchers interpreted a main purpose of the encircling perimeter drainage channel as a boundary 'moat' separating sacred and secular urban areas of the city [1,2]; the perimeter drainage channel additionally served as a vital part of a complex network of supply and drainage channels together with aquifer drainage to promote rapid post-rainy season soil dry-out with health benefits to city inhabitants. Excess water collecting into the perimeter drainage channel from aquifer drainage, rainy season runoff and flow to and from surface and subterranean canals then rapidly exited through a connecting canal to the Tiwanaku River connection to Lake Titicaca thus forming an integrated hydraulic/hydrological network designed to perform the (1) to (6) functions listed above.

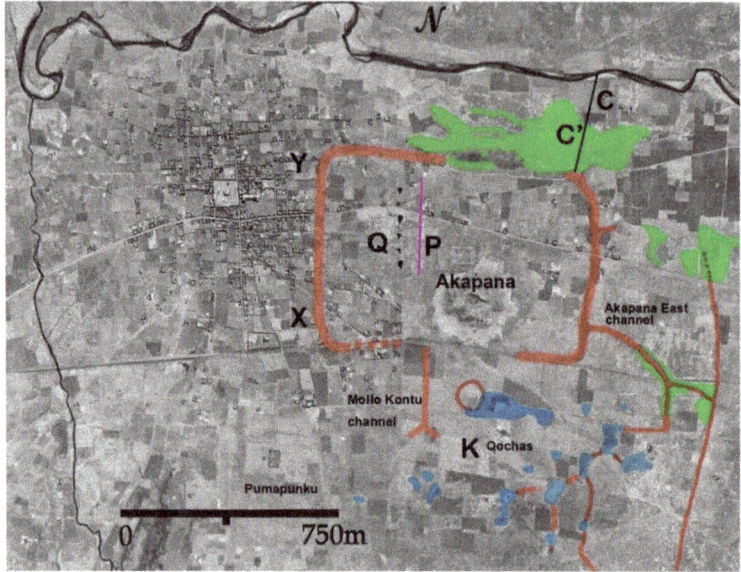

Figure 3. Details of the (red) intersection path of the Mollo Kontu M channel with the southern arm of the perimeter drainage channel. Note that excess perimeter drainage channel water is drained to the Tiwanaku River through channel C and the floodplain agricultural (green) area C′. The Akapana East Channel shown supplies water to qocha agricultural and (green) pasturage areas to the southwest and served to drain excessive water to the perimeter drainage channel to maintain required moisture levels for agriculture and pasturage.

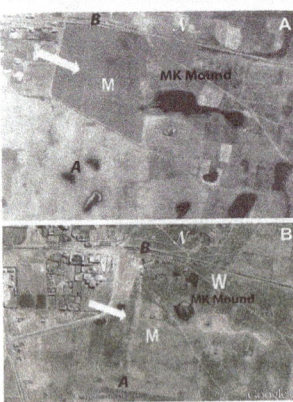

Figure 4. The Mollo Kontu M channel and surface features from Google Earth satellite imagery used to compose Figure 1a.

Shown in Figures 2 and 3 are the 1930s aerial photographs of the main ceremonial center enclosed by the perimeter drainage channel; Figure 4 is a recent Google aerial photograph of the same area.

Of major interest is the trace of a canal not previously noted in research studies—the Mollo Kontu channel (Figures 1a and 2, Figures 3 and 4) is now designated the M channel. The M channel segment shown in Figures 2–4 aerial photographs is supplied through a stone-lined transfer channel constructed through a marsh region supplied from springs and reservoirs located at the base of the Corocoro Mountain range to the south of the ceremonial center. The newly discovered M channel adds the missing link to the design function of the urban water system conceived, designed and implemented by Tiwanaku hydraulic engineers. With the discovery of this canal and its function, the ingenious water supply and distribution system of urban Tiwanaku, new insights about the water engineering knowledge base of a pre-Columbian society are revealed for the first time in report sections to follow. Figure 5 indicates the presence of one (P) of the two subterranean canals (P and Q) whose location is given in Figure 1a,b. This discovery adds a third dimension to the intricate water supply and distribution network not previously known. The totality of this water network's design and use and the water engineering involved is described in sections to follow.

Figure 5. Photo and plan view of the excavated portion of the subterranean channel P.

Figure 5 is illustrative of one of the dual subterranean water channels located under structures within the ceremonial center bounded by the perimeter drainage channel. These subterranean channels are designated the P and Q channels with their positions illustrated in Figure 1a,b. Given the preliminary details of the ceremonial area and its encircling perimeter drainage channel, the network of surface canals and subterranean channels connected to the perimeter drainage channel is illustrated in the Figure 1a model together with detail of ceremonial structures within the perimeter drainage channel shown in Figure 1b. The FLOW-3D CFD model consists of several million grid cells necessary to preserve accuracy for both surface water and internal aquifer water flows. Canal and channel water flows are characterized by a k-ε turbulence model; aquifer flow physical properties are given in a subsequent section. Transient calculations initiate from initial conditions on water flow rates at canal origin locations and initial aquifer fluid fraction (ff) saturation values within the phreatic evaporation layer surface and groundwater layers. Figures 6–10 represent sample CFD calculation results of transient aquifer and ground surface fluid fraction levels occurring during water transfer processes and illustrate aspects of the urban water system in operation. Boundary conditions are estimated canal flow rates at canal origin locations on the computational grid boundaries.

During the dry season, continued aquifer seepage from rainy season infiltrated rainwater and flow from the M channel recharged the groundwater to maintain and stabilize its height through seasonal changes. During the rainy season, excess runoff and aquifer seepage water was directed into the perimeter drainage channel. With the depth of the perimeter drainage channel set at approximately the height of the stabilized groundwater profile, the saturated perimeter drainage channel bed limited further internal seepage past the perimeter drainage channel bottom and thus transferred arrival water to dispersal areas by the C channel to the Tiwanaku River and the C' farming area (Figure 1a). Water directed to the Tiwanaku River by transfer from the perimeter drainage channel then flowed directly into Lake Titicaca. Figure 1b illustrates the major sites within the perimeter drainage channel boundary shown in Figure 1a; although two subterranean channels P and Q are known through excavation, more subterranean channels likely exist as several surface channels with major structures have slopes that lead water away from P and Q drainage exits.

Figure 1d illustrates two canals below the floor of the La Karaña complex (Figure 1a) that extend to and drain into channels P and Q. The extension of additional perimeter drainage channels surrounding three sides of the base of the Akapana are as yet unexcavated but must have a channel path either on the ground surface or a subterranean channel (or both) connection to the perimeter drainage channel. Of interest is a large highly sloped channel that runs from the top platform of the Akapana down the pyramid's side to the pyramid's base; the slab-covered top platform part of this channel is shown in Figure 1e.

In summary, one additional effect of the stabilized groundwater level through rainy and dry seasons was to maintain the bearing strength of soil under large monuments within the ceremonial center to limit structural distortions [3]. Water accumulating in the perimeter drainage channel from aquifer drainage and channeled spring water flow provided flow through dual subterranean channels P and Q (Figure 1a,b) to flush human waste delivered into the subterranean channels from elite compound structures to maintain hygienic conditions in the compounds. Figure 1c illustrates a section of the P canal below the Putini floor.

The multi-faceted hydrological aspects of the perimeter drainage channel served city environmental and hygienic conditions through rapid soil drying in city housing areas while promoting structural stability for the site's many monuments as well as aiding in rainy season drainage from the Semisubterranean Temple floor (F, Figure 1a,b). While groundwater control mastery is apparent in the urban setting, additional research on Tiwanaku raised-field agriculture [4–20] indicates similar advances in use of groundwater control technology in urban settings.

To demonstrate the seasonal interaction of surface and aquifer water flows more fully, the porous media aquifer CFD model (Figure 1a) is utilized in later discussions using CFD analysis to demonstrate the perimeter drainage channel's role as a hydrological control element vital to the city's sustainability during wet and dry seasons.

3. Settlement History

The ancient city of Tiwanaku, capital of a vast South American empire, has been the subject of research starting from early 20th century scholars that continues to the present day [21–30]. The city, located at the southern edge of the Lake Titicaca Basin in the south-central portion of the South American Andes at an altiplano altitude of ~4000 m.a.s.l. incorporated an elite area bounded by an encompassing perimeter drainage channel that enclosed temple complexes, palace architecture and the seven-stepped monumental Akapana pyramid (Figure 1b) designed to serve ceremonial sacrifice functions and provide rooms on the top surface for special ceremonies. Recent excavation of one of the top rooms indicated a large collection of llama bones used in ceremonial functions. Outside of the center lay a vast domain of secular urban housing structures. An intricate network of canals acting in conjunction with the perimeter drainage channel performed hydrological functions that included rapid ground drainage during both wet and dry seasons to promote health advantages for the city's 20,000 to 40,000 inhabitants as well as flood defense to preserve the ritual center and surrounding urban structures. The management of water systems within the city demonstrates hydrologic engineering expertise consistent with that found in Tiwanaku's raised field agriculture and demonstrates Tiwanaku hydrologic engineering mastery. The 1930s aerial photographs provide data to interpret the extent of, and insight into, the hydrologic function of the perimeter drainage channel. The early photographs reveal traces of the perimeter drainage channel's north and south arms in addition to the Mollo Kontu M channel as well as traces of the support structures within the ceremonial center (Figures 1a and 2, Figures 3 and 4). Additional channels intersecting the southern part of the perimeter drainage channel southern arm are indicated in Figure 1a. The perimeter drainage channel collected flow from adjacent channels and canals, rainfall runoff and infiltrated rainfall seepage from the saturated, near-surface phreatic layer of the aquifer as well as from the deep groundwater aquifer as the perimeter drainage channel depth intersected to top portion of the deep groundwater layer to transfer water into the nearby Tiwanaku River during the rainy season to prevent deep groundwater recharge. During the dry season, continued phreatic aquifer seepage into the perimeter drainage channel plus water from the intersecting M channel maintained the deep groundwater aquifer level relatively constant while surface evaporation and recession of the near-surface phreatic aquifer served to rapidly dry the ground surface promoting health and livability benefits for city inhabitants. The design intent of the builders of the perimeter drainage channel thus envisioned control of the deep groundwater level through wet and dry seasons to maintain the physical integrity of monumental structures by preventing the dry-out settling of the deep aquifer soils underlying the main ceremonial core as water continually occupied aquifer pore spaces. Given a stable upper boundary of the deep groundwater layer throughout the year, physical strength properties of foundation soils were maintained thus limiting structural distortion and settling of the massive platforms of the Kalasasaya and Akapana pyramid (R, Figure 1a,b) within the ceremonial center. Additionally, with the stable groundwater layer well below the floor of the Semisubterranean Temple (F, Figure 1a), rainy season drainage from the site floor into the aquifer region above the deep groundwater level was facilitated. This excess water then percolated toward the perimeter drainage channel's sidewalls for delivery to the Tiwanaku River. Thus, beyond the perimeter drainage channel's role in creating a ritual and social boundary between the elite residence ceremonial center and secular residential city districts, its engineering design contributed many practical benefits to living conditions for city residents throughout wet and dry seasons.

4. Tiwanaku Hydraulic Analysis

To demonstrate the perimeter drainage channel's hydrologic functions, multiple data assemblages used to construct a CFD hydraulic/hydrological model (Figure 1a) include results of archaeological mapping and excavation [31–33], Google Earth imagery and the aerial photos taken over the site of Tiwanaku. These aerial photos reveal the site decades before modern urbanization and monument reconstruction began and were taken at a time of year when many features held water thus providing a clear view of Tiwanaku's hydrological features. From these 90-year-old photographs, the outline of the perimeter drainage channel is shown in Figure 2 as the dark encircling boundary to the ceremonial center. The curvature of the drainage canal V-D-W-X shown in Figure 1a is derived from earlier observations of the channel made before years of erosion and soil deposition infilling that continues to the present day Previous explorers of decades past listed in [13,14,17,18] provided the foundation for current studies of the perimeter drainage channel.

The east drainage canal arm (denoted 'Moat' in the east arm in Figure 2) averages 5 to 6 m deep and ranges 18 to 28 m in top width. Subterranean canals originating from the perimeter drainage channel's south arm were drainage conduits for Tiwanaku's monumental and elite residential structures [2,34–38]. Since the south arm of the perimeter drainage channel is shallower in depth than the north arm as determined by ground contour measurements, a fraction of the water that accumulated in this arm flowed down-slope through the perimeter drainage channel's east and west arms toward the Tiwanaku River while a portion of accumulated water in the south arm flowed into the dual subterranean P and Q channels (Figures 1a and 5) underlying the ceremonial center. Given the two-degree declination slope of the subterranean channels, water accumulating in the perimeter drainage channel's bottom during the wet and dry seasons provided flush water cleaning for the Putuni palace's waste removal/drainage facilities (Figure 1b). North of Tiwanaku's monumental center, the shallow alluvial plain drops sharply downward toward the Tiwanaku River's marshy floodplain. One portion of the east arm of the canal turns west and disappears into the floodplain (C', Figure 2) while canal C continues north toward the Tiwanaku River. One portion of the west perimeter drainage channel arm led into the marsh north of the Kalasasaya Platform (Figures 1a and 3) while an ancillary arm continued northeast toward the river. The north portion of the perimeter drainage channel divided into several branch canals that intersected the floodplain and drained accumulated water from the north arm of the perimeter drainage channel. Water not directly shunted to the Tiwanaku River through canal C (Figures 1a and 5) drained water in the floodplain's aquifer into the Tiwanaku River. The floodplain area served as a nearby productive agricultural area for the urban center.

5. The Perimeter Drainage Channel in the Urban Hydrological Network

Where the groundwater surface emerged from depressed land areas, springs formed. Several canals in the southern portion of Tiwanaku were engineered to utilize the canal's water input. The westernmost Choquepacha area's canal [36] is derived from a natural spring on a bluff southwest of the Pumapunku complex (Figure 1b). The spring was fitted with a reservoir basin that included several incised stones carved to convey water. Combined with the output of an adjacent stream that drained the marshy area, the Choquepacha area supported extensive terrain amenable to pastoral grazing and farming immediately to the west of the Tiwanaku urban area. Other features relate directly to the hydrological function of the perimeter drainage channel. The first feature is the north–south Mollo Kontu region M canal that supplied water from springs and reservoirs originating from the southwest portion of the site near the Pumapunku complex into the southwest portion of the perimeter drainage channel (Figure 1(S-M), Figure 2(M), Figure 3(M) and Figure 4(M)). The second feature is an interlinked cluster of sunken basins (qochas) that occupied the southeast portion of the site (Figures 1a and 2). Qochas are pits excavated into the aquifer layer that capture and store rainwater and serve to expand planting surfaces and pasturage while creating micro-lacustrine environments that attract waterfowl [37]. Figure 2 depicts a series of canals dendritically linking the qochas to one another with a branch connecting to the Akapana East canal (L, Figure 1a) that drained into the east arm of the perimeter drainage channel. The third major feature is a long, narrow, outer canal (J) on the east side of Tiwanaku (Figures 1a, 2 and 3). While the role of this canal is unclear, its southern portion is straight and follows an alignment that mirrors that of the Pumapunku complex to the west; its northern portion shifts course and bounds the east edge of the site. The east canal (L, Figure 1a) links with the perimeter drainage channel (Figure 1a (W-D-V-X)), Figure 2) by connector canal I and indicates that the outer canal was part of an encompassing urban hydraulic network. The areas immediately east of the perimeter drainage channel contain Tiwanaku's residential sectors that include Ch'ijiJawira, a barrio of ceramic producers that depended on a constant water supply [39]. Immediately east of the Ch'ijiJawira sector is a low brackish marsh; from this marsh, the outer canal (J) provided fresh water from springs for Tiwanaku's easternmost residential sectors and drainage of excessive canal flow during the rainy season.

The east and west arms of the perimeter drainage channel directed water around the monumental complex area toward the Tiwanaku River to the north (Figure 1A'-B' and Figure 2). The C' floodplain was an integral part of Tiwanaku's larger hydraulic network that served to facilitate drainage of both groundwater seeping from the perimeter drainage channel arm D-V (Figure 1a) and rainwater runoff during rainy season peaks. Intricate surface canals and dual subterranean stone-slab constructed canals (Figure 1a,c,d and Figure 5) provided additional drainage and water transfer within the elite compound area bounded by the perimeter drainage channel. The elaboration of surface canals on the interior floor of the Semisubterranean Temple (F, Figure 1a) and areas outside the Kalasasaya (G, Figure 1a) indicate a drainage connection to either (or both) the perimeter drainage channel and the subterranean channel P (Figure 1a,b); additional drainage by seepage into the stabilized low groundwater layer below the temple floor helped to keep the temple floor dry throughout the rainy season. The Akapana pyramid (R, Figure 1a,b) incorporated an intricate, stone-lined canal that routed water from the uppermost level down through successively lower platforms and finally out through several portals in its basal terrace (Figure 1e). Water delivered to several open surface basins draining into vertical pipes (and/or surface channels leading to the perimeter drainage channel) conveyed water into the subterranean channels (Figure 1c,d) then into the perimeter drainage channel arm V, Figure 1a. Subterranean channel P (Figure 1a,b and Figure 5) provided flushing water to remove human waste from the Putini residential compound bounded by the perimeter drainage channel for conveyance to the Tiwanaku River. Water was temporarily pooled in the sunken courtyards near platform monuments rendering them lakes for ritual events and reservoirs for controlled water distribution.

Excavations between Putuni and Kerikala complexes (Figure 1b) indicated structures within the ceremonial core region that articulated with Tiwanaku's subterranean drainage network. This area housed high status groups until, at approximately AD 800, the construction of the Putuni palace repurposed the space to support recurring state-sponsored ceremonies [40–43]. Located~2.5 m below the current ground surface, subterranean channel P (Figures 1 and 5) consisted of sandstone slab masonry with vertical side slabs approximately 1. 0 m high and horizontal slabs about 0.8–0.9 m in width. Several vertical pipes consisting of multiple stacked, perforated stone disks conducted surface water from features within the Putini into the lower subterranean channel P (Figures 1a and 5). Water from the perimeter drainage channel's southern arm V supplemented by water from canals L and M (Figure 1a) together with seepage water from both phreatic and top portions of the deep aquifer was used to flush waste water through subterranean channels P, Q located in the west portion of the monumental core.

6. Water Management at Tiwanaku

To demonstrate insights related to the hydrological function of the perimeter drainage channel, use of CFD is made for cases that address seasonal variability in water input. Here the equations were numerically solved by finite difference methods [44] governing aquifer percolation [45,46] to show transient water transfer within the aquifer for two seasonal water availability cases. Case 1 considers effects existing at the termination of a rainy season on Tiwanaku's canal systems and city open surface areas. The rainy season in the south-central Andean altiplano generally runs from November through March. Case 2 considers effects of limited water input from springs and aquifer seepage into the deep groundwater layer during the April through October dry season. Data from aerial photographs, Google Earth imagery, contour maps and ground survey provided the basis for the CFD computational model (Figure 1a) to demonstrate hydrological features of the perimeter drainage channel and its encompassing hydrological network. A porous soil model of the subsurface aquifer is used to demonstrate hydrological responses of the canal network and perimeter drainage channel for the two cases. The Figure 1a CFD model surface and subterranean features are on the same scale as Figures 2–4 and represent best estimate water supply and distribution network canal paths inferred from photographic and ground survey data. The canal inlets shown (J, N, O Figure 1a) are sourced by canalized Corocoro springs and reservoirs located south of the modeled area as are canals (S-M) leading from the Pumapunku area. Key monument architectural and hydrologic features are:

A′–B′: the Tiwanaku River, flow direction A′ to B′

C: perimeter drainage channel to A′-B′ Tiwanaku River

C': floodplain drainage and agricultural complex supplied from the perimeter drainage channel

V: arm

E: La Karaña residential complex

F: Semisubterranean Temple

G: Kalasasaya Platform

H: Putuni Palace

I: Connecting channel between canals L-K-L and J

K: multiple interconnected qocha region supplied by canal N arm D

L: Akapana East canal, which drained qocha region K toward perimeter drainage channel

M: Mollo Kontu canal linking supply canals O and S to perimeter drainage channel arm W

N: Supply canal to qocha region K

O: Connecting canal to canal M

P and Q: subterranean channel pair with declination slope of two degrees to the Tiwanaku River; channels P and Q run underneath the Putuni Palace H (Figure 1b)

R: Akapana seven-stepped truncated pyramid

S: branch canal to Mollo Kontu canal M

T: lateral transverse canal to L; shunt canal I to canal J and/or canal drainage to the perimeter drainage channel U: Mollo Kontu monument

Z: the Kalasasaya compound—a connection to the perimeter drainage channel indicated in Figures 1a and 6–10.

W-D-V-X: the perimeter drainage channel circuit around the monumental core of Tiwanaku; original depth of the channel estimated at~5–6 m at location D.

Y: drainage canal from V to the Tiwanaku River (from the 1930's aerial photographic source); its inclusion in the model has a minor drainage effect compared to drainage features C, C' P, Q originating from the perimeter drainage channel's Z canal below the west side of the Akapana (R, Figure 1a) draining toward drainage canal segment D. (Figure 1a).

The CFD model is composed of a porous medium aquifer duplicating soil material properties (porosity, permeability) found at the site through which aquifer water percolates. The CFD model incorporates both the east-to-west ground slope declination and a south-to-north declination observed from field measurements. The momentum resistance to flow in the porous medium representation of an aquifer [46] is expressed as a vector drag term F_d u where F_d is the porous media drag coefficient and u the velocity vector u = q_x i + q_y j + q_z k with q_x, q_y, q_z velocity components in the i, j, k (x, y, z) coordinate directions (Figure 1a). The permeability k is defined as k = V_f μ/ρ F_d where V_f is the volume fraction (open volume between soil particles/total volume), μ is the water viscosity and ρ the water density. For the present analysis, k is on the order of ~10^{-11} cm^2 based upon the site soil type [43] within the model area excepting elite monumental paved areas for which k is on the order of ~10^{-5} cm^2. For model area soils, 0.43 < V_f < 0.54. Based on these estimates, the average drag coefficient F_d is estimated to be ~0.80. While deviations from this value occur due to varying soil properties with depth and location, flow delivery rates from the saturated part of the aquifer to the perimeter drainage channel's seepage surface (defined as the exposed interior wall soil surface of the perimeter drainage channel exposed to the atmosphere) will be affected but calculations will nevertheless demonstrate qualitative conclusions regarding the perimeter drainage channel's function. In the CFD model, the deep groundwater layer is composed of saturated soil and is stabilized throughout the year at ~5 to 6 m below the ground surface as well probe data indicate. The saturated phreatic aquifer layer is assumed to lie above the deep groundwater surface for Case 1 calculations indicative of an intense, long duration rainfall period. The bottom depth of the perimeter drainage channel intersects the upper portion of the deep groundwater layer in the Figure 1a CFD model and the capillary fringe zone and saturated phreatic top surface water layers provide seepage water into the perimeter drainage channel together with runoff water and canal water supplied by springs and reservoirs south of the city (Figure 1a). For a less intense rainfall period, a capillary fringe zone extends upward from the deep groundwater zone to intersect the bottom reaches of a surface saturated phreatic layer; the contact region size depends upon the amount of intercepted rainfall. Thus, deep aquifer recharge can occur when the surface phreatic layer extends sufficiently downward to penetrate the groundwater capillary fringe zone during long duration rainy periods. For minimal rainfall, surface phreatic layer is considered a small depth evaporation zone that vanishes in depth as the dry season continues. Again, when seasonal rainfall is intense and of long duration, the phreatic and deep groundwater layers merge; for this case, aquifer seepage into the saturated perimeter drainage channel bottom cannot occur and excess drainage water is rapidly shunted to the nearby Tiwanaku River. This effect limits the height excursion of the deep groundwater layer to the base depth of the perimeter drainage channel.

For Case 1 analysis, the post-rainy season phreatic layer is saturated and lies above the saturated deep groundwater layer; as no further infiltrated rainwater can be absorbed into the deep saturated groundwater layer's bottom surface, water accumulation occurs by aquifer seepage from the perimeter drainage channel walls. The aquifer drainage flow into the perimeter drainage channel then flows out to the Tiwanaku River through channel C and seepage to the C' farming area. As the dry season progresses, surface evaporation shrinks the phreatic layer upward toward the ground surface and soil drying occurs to a depth enhanced by aquifer drainage. In 1000–1400 AD times of extended drought, the phreatic top layer and ultimately the deep groundwater layer contracted leading to soil dry-out conditions to a large depth. This climate condition is the basis for the extended drought reason for collapse of Tiwanaku's raised-field agricultural systems in the AD 1000–1100 time period (Figure 17) that ultimately led to Tiwanaku's demise [47–50].

7. Case 1—Post-Rainy Season Ground Saturation Conditions

Case 1 examines post-rainfall conditions typical of the end of the altiplano rainy season characterized by phreatic zone saturation and continuous water flow through canals O, S, N, M and J from Corocoro springs and reservoirs (Figure 1a). Aquifer seepage to the bottom of the perimeter drainage channel from the saturated phreatic layer is transferred to perimeter drainage channel arms D, V and W to X-Y and then to the Tiwanaku River (A'-B', Figure 1a) and ultimately to Lake Titicaca as all canals and channels have a down-slope toward the river. Additional seepage occurs from the top reaches of the deep groundwater layer into the perimeter drainage channel. Water from the perimeter drainage channel's east and west arms then led to the Tiwanaku River through the C canal branch and seepage from to C' area. Water arriving into inlet N (Figure 1a) was conducted by canals K and L into either (or both) canals D and then from I to J. A summary of rainy season water inflows/outflows from a representative section of the perimeter drainage channel is shown in Figure 11 and indicates seepage from the top surface saturated aquifer region together with runoff from the saturated soil surface collecting at the saturated drainage canal bottom then transferring down-slope to the Tiwanaku River (A'-B' in Figure 1a). For dry season conditions (subsequently discussed in a later section), Figure 12 summarizes continued seepage flows from the vadose near-surface aquifer region and input flows channeled from the Corocoro spring/reservoir region by the M channel deposited on to the saturated perimeter drainage channel bottom then directed down-slope to the Tiwanaku River. For both cases, additional drainage from the eastern branch of the perimeter drainage channel is provided by subterranean channels P and Q (Figure 1a) directly to the Tiwanaku River.

In figures to follow, the (red) fluid fraction ff = 1 indicates aquifer saturation; ff = 0 indicates no water content to dry aquifer soil; intermediate ff values indicate intermediate levels of water content in aquifer soils. Numerical solutions of equations governing saturated aquifer and surface/subterranean canal flows give a picture of transient water transfers to and from the perimeter drainage channel from aquifer seepage and canal flows given estimates of flow rates based on supply flow rates in canals. Given Case 1's post-rainy season conditions, surface runoff has been largely collected into the perimeter drainage channel and transferred to the Tiwanaku River; further water transfer to the bottom of the perimeter drainage channel is from aquifer seepage and adjacent canal M water flow input—a fraction of this water supply goes into subterranean channels P and Q with accumulated water in the perimeter discharge channel's V arm discharged into the Tiwanaku River. Figures 6 and 7 show a time progression of water seepage from the perimeter drainage channel's open surface area and progressive surface drying as the phreatic layer deflects downward due to drainage into the perimeter drainage channel.

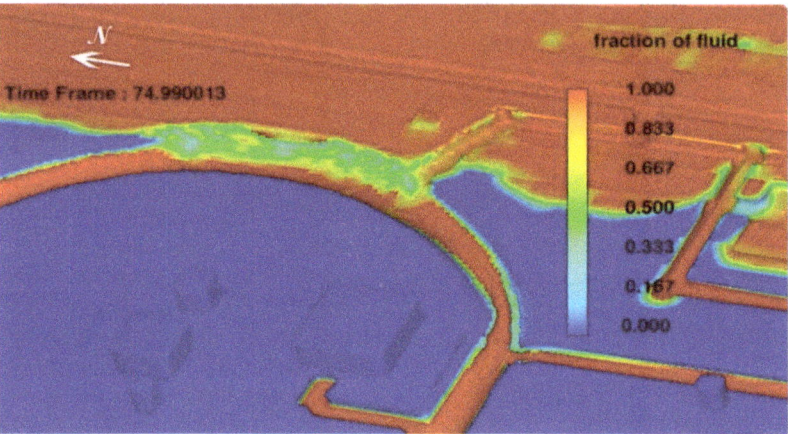

Figure 6. Post rainy season fluid fraction detail from FLOW-3D calculations of the east arm of the perimeter drainage channel showing the perimeter drainage channel's surface walls conducting seepage water to the saturated bottom of the perimeter drainage channel and the start of progressive surface drying for Case 1 conditions.

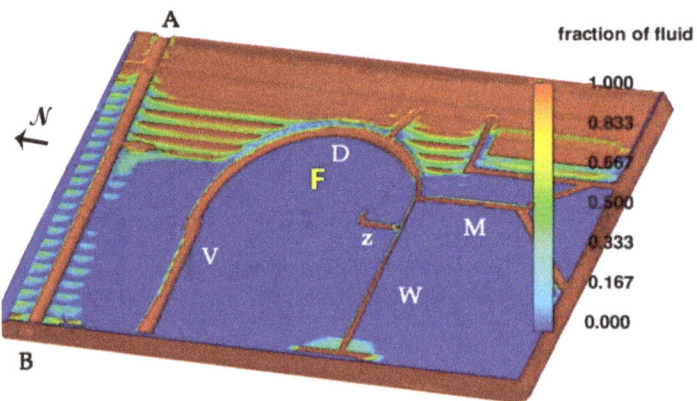

Figure 7. Later time fluid fraction surface drying achieved by aquifer seepage and surface evaporation at the end of the rainy season; note that channel M continues to provide water flow to subterranean P and Q channels and the perimeter channel bottom. Note low values of fluid fraction starting eastward on the drying ground surface for Case 1 conditions as the dry season initiates.

Rapid water removal from the perimeter drainage channel via channels C and C' to the Tiwanaku River (Figure 1a) limited water transfer from the phreatic aquifer to the deep groundwater layer causing deep groundwater level stabilization. The Akapana monumental internal core experienced limited rainfall infiltration due to extensive terrace and side wall paving and compound roofing that ultimately promoted runoff into the perimeter drainage channel aided by the exterior Akapana sloped channel (Figure 1e) originating from the Akapana top surface. Water that managed to infiltrate between paved areas of the Akapana then drained into the interior of the Akapana where it reemerged from base openings to join drainage canal and subterranean channel extensions P and Q (Figures 1a, 5 and 8) that directed water to the Tiwanaku River through C' drainage and C, channels. As the perimeter drainage channel depth extended to the top fringe of the deep groundwater layer, water drainage in the rainy period and water addition during the dry period helped to stabilize the deep groundwater level through season changes.

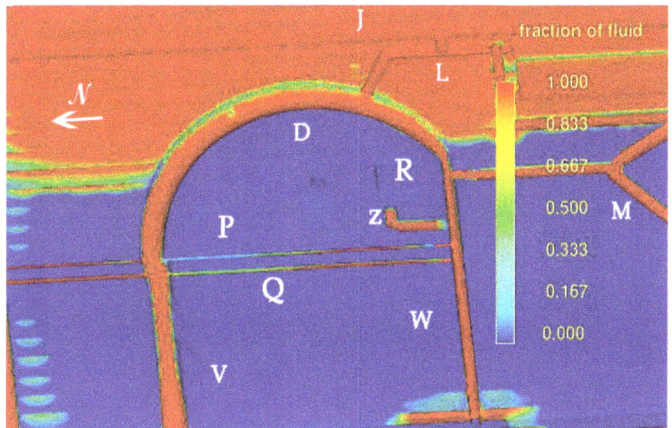

Figure 8. Case 1 fluid fraction results at P-Q depth from the ground surface; water input from channel M to subterranean channels P and Q flush wastewater from the Putuni Palace complex to drainage canal arm V then on to the Tiwanaku River A'-B' and Lake Titicaca. Z indicates a water channel connection from the Kalasasaya Platform to the perimeter drainage channel.

The location of the Semisubterranean Temple floor (F, Figure 1a) above the stabilized deep groundwater level and its nearness to the perimeter drainage channel helped to promote a dry floor through seasonal changes. Rainfall accumulating on the temple floor infiltrated into the phreatic layer then drained to the nearby perimeter drainage channel and groundwater layer. Fluid fraction results at the inner face of the perimeter drainage channel bounding the ceremonial center confirm runoff and aquifer seepage was minimal from what little infiltrated rainwater existed in this largely paved and roofed elite area. What little infiltrated water was conducted to the saturated perimeter drainage channel's bottom and quickly removed by canals C and field area C' aquifer water transfer to the Tiwanaku River. From the paved elite areas, rainy season rainfall runoff constituted a major water contribution to the perimeter drainage channel. Figure 8 shows the water transport in subterranean channels P and Q in the dry season (Case 2). Channel P lies below the floor of the Putuni palace; channel Q lies at the same depth as P but ~10 m west of P. Vertical pipes connected drainage areas in the Putuni courtyard and palace to canal P with collected water directed toward the V arm of the perimeter drainage channel (Figure 1a–c). The P and Q subterranean channels required a constant input of flowing water from canal M and aquifer seepage water into the perimeter drainage channel arm W to maintain dry elite residential area hygienic conditions. As the P, Q channels, the C canal, the C' area and the perimeter drainage channel bottom all sloped downhill toward the Tiwanaku River, water flow from the perimeter drainage channel arm W directed water and waste solids into the Tiwanaku River.

8. Case 2—Dry Season Initiation

Case 2 considers the perimeter drainage channel function under dry season initiation conditions (zero rainfall and continuous, but limited, water supply from Corocoro springs and reservoirs into surface canals N, O, S and M as well as regions K and C'. Figure 9 indicates that water supply from the M canal plus aquifer seepage continues to supply subterranean P and Q channels to flush human waste from the Putuni Palace compound structures during dry season initiation. Figure 10 shows the situation during the late part of the dry season—only limited aquifer seepage and water from the M canal constitutes the near total water supply into the perimeter drainage channel.

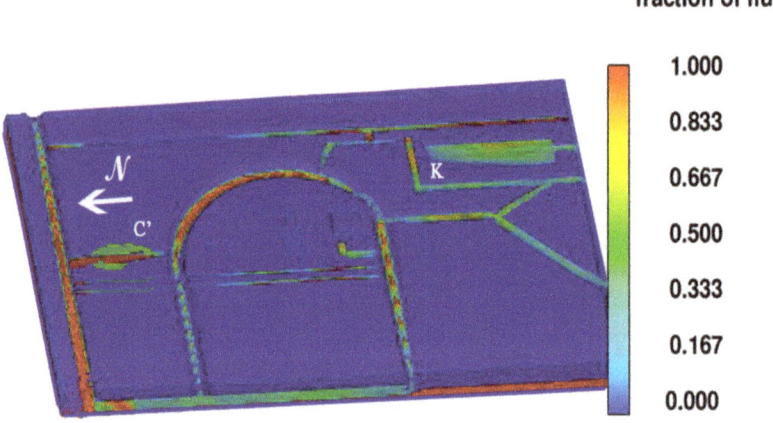

Figure 9. Dry season (Case 2) fluid fraction results on a plane below the ground surface; moisture levels in qocha region K and depressed area C' indicate sustainable pasturage and agriculture due to contact with the deep-water table and discharge from the V arm of the perimeter drainage channel. Water input from canal M delivers water to the perimeter drainage channel that enters subterranean channels P and Q.

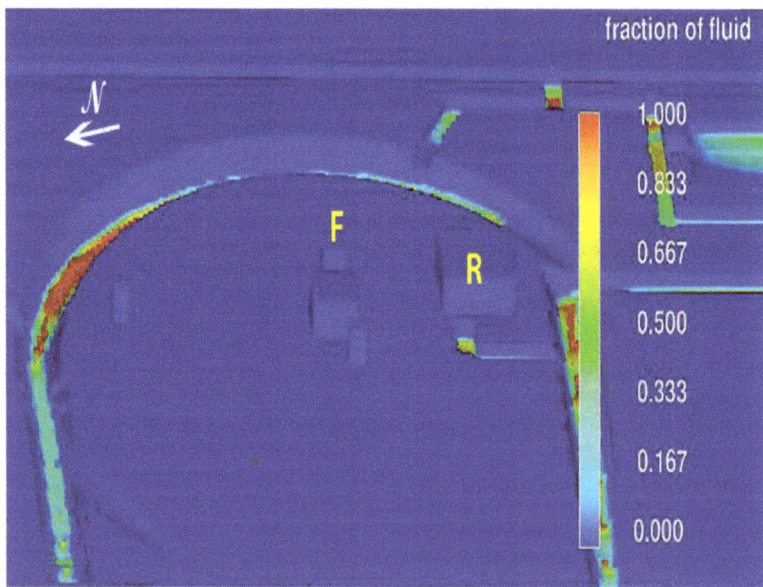

Figure 10. Late dry season (Case 2) fluid fraction results for the east arm of the perimeter drainage channel indicating dry season decreased seepage water into the perimeter drainage channel and extensive surface drying.

The C' and K regions (Figure 1a) remained functional due to their depth penetration into the receding phreatic zone water level indicating agriculture and pasturage were possible during the dry season. By transfer of seepage water and M channel water, the deep-water table remained stabilized throughout the dry season. Figure 10 indicates the situation well into the dry season with dry ground conditions and aquifer seepage from the perimeter drainage channel walls minimal; the ground surface is dry (fluid fraction approaches zero) and the W arm of the perimeter drainage channel indicates water supply mainly from the M channel. Figure 12 summarizes the water transfer mechanisms associated with dry season operation. Subterranean channels P and Q have continued water transport and flushing activity from perimeter drainage channel segment W as the dry season progresses with the M channel providing water supply during the dry season.

9. Newly Discovered Features of the Perimeter Drainage Channel

During the construction of the platform base of the Akapana pyramid, the phreatic aquifer layer was compressed by the heavy construction base's rock and gravel fill within stone compartments. As a result of the compression, water was expelled from the pyramid base aquifer into the nearby perimeter drainage channel. As additional heavy platforms were added (seven total) and the structure weight increased, further consolidation and compression of the aquifer below the pyramid resulted in reduction in aquifer porosity. While some rainfall infiltration into the increasingly consolidated foundation base soil occurred, less water was available due to low foundation soil porosity and increased aquifer drainage into the nearby perimeter drainage channel. As platforms were added and the compressive structural weight increased, a consolidated foundation impervious to water infiltration was created ensuring further minimal structural deflection and distortion. It is likely that Tiwanaku city planners included the creation of the perimeter drainage channel contemporary with construction of heavy ceremonial core region structures to promote monument stability. The Akapana to this day still retains its structural integrity without settling distortion as a testament to this original planning.

10. Rainy and Dry Season Groundwater Profiles

Figure 8 shows a constant depth transect through location D (Figure 1a) that indicates a fluid fraction of unity (ff = 1) consistent with ground saturation during the rainy season. As the rainy season concluded, seepage from the perimeter drainage channel side walls from adjacent saturated soil areas accelerated ground surface drying. Figure 11 summarizes all water supply and drainage paths relevant to maintain the deep groundwater level constant during the rainy season; Figure 12 summarizes all water supply and drainage paths during the dry season. Figures 11 and 12 summarize drainage and water supply conditions necessary for deep groundwater level stabilization during seasonal rainfall change. Figure 13 indicates a y plane transect through the Figure 1a CFD model—under heavy rainfall conditions, the aquifer is completely saturated down to the deep groundwater level as indicated by the fluid fraction ff = 1 as the CFD calculation verifies. With the onset of the dry season, the surface aquifer region contracts due to seepage into perimeter drainage channel together with surface evaporation while the deep aquifer level remains constant.

Additional water arrives into the Tiwanaku urban area by percolation from infiltrated rainfall into areas far to the east of the site. Given the slow percolation rate of water through an aquifer, intercepted rainfall originating from past decades arriving to the site from distant sources is a further contribution to groundwater level maintenance although the rate of water delivery is not constant due to the randomness of post year climate events that influence rainfall rates and amounts of water infiltrated into groundwater. Under severe long-term drought conditions in the 10–11 century AD time period, the Titicaca Lake level dropped severely [43,44] affecting the lowering of the groundwater level close to the lake edge; this effect then reduces the water level in raised-field swales and severely contracts agricultural production [45–48].

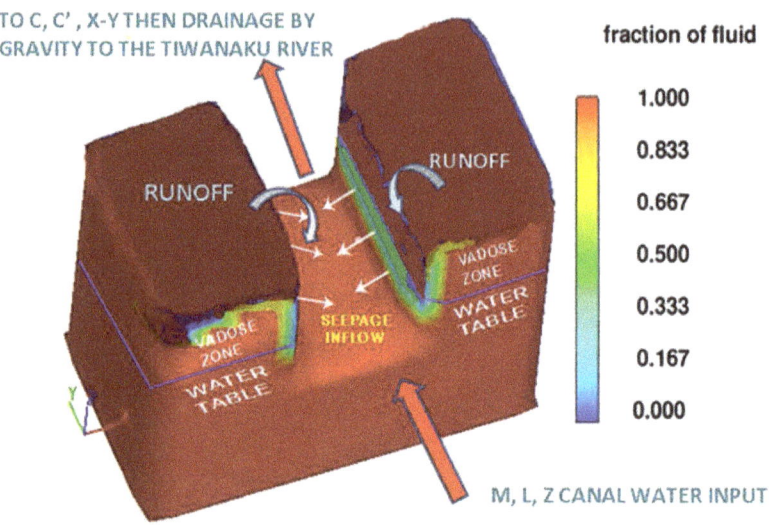

Figure 11. Summary fluid fraction diagram on water input/output flows on a typical perimeter drainage channel section near Figure 1a, and represent typical flow conditions near the end of the rainy season.

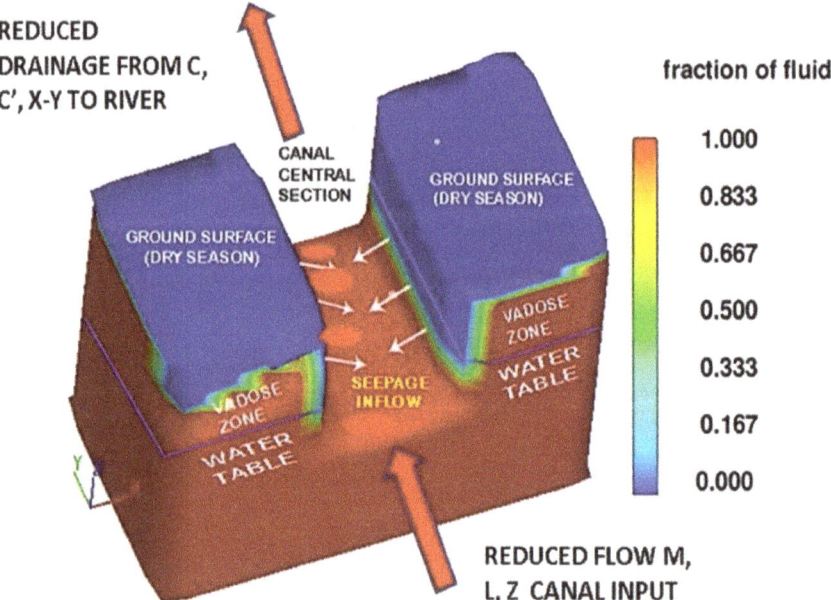

Figure 12. Dry season (Case 2) fluid fraction results for the monumental center with decreased water supply from spring-supplied canals; rainfall infiltration and seepage limited by large paved and roofed areas of the ceremonial center.

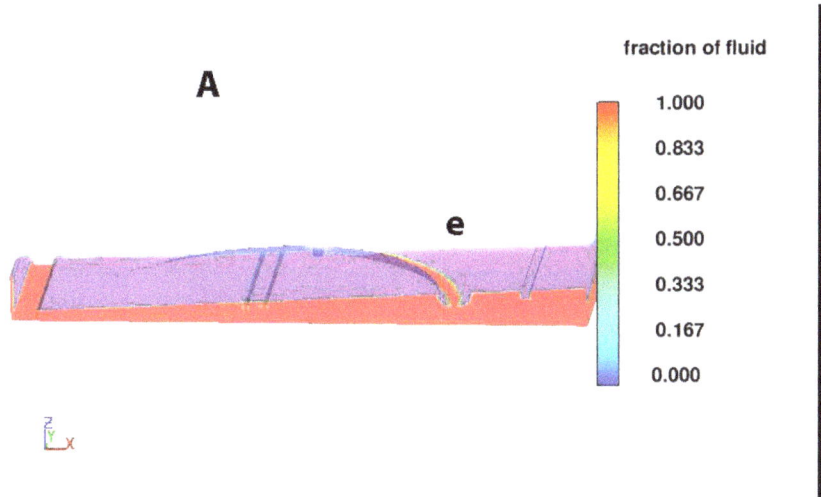

Figure 13. Constant y transect through the D drainage channel (Figure 1a) of groundwater profile for rainy season ground saturation conditions. The ff = 1 fluid fraction condition indicates layer aquifer saturation and bottom saturation of the perimeter drainage channel bed proceeding from design depth of the perimeter drainage channel.

The perimeter drainage channel bottom (e) depth is designed to intersect the top of the seasonally stabilized groundwater layer. In a severe rainy season, infiltrated water penetration may extend to the depth of the deep groundwater layer; in a normal rainy season, ground saturation extends the depth of the near-surface evaporation layer. For normal rainy season conditions, water infiltration drainage and surface runoff into the perimeter drainage channel proceeds to drain the evaporation layer; for normal dry season conditions, limited evaporation layer drainage continues but now Corocoro spring and reservoir water channeled into the perimeter drainage channel is added to maintain the deep groundwater level. Note that in the rainy season, Corocoro spring water continues to flow into the perimeter drainage channel but the excess beyond that to maintain the deep groundwater height is shunted into both the subterranean channels P and Q and the perimeter drainage channel to discharge into the Tiwanaku River. Figure 12 indicates that the role of water input from M channel prevents further recession of the deep groundwater layer as the dry season progresses. Thus, the intersection of the perimeter drainage channel bottom with the deep groundwater layer provides groundwater stabilization that underlies the conclusions of the prior sections. Although a case has been made for large monument foundation stability and its relation to the perimeter drainage channel, this result may have been fortuitous as knowledge of aquifer dynamics under compressive forces known to Tiwanaku engineers is as yet subjective with the present case the only known example to draw from.

11. Hydrologic Applications Exterior to City Precincts: Further Examples of Tiwanaku

Mastery of Groundwater Science

Research conducted on Tiwanaku's raised-field agricultural systems in the Pampa Koani region and water systems under Tiwanaku influence on the northwest regions of Lake Titicaca have indicated use of advanced hydrological methodology underlying crop sustainability and yield improvement [51–53]. Raised-fields are described as trenches dug to penetrate the water table by about 1.0 m with excavated soil piled up to form planting surface berms. Typical aerial views of berm geometry and placement are shown in Figures 14 and 15. Among the advances in agricultural science is the use of solar heat transfer technology to limit crop destruction by freezing during cold altiplano nights [4] as well as hydrological and hydraulic control mechanisms providing groundwater height control to stabilize raised-field swale water height through seasonal changes in water availability. Additionally, raised-field technology has been shown to be the most efficient design choice to limit short term drought effects on crop yield—this is due to continual groundwater supply from intercepted rainfall over vast eastern collection areas continually flowing toward the Lake Titicaca basin. Further analysis [54] demonstrates that Tiwanaku raised-field berm design is optimum to yield the maximum agricultural output per unit land area. Analysis of groundwater control mechanisms in the Pajiri agricultural area [52] reveals different berm heights and swale water depths appropriate to different crop types. These observations coupled with management of different nutrient chemical compositions of swale water from different springs and river sources necessary for maximum growth of different crop types point to an advanced agricultural science used by Tiwanaku water engineers to maximize and sustain crop yields in the Tiwanaku heartland.

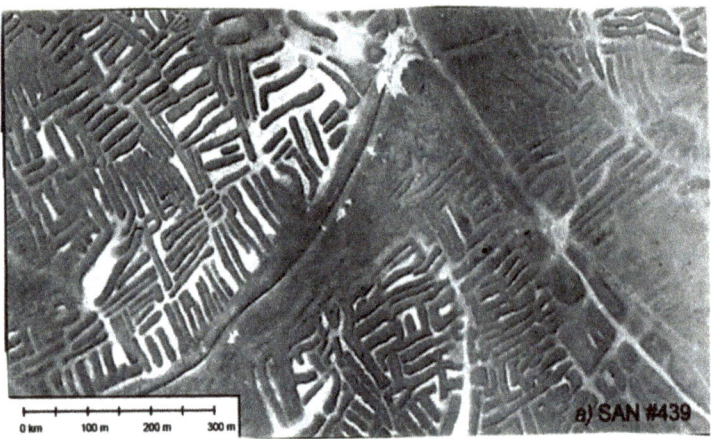

Figure 14. Tarraco raised-field aerial view.

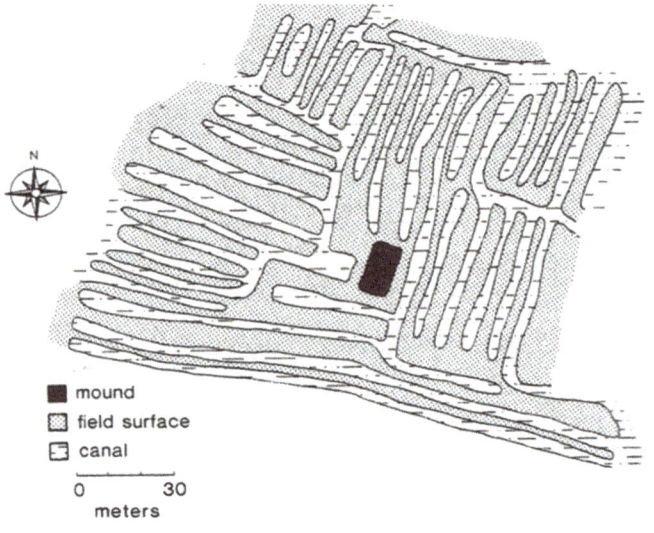

Figure 15. Lakaya sector raised-field geometry in the Pampa Koani system.

In the Pampa Koani and northwest Tarraco regions of Lake Titicaca, different patterns of raised-field lengths, widths, orientations and berm heights are frequently inserted within more regular patterns (Figures 14 and 15) with each pattern is appropriate for the water needs of different crop types [49]. Excavation of raised-field berms in the Pampa Koani area indicated stone base lining and clay layers to limit cold capillary water transfer from deep groundwater regions into berm interior regions; capillary water transfer to the berm interior region is mainly provided from swale water.

Due to higher swale water temperature from solar radiation input [4] the additional storage heat to berm interior regions limits berm outer surface convection and radiation heat withdrawal during cold altiplano nights to prevent freezing damage to crop root systems. The latent heat removal for water to ice transition within berm interiors during cold altiplano nights is limited by additional heat transfer from elevated temperature swale water heat transfer into berm interiors Examination of early Tarraco raised-field berm patterns [54] in northernmost regions of Lake Titicaca and raised-fields in the Pampa Koani region north of urban Tiwanaku reveals an average berm shape consistency. Figures 14 and 15 show that swales are interconnected leading to a continuous water path surrounding berms. When a typical berm is described as an elongated ellipse with major axis a and minor axis b, then the a/b ratio from 10 to 15 appears to characterize the average of berm geometries. This ratio for an elongated ellipse (a >> b) is significant in that the ellipse perimeter is a maximum for the given berm surface area (π a b) for this class of ellipse. This indicates that the average berm pattern configuration yields the maximum wetted berm perimeter [54] and thus requires a minimum of interconnected swale widths to provide capillary water transfer to narrow berms. Thus, the berms can be placed closer together to maximize agricultural surface per unit field area. Here the narrow berms provide an easy path for elevated temperature capillary water to reach berm interiors. This, in turn, reduces the exposed water surface area of the interconnected swales reducing evaporation loss that helps to locally maintain a constant groundwater profile to maintain swale water height. The net effect is that a greater number of closely spaced berms can be watered properly per unit field area to maintain the crop freezing defense while the greater area under cultivation produces more agricultural yield per unit field area. These advantages are a key indicator of an advanced agricultural science being employed to protect and increase the yields of raised-field agricultural systems. Thus, the a/b ratio of individual berms contains important information related to Tiwanaku water engineering practice as their design incorporates a level of optimization to limit swale water area to reduce evaporation losses and maximize the farming area per unit field area. Further analysis results [54] show that a/b ratios from 10 to 15 are an optimum berm design to yield the maximum wetted boundary perimeter for berms.

The conclusion that applies for the Tarraco raised-field geometry also applies in the Pampa Koani region as Figures 14 and 15 indicate similar use of a technology to maximize agricultural land area per unit field area. Although regional differences exist in raised-field designs at different locations built at different times and the Tarraco system design reflects different groundwater water availability, ambient air temperatures and crop types than those for the Tiwanaku Pampa Koani raised-field design, both systems reflect knowledge of berm designs to maximize agricultural production per unit field area.

12. Retrospectives on Tiwanaku Societal Structure

Ceramics and other objects in Tiwanaku style are widely distributed throughout the south-central Andes from the southern coastal valleys of Peru and Chile to the lowland eastern slopes of the Andes. The distribution of Tiwanaku cultural artifacts exhibit stylistic variations with different types and quantities of Tiwanaku style materials occurring in different regions. The governance principles operational in the Tiwanaku polity that led to the distribution of cultural material provides insight about the social structure of the Tiwanaku polity and their expansionist policies. Current theories related to Tiwanaku territorial and agricultural expansion are broadly summarized as:

(1) The distribution of high-status cultural artifacts results from Tiwanaku imperial expansion outside the Titicaca Basin with colonies and conquest aimed at lowland, highland and tropical base resource extraction.

(2) The widespread distribution of cultural materials results from the growth of an archipelago system where discontinuous territorial and ecological niches were exploited through placed colonies.
(3) Tiwanaku style materials spread spatially through trading networks headed by the Tiwanaku administrative polity.
(4) Tiwanaku expansion was ideological and/or ritual in nature devoid of political control from the central government of Tiwanaku. Tiwanaku expansion occurred as a combination of these paths and is characterized by a cultural rather than military expansion policy.

From the analysis thus far detailed and its relevance to the characteristics of Tiwanaku expansion policy, the increase and diversity of agricultural resources available to the Tiwanaku urban center, it is apparent that technical transfer of Tiwanaku water technologies to outlying areas and societies provided an additional economic and social inducement to associate with the central Tiwanaku polity. For the Tiwanaku urban center, agricultural systems incorporated groundwater-based raised-fields, rainfall supplied terraces, spring and groundwater-supplied raised-field areas, qochas and urban canal-supplied agricultural basins (K, Figure 1a). For sites distant from the Tiwanaku heartland that exploited different ecological conditions for farming, typical agricultural systems incorporated lowland valleys with river supplied canal irrigation systems, channeled spring-supplied agricultural fields and a variety of field systems located at different altitudes with different soil, water supply and temperature conditions that permitted a different range of crop types not possible to cultivate at the high altiplano elevation of urban Tiwanaku. Individual satellite sites required mastery of different hydrological regimes for irrigation. From the analysis of Tiwanaku urban and raised-field hydraulic/hydrological knowledge aspects of this technical base were likely exported to remote sites to maximize food production to economically justify the effort to maintain extensive trade and food import supply networks. The presence of cultural artifacts at many Tiwanaku colonies and subject areas verifies that association with central Tiwanaku had occurred based upon mutual economic benefits for all concerned. As some site areas were already occupied by different societies, some cooperative and others not accommodating an intrusion from a dominant competing society for use of limited water and land resources, the existing agricultural technology at distant areas outside of direct influence from the Tiwanaku urban core may not have initially generated sufficient surplus to interest incorporation into the Tiwanaku archipelago. However, by incorporation of advanced agrotechnical knowledge from urban Tiwanaku's hydraulic/hydrology experts, the agricultural output could be increased to justify mutually beneficial import/export status and cooperation between independent societies and the Tiwanaku core administrative region. Thus, the establishment and economic success of satellite archipelago sites appears influenced from a central authority based in the Tiwanaku urban center that included advice and council of technical experts versed in hydraulic and hydrological matters. This observation is best stated [55] as the challenge of new technologies: " ... certain individuals were probably empowered by technological or knowledge status and decisions made regarding the adoption, invention and use of certain technologies must have been made by technocratic and expert-centered individuals..." Thus, technical knowledge was a valuable export item. While trade in sumptuary goods to outlying societies was prevalent in the Tiwanaku sphere of influence and served to expand Tiwanaku influence and cultural traits into outlying areas, the exportation of agricultural knowledge had value to outlying societies that promoted economic benefits of association with the Tiwanaku urban core. While export of ally groups and technical experts from the Tiwanaku urban core experienced in agricultural production was manifest, the conversion to full agricultural potential of new resource-rich areas, given the different ecological conditions and challenges than those existing in the Tiwanaku urban core altiplano heartland, required indigenous creativity and invention in hydraulic and hydrological science. Given the sophistication of the urban Tiwanaku water system thus far described, it is clear that engineering creativity was a major focus of the Tiwanaku administration. In this respect, the many surface and subterranean canals that were found associated with the urban Tiwanaku perimeter drainage channel as well as water control canals in the outlying Pampa Koani raised-field heartland contain a comprehensive technology base that would be of vital use in the Tiwanaku-influenced

Moquegua Valley canal irrigated areas of Omo and Chen Chen. While some researchers claim that that individual farming communities could invent optimum agricultural field systems in a bottom-up manner without the need of a central controlling administration overview group, they underestimate the technical complexity involved in surface and groundwater control related to agricultural production over vast raised-field areas together with city water control engineering. Thus, a central planning (top-down) administration able to collect and invent farming methodologies and urban water control systems for use in areas under its direct control and manage the labor force for implementation of complex water technologies was vital to the expansion, development and integration of satellite areas to the Tiwanaku Empire.

The subterranean channels P and Q (Figure 1a,b) are examples of advanced technology applied to this end. The ~25,000 hectares of Pampa Koani raised-fields required expertise in local groundwater height control by means of a canal supply and drainage network operational over vast areas to provide tailored agricultural berm moisture levels and different berm geometries for different crop types. The agricultural system at the outlying site of Pajchiri is a prime example of water control mechanisms of this type developed for specialty crops. Thus, extrapolating from the bottom-up agricultural success at the local level by small independent allyu kin groups exploiting small farming areas to what is required to reliably support food supply for an urban Tiwanaku population of 20,000–40,000 through seasonal variations in groundwater and swale water level water, the management structure must logically incorporate a top-down overview structure [56] capable of assigning and relocating local and satellite agricultural zones to maintain a constant city food supply throughout seasonal weather changes. Despite this capability in agricultural technology, during the last stages of the 10–11th century drought, the groundwater level coupled with declining lake levels [5] forced abandonment of near lake edge raised-field agriculture as groundwater levels declined below swale bottom levels. This caused agriculture to move to the outer fringes of Pampa Koani, as noted by [54], where the water table remained high in swales far from Lake Titicaca due to incoming intercepted groundwater from distant sources and earlier rainfall events. This drop in agricultural output from major field system areas then led to city population dispersal to sustainable farming zones by different segments of the city population and the ultimate decline of the Tiwanaku Empire.

Summarizing, an overview of vast agricultural land and water management to ensure successful agricultural yields required knowledge of optimum berm designs and groundwater height control only possible from a top-down overview perspective of land, water and labor management for vast areas under their control. While other sites had value for imports of non-agricultural resources, sites with the potential for optimization of agricultural resources could be improved by optimization technologies for raised-fields and other agricultural methodologies at coastal valley sites to improve the economic basis for agricultural imports. Clearly optimization of river/spring source irrigation, raised-field agriculture and control of urban water supplies for hygienic advantage to the city population demonstrated that exported Tiwanaku oversight to apply engineering methodology to optimize food production and city living benefits would of advantage for candidate sites to associate with the Tiwanaku hierarchy and share the mutual benefits of association. It would be expected that this oversight activity was applied to rate potential archipelago satellite sites for incorporation given that the economic burden of long-distance transport of perishable goods to urban Tiwanaku. Within the Tiwanaku governmental structure were religious rites, rituals and ceremonies elaborated with elaborate ceremonial, royal and administrative architecture to provide the religious accompaniment to the worldly success of their agroscience, both locally and distant from the urban core of Tiwanaku. Thus, aspects of all the above (1) to (4) categories provided the basis and rationale for Tiwanaku expansion from its heartland—this was only made technically and economically possible with the underlying centrally planned agrotechnical base provided by a top-down corporate management structure at urban Tiwanaku. Thus, the Tiwanaku corporate structure provided the success basis for satellite trade networks in agricultural goods together with the export of cultural traits and artifacts from the urban center of Tiwanaku to cement cultural ties back to the homeland source as observed from the archaeological record. A further argument for a top-down Tiwanaku management structure can be posited on an economic advantage basis. From similitude analysis methods [54] a mathematical model of two competing ally groups is considered: the first arrival group (1) sets up localized, near lake raised-field agriculture supplied with a local spring-fed canal system; the second group (2) arriving at a later time sets up available outlying raised-field system with long canal lengths to irrigate their distant fields. The first group clearly has an economic advantage due to better water access (shorter canals) requiring less labor to tend to the agricultural land. Analysis [54] shows a computable economic advantage for both (1) and (2) groups to combine land and water resources by use of a newly designed canal irrigation network that more effectively irrigates both land areas and reduces labor input from both groups to maintain the combined agricultural land area while raising the agricultural output of the combined land area. The advantages to both groups to combine their resources under collective management that demonstrates economic advantages to both groups serves to promote top-down oversight of the combined raised-field area to the advantage of all participating ally groups through a governing organization that provides direction and oversight on project activity. Here the formalism of the similitude methodology [54] permits a calculation of the increase in food production through collective top-down management oversight compared to the bottom-up system of disconnected groups managing localized field plots.

While certain researchers suggest less centralization and more local autonomy in the Tiwanaku core region as opposed to other archaeologist's vision of a highly centralized state-directed agrarian production, the present discussion demonstrates that the agricultural engineering base together with knowledge of urban Tiwanaku water control design and operation essentially defines the success of the Tiwanaku society. Thus, massive public reclamation and construction projects requiring a large and coordinated labor force supported by an advanced technology base much in the same way that modern progressive societies function appears to verify top-down management directing complex high technology projects. As to the demise of Tiwanaku colonies located in the Moquegua Valley, collapse dates are consistent with, or follow somewhat, the final collapse dates of Tiwanaku urban complexes. As detailed by Sharatt et al. [57], evidence of Moquegua colonies persisted into Ilo–Tumilaca–Cabuza coastal phases and highland Tumilaca Phases l past~1000 AD, indicating in many cases the extension of some of the Tiwanaku city traditions and stylistic practices in textile and ceramic designs. As the slow development of altiplano drought initiates in the 10th century AD, the rainfall runoff-based canal agriculture of Moquegua Valley colonies invariably responded to rainfall runoff decrease in vulnerable valley rivers challenging the continuity of their irrigation agricultural field systems. This leads to ultimate population contraction of Moquegua societies. The establishment of Moquegua Valley Estuquiña highland valley society at higher altitudes with higher rainfall levels is a natural survival consequence compared to valley societies dependent on river-sourced agriculture. As groundwater decline for the altiplano Tiwanaku is a slow process due to recharge from distant infiltrated rainwater sources continually flowing through the aquifer toward Lake Titicaca, slow groundwater level decline permits longer continuation of raised-field agriculture in outlying regions of the Pampa Koani area well past that of rainfall runoff river supplied agricultural system of the outlying Tiwanaku Moquegua colonies. Thus, it is expected that the colonies also ultimately diminish in size due to drought but at a different rate than hardier raised-field systems of highland Tiwanaku due to their different agricultural water supply means. To assign the Tiwanaku societal collapse to socio-political mechanisms would likely reflect the catalytic effects of drought-induced agricultural contraction on the sustainability of a society. The Tiwanaku collapse appears to be a slow process over decades as the near lake raised-fields decline first as the Titicaca lake level subsides due to rainfall decrease. Agriculture continues at a minor level at more distant raised-fields where groundwater decline lags that of near lake fields.

13. Visions of the Last Days of the City of Tiwanaku

The final stages of urban Tiwanaku due to extended drought are described in [44]. Establishment of extended drought conditions that led to the gradual demise of urban Tiwanaku and its associated raised-field systems is well substantiated through geophysical means originating from ice core data.

Figure 16 summarizes the drought decline by integration of the nine year moving averages of ice cap thickness derived from [49,50,58] references. As drought initiates, yearly rainfall declines leading to smaller distances between successive ice deposit layers. Essentially the drought slowly lowered the water table supporting raised-field agriculture for the Tiwanaku city's 20,000 to 40,000 inhabitants; additionally, nonexistent water levels in raised-field swales promoted loss of any surviving crops to freezing events and water unavailability to plant root systems. Given the vast area devoted to raised-field agriculture, restoration of the fields by excavating swale depths to penetrate the declining groundwater level together with lowering field system berm heights to accommodate crops with root system depths necessary for plant growth proved to be an impossible task given the vast labor requirements to perform these tasks. Evidence of use of raised-field systems remote from the edge of Lake Titicaca where the groundwater height remained high exist requiring relocation of elements of city population to distant areas from the city. The presence of scattered qocha farming pits excavated to groundwater phreatic levels located distant from the city center indicated population fragmentation in order to conduct localized survival farming. New information pertinent to the last days of Tiwanaku city life [57,59] is available from the use of multiple stable isotope methods involving analysis of skeletal remains dating from the ~1100 AD time period which corresponds to city abandonment dates at the contemporary site of Wari. Noted are dietary changes from previous norms experienced by city population as drought intensified: these changes include absence of fish from the diet and no reported instances of child remains incorporating nutrients from fish or marine sources. This later observation may represent partitioning high nutrient food types to the most productive society members capable of generating food resources in emergency situations. As population decline and migration continued in this time period, specialized industries randomly lost key members necessary to sustain the group's function effectively; hence the loss of skilled fisherman can diminish the amount of fish available from the lake source. From modern observation of villages' use of lake resources, small minnows can be gathered from the near shoreline by nets which provide a protein source for site inhabitants.

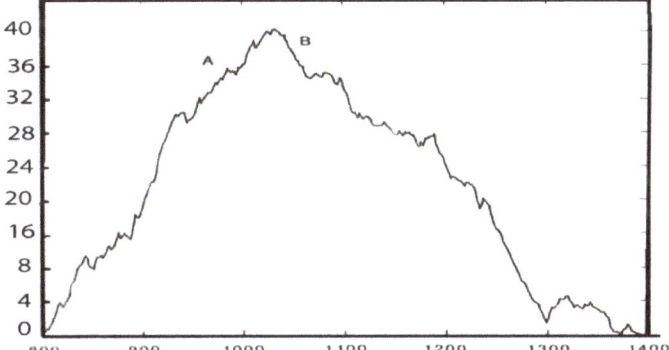

Figure 16. In the time period AD 800–1400, the 9 year moving average of Huascaran Mountain ice core yearly deposit layer thickness begins a decline indicating the start of extended severe drought conditions. Ordinate scale is in centimeters [49,50].

However, as lowered lake levels resulted from extended drought conditions, limited access to shallow shoreline depths together with increased salinity that affected fish stocks, marine resources by lake fishing and shoreline collection was likely reduced from previous norms. Results from Miller et al. [60] indicate the substantial presence of maize as a food source in the ~1000 AD time period—this indicates a likely increase in importation from different satellite areas where this crop could be successively raised and transported in dried kernel form. Throughout the existence of Tiwanaku, maize importation constituted a large fraction of the population's diet and source of chicha for celebratory, social binding rituals. Although the totality of effects on population decrease and dispersion are yet to be brought forward under extended drought conditions, the use of stable isotope methods opens new paths to understand the final days of Tiwanaku city closure. Recent studies [59] detail the societal collapse of a contemporary Middle Horizon Wari site due to food shortages in the same time period; here dietary shifts, associated with the extended AD 10–11th century drought, assign limited food resources to productive society members and leadership individuals capable of sustaining and guiding the society into a recovery period. As both highland societies are contemporary and experience the same drought period conditions, it is of interest to note similar responses to protect vital members of their societies. Figure 17 indicates that severe drought conditions at AD 1000–1100 and AD 600–700 altered the survival fate of different societies—in some cases, certain societies survived by altering their farming methodologies (or conquest of other societies with significant land and water resources) while other societies disappear from the archaeological record (Figure 17), of interest is the Medieval Warming Period post-drought recovery period in the 13–14th century AD period (Figure 17) that led to Inka expansion and control over vast areas of Peru and Ecuador with no state level polities left to contest their dominance.

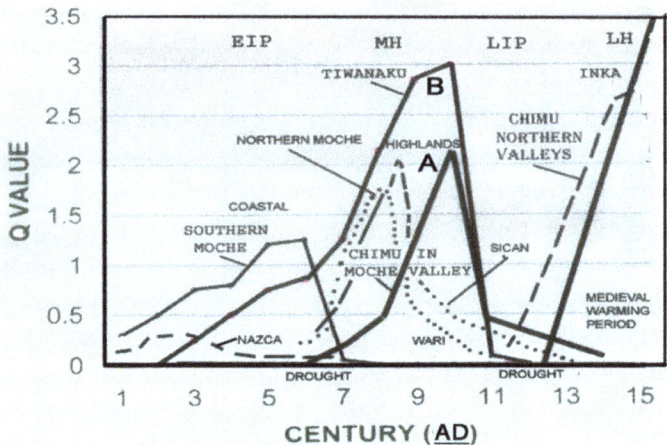

Figure 17. The AD 10–11th century drought led to societal decline for both the highland Tiwanaku and Wari polities as well as for Peruvian north coast societies. Occurrences of El Niño, La Niña and other ENSO climate change effects on continuity of major Andean societies constitute a vital part of understanding Andean prehistory.

14. Conclusions

CFD results suggest that the perimeter drainage channel accelerated Tiwanaku's ceremonial center and the surrounding urban areas seasonal dryness throughout the year's seasonal changes in rainfall promoting the city's hygienic benefits. For example, reduction of dampness in indoor habitable structures limits the occurrence of many respiratory and mould-borne diseases [61]. In the rainy season, the deep groundwater level was stabilized by runoff and aquifer drainage into the perimeter drainage channel. In the dry season, additional seepage from the aquifer and the M channel flow kept the deep groundwater boundary from subsiding. The resulting stabilization of the deep groundwater level prevented the settling of monumental structures in the ceremonial core that originated from the design feature of the perimeter drainage channel's depth intersection with the top fringe of the deep groundwater layer and introduced the possibility of water resource availability during local drought periods. The stable groundwater level promoted the existence of wells and qochas for localized water supplies (Figure 2) to urban districts and inter-city agricultural zones. Subterranean channels P and Q largely served the hygienic requirements of the Putuni and Kerikala structures by providing continuous water flow from perimeter drainage channel aquifer seepage water and the M channel water arriving into perimeter drainage channel arm W. Each major monumental structure maintained an intricate drainage system that simultaneously served practical and symbolic purposes as exemplified by the Akapana's elaborate drainage network that limited rainfall infiltration into its compartmentalized earth-fill interior to preserve its structural integrity. Canals O and N south of the perimeter drainage channel directed water to the qocha complex K for inter-city agricultural and pasturage purposes (Figure 2). Rainy season runoff water that washed into canals N, O and S exceeding their carrying capacity was diverted into canals L, I and J leading to the Tiwanaku River, thus protecting urban regions from canal overflow flooding. In totality, the perimeter drainage channel was the linchpin of an intricate hydraulic network that controlled surface and aquifer flows as rainfall amounts varied from rainy to dry seasons.

Analysis of the perimeter drainage channel's hydrological function indicates that qochas and wetland systems C' and K were an integral feature of urban Tiwanaku. Interlinked by canals fed by Corocoro springs and reservoirs, qocha clusters C' and K occupied massive portions of the city and likely supported camelid herds and caravans brought to the center during key social gatherings. Recent excavations in adjacent Mollo Kontu residential compounds support the hypothesis that llama and alpaca herds were important in this part of Tiwanaku and were served by C' and K qocha pasturage areas. Raised-field and qocha systems that occupied the edges of some of the city's canals are evident from aerial photos of the edges of canals I, J, L and C to support localized in-city agriculture and pasturage. The C' floodplain at the south edge of, and several meters below, the main portion of the city area of Tiwanaku supported an extensive cluster of integrated raised-field networks and qochas to support additional intra-city agriculture and pasturage.

Prior studies focused on Tiwanaku's hinterland demonstrated an understanding of hydrologic principles to develop intensive raised-field farming systems. Present research indicates that the urban center of Tiwanaku incorporated a complementary intricate hydrological network focused on the perimeter drainage channel that effectively managed seasonal water variations through surface canals, subsurface channels and aquifer drainage manipulation. CFD results detail many practical hydrological features of the perimeter drainage channel related to environmental and population livability concerns—these include rapid drying of subsurface soils surrounding elite ceremonial and secular housing districts to limit soil dampness and its negative health effects on the city's population. The perimeter drainage channel further supplied water to flush the subterranean channel network underlying the elite ceremonial core region to transfer human waste material to the nearby Tiwanaku River. These health-related features and remarkable plumbing features are the first reported for any Andean pre-Columbian city. Tiwanaku city planners demonstrated an extraordinary level of knowledge regarding hydrologic and structural maintenance principles based upon surface and groundwater manipulation to maintain high livability standards for their population under harsh altiplano environmental conditions. Building on prior studies of the groundwater-based raised-field systems that supported agriculture for the large population of Tiwanaku, analysis results demonstrate that knowledge of surface and groundwater flows within urban Tiwanaku merit further consideration in assessing New World engineering science.

The raised-field technology devised by the ancient Tiwanaku has been brought back to life once again after ~1000 years of raised-field abandonment by inhabitants of local altiplano villages many of whom participated in the original NSF Proyecto WilaJawira project. Restoration of segments the ancient raised-fields followed by planting native crops resulted in yields~5X over that of their current agricultural practice. The lesson here is that reexamination and restoration of agricultural methodologies of societies in past millennia may have great benefits for third world societies with large labor resources but who have limited access to modern machinery and chemical fertilizer supplements that are beyond their means to acquire.

Funding: No external funding requested by author for project participation.

Acknowledgments: Report is based upon six years of participation in the original NSF funded Proyecto WilaJawira project in Bolivia.

Conflicts of Interest: The author declares no conflict of interest.

References

1. Kolata, A.L. Agricultural Foundations of the Tiwanaku State. *Am. Antiq.* **1986**, *1*, 748–762. [CrossRef]
2. Kolata, A.L. The Technology and Organization of Agricultural Production in the Tiwanaku State. *Lat. Am. Antiq.* **1991**, *2*, 99–125. [CrossRef]
3. Peck, R.; Hanson, W.; Thornton, T. *Foundation Engineering*; John Wiley and Sons: New York, NY, USA, 1974.
4. Kolata, A.L.; Ortloff, C.R. Thermal Analysis of the Tiwanaku Raised-Field Systems in the Lake Titicaca Basin of Bolivia. *J. Archaeol. Sci.* **1989**, *16*, 233–263. [CrossRef]
5. Kolata, A.L.; Ortloff, C.R. Agroecological Perspectives on the Decline of the Tiwanaku State. In *Tiwanaku and its Hinterland: Archaeology and Paleoecology of an Andean Civilization*; Kolata, A.L., Ed.; Smithsonian Institution Press: Washington, DC, USA, 1996.
6. Ortloff, C.R. Engineering Aspects of Groundwater Controlled Agriculture in the PreColumbian Tiwanaku State of Bolivia in the Period 400-1000 AD. In *Tiwanaku and its Hinterland: Archaeology and Paleoecology of an Andean Civilization*; Kolata, A.L., Ed.; Smithsonian Institution Press: Washington, DC, USA, 1996.
7. Ortloff, C.R. Engineering Aspects of Tiwanaku Groundwater Controlled Agriculture. In *Tiwanaku and its Hinterland: Archaeology and Paleoecology of an Andean Civilization*; Kolata, A.L., Ed.; Smithsonian Institution Press: Washington, DC, USA, 1996; Volume 1.
8. Kolata, A.L.; Ortloff, C.R. Tiwanaku Raised-field Agriculture in the Titicaca Basin of Bolivia. In *Tiwanaku and its Hinterland: Archaeology and Paleoecology of an Andean Civilization*; Kolata, A.L., Ed.; Smithsonian Institution Press: Washington, DC, USA, 1996.
9. Binford, M.W.; Brenner, M.; Leyden, B.W. Paleoecology and Tiwanaku Agroecosystems. In *Tiwanaku and its Hinterland: Archaeology and Paleoecology of an Andean Civilization*; Kolata, A.L., Ed.; Smithsonian Institution Press: Washington, DC, USA, 1996.
10. Bruno, M. Beyond Raised-fields: Exploring Farming Practices and Processes in the Ancient Lake Titicaca Basin in the Andes. *Am. Anthropol.* **2014**, *116*, 1–16. [CrossRef]
11. Ortloff, C.R. *Water Engineering in the Ancient World: Archaeological and Climate Perspectives on Societies of Ancient South America, the Middle East and South-East Asia*; Oxford University Press: Oxford, UK, 2010.
12. Ortloff, C.R. New Discoveries and Perspectives on Water Management and State Structure at AD 300–1100 Tiwanaku's Urban Center (Bolivia). *MOJ Civ. Eng.* **2016**, *1*, 57–66.
13. Créqui-Montfort, G. *Fouilles de la Mission Scientifique Française à Tiahuanaco. Ses Recherches Archéologiques et Ethnographiques en Bolivie, au Chili et dans la République Argentine*; International Congress of Americanists: Stuttgart, Germany, 1906; pp. 531–550.
14. Bandelier, A. The Ruins at Tiahuanaco. *Proc. Am. Antiqu. Soc.* **1911**, *21*. Available online: https://www.americanantiquarian.org/proceedings/44817263.pdf (accessed on 1 December 2020).
15. Denevan, W. *Cultivated Landscapes of Native Amazonia and the Andes*; Oxford University Press: Oxford, UK, 2001.
16. Means, P. *Ancient Civilizations of the Andes*; C. Scribner's Sons Publishers: New York, NY, USA, 1931.
17. Bennett, W.C. Excavations at Tiahuanaco. *Anthropol. Pap. Am. Mus. Nat. Hist.* **1934**, *34*, 354–359.
18. Posnansky, A. *Tihuanacu: The Cradle of American Man, Volumes. I & II*; J. Augustin Publishers: New York, NY, USA, 1945.
19. Ponce Sangines, C. *Informe de Labores. La Paz: Centro de Investigaciones Arqueologicas en Tiwanaku: Espacio, Tiempo y Cultura: Ensayo de Sintesis Arqueológica*; Academía Nacional de Ciencias Publication: La Paz, Bolivia, 1961.
20. Ponce Sanguines, C. *Descripción Sumaria del Templete Semisuterraneo de Tiwanaku*; Juventud Publicaciones: La Paz, Bolivia, 2009.
21. Kolata, A.L. *The Tiwanaku: Portrait of an Andean Civilization*; Wiley: Hoboken, NJ, USA, 1993.
22. Kolata, A.L. Tiwanaku Ceremonial Architecture and Urban Organization. In *Tiwanaku and Its Hinterland: Archaeology and Paleoecology of an Andean Civilization*; Kolata, A.L., Ed.; Smithsonian Institution Press: Washington, DC, USA, 2003; Volume 2, pp. 175–201.

23. Ortloff, C.R. Hydraulic Engineering in Ancient Peru and Bolivia. In *Encyclopaedia of the History of Science, Technology and Medicine in Non-Western Cultures*; Springer Publications: Heidelberg, Germany, 2014.
24. Janusek, J. *Ancient Tiwanaku*; Cambridge University Press: Cambridge, UK, 2008.
25. Kolata, A.L. Tiwanaku and its Hinterland. In *Archaeology and Paleoecology of an Andean Civilization*; Kolata, A.L., Ed.; Smithsonian Institution Press: Washington, DC, USA, 1996.
26. Erickson, C. Lake Titicaca Basin: A Precolumbian Built Landscape. In *Imperfect Balance: Landscape Transformations in the Precolumbian Americas*; Lentz, D., Ed.; Columbia University Press: New York, NY, USA, 2000; pp. 311–356.
27. Ortloff, C.R.; Janusek, J. Water Management and Hydrological Engineering at 300 BCE-1100 CE Precolumbian Tiwanaku (Bolivia). *J. Archaeolog. Sci.* **2014**, *44*, 91–97. [CrossRef]
28. Janusek, J. The Changing Face of Tiwanaku Residential Life: State and Social Identity in an Andean City. In *Tiwanaku and Its Hinterland: Archaeology and Paleoecology of an Andean Civilization*; Kolata, A.L., Ed.; Smithsonian Institution Press: Washington, DC, USA, 2003.
29. Janusek, J. Craft and Local Power: Embedded Specialization in Tiwanaku Cities. *Lat. Am. Antiq.* **1999**, *10*, 107–131. [CrossRef]
30. Ortloff, C.R. Groundwater Management at the 300 BCE-CE 1100 Precolumbian City of Tiwanaku (Bolivia). *Hydrol. Curr. Res.* **2014**, *5*, 1–7.
31. Couture, N. The Construction of Power: Monumental Space and an Elite Residence in Tiwanaku, Bolivia. Ph.D. Disertation, Department of Anthropology, University of Chicago, Chicago, IL, USA, 2002.
32. Janusek, J.; Earnest, H. Excavations in the Putuni: The 1988 Season. In *Tiwanaku and its Hinterland, Report submitted to the National Science Foundation and the National Endowment for the Humanities*; Kolata, A.L., Ed.; University of Chicago Press: Chicago, IL, USA, 2009.
33. Bentley, N. The Tiwanaku of A.F. Bandelier. In *Advances in Titicaca Basin Archaeology*; Vranich, A., Levine, A., Eds.; Cotsen Institute of Archaeology, University of California: Los Angeles, CA, USA, 2013.
34. Couture, N.; Sampeck, K. Putuni: A History of Palace Architecture in Tiwanaku. In *Tiwanaku and its Hinterland: Archaeology and Paleoecology of an Andean Civilization*; Kolata, A.L., Ed.; Smithsonian Institution Press: Washington, DC, USA, 2003.
35. Ortloff, C.R.; Janusek, J. *Water Management at BCE 300- CE 1100 Tiwanaku (Bolivia): The Perimeter Canal and its Hydrological Features*; Encyclopaedia of the History of Science, Technology and Medicine in Non-Western Civilizations; Springer Publications: Berlin/Heidelberg, Germany, 2015.
36. Bruno, M. Sacred Springs: Preliminary Investigation of the Choquepacha Spring/Fountain, Tiwanaku, Bolivia. In Proceedings of the 65th Annual Meeting of the Society for American Archaeology, Philadelphia, PA, USA, 5–9 April 2000.
37. Vallières, C. Taste of Tiwanaku: Daily Life in an Ancient Andean Urban Center as Seen through Cuisine. Unpublished Ph.D. Thesis, Department of Anthropology, McGill University, Montreal, QC, Canada, November 2012.
38. Rivera, C.; Ch'iji, J. Evidencias sobre la Producción de Cerámica en Tiwanaku. Licenciatura Thesis, Universidad Mayor de San Andrés, La Paz, Bolívia, 1994.
39. Rivera, C.; Ch'iji, J. A Case of Ceramic Specialization in the Tiwankau Urban Periphery. In *Tiwanaku and its Hinterland: Archaeology and Paleoecology of an Andean Civilization*; Kolata, A.L., Ed.; Smithsonian Institution Press: Washington, DC, USA, 2003.
40. Browman, D. Political Institutional Factors contributing to the Integration of the Tiwanaku State. In *Emergence and Change in Early Urban Societies*; Manzanilla, L., Ed.; Plenum Press: New York, NY, USA, 1997; Chapter 9.
41. Janusek, J. *Identity and Power in the Ancient Andes*; Routledge Press: London, UK, 2004.
42. Doig, F.K. *Manual de Arqueologia Peruana*; Ediciones Peisa Publicaciones: Lima, Peru, 1973.
43. McAndrews, J.; Albarracin-Jordan, J.; Bermann, M.P. Regional Settlement Patterns of the Tiwanaku Valley of Bolivia. *J. Field Archaeol.* **1997**, *24*, 67–83.
44. Flow Science. *FLOW-3D User Manual V. 9.3*; Flow Science Inc.: Santa Fe, NM, USA, 2019.
45. Bear, J. *Dynamics of Fluids in Porous Media*; Ch. 4, Sections 4.5 to 4.7; Dover Publications: New York, MY, USA, 1972.
46. Freeze, R.A.; Cherry, J.A. *Groundwater*; Prentice-Hall, Inc.: Englewood Cliffs, NJ, USA, 1979.
47. Ortloff, C.R.; Kolata, A.L. Climate and Collapse: Agro-ecological Perspectives on the Decline of the Tiwanaku State. *J. Archaeol. Sci.* **1993**, *16*, 513–542. [CrossRef]

48. Binford, M.W.; Kolata, A.L.; Brenner, M.; Janusek, J.W.; Seddon, M.T.; Abbott, M.; Curtis, J.H. Brenner Climate Variation and the Rise and Fall of an Andean Civilization. *Quat. Res.* **1997**, *47*, 235–248. [CrossRef]
49. Thompson, L.G.; Mosley-Thompson, E.; Davis, M.E.; Lin, P.N.; Henderson, K.A.; Cole-Dai, J.; Bolzan, J.F.; Liu, K.B. Late Glacial Stage and Holocene Tropical Ice Core Records from Huascaran, Peru. *Science* **1995**, *269*, 46–50. [CrossRef] [PubMed]
50. Thompson, L.G.; Davis, M.E.; Mosley-Thompson, E. Glacial Records of Global Climate: A 1500-year Tropical Ice Core Record of Climate. *Hum. Ecol.* **1994**, *22*, 83–95. [CrossRef]
51. Goldstein, P. *Andean Diaspora: The Tiwanaku Colonies and the Origins of South American Empire*; University of Florida Press: Gainesville, FL, USA, 2005.
52. Stanish, C.K.; Frye, E.; de la Vega, M. Sneddon Tiwanaku Expansion into the Western Titicaca Basin, Peru. In *Advanced in Titicaca Basin Archaeology-1*; Stanish, C., Cohen, A., Aldenderfer, M., Eds.; Cotsen Institute of Archaeology: Los Angeles, CA, USA, 2008; pp. 103–114.
53. Henderson, M. The Ancient Raised-Fields of the Taraco Region of the Northern Lake Titicaca Region. In *Advances in Titicaca Basin Archaeology-III*; Vranich, A., Clarich, E.A., Stanish, C., Eds.; Memoirs of the Museum of Anthropology: Ann Arbor, MI, USA, 2012.
54. Ortloff, C.R. *The Hydraulic State: Science and Society in the Ancient World*; Routledge Press: London, UK, 2020.
55. Dillehay, T. (Ed.) Foundational Understandings. In *Where the Land Meets the Sea*; University of Texas Press: Austin, TX, USA, 2017.
56. Janusek, J.W.; Kolata, A.L. Top-down or Bottom-up Rural Settlement and Raised-Field Agriculture in the Lake Titicaca Basin, Bolivia. *J. Anthropol. Archaeol.* **2004**, *23*, 404–430. [CrossRef]
57. Sharratt, N. Tiwanaku's Legacy: A Chronological Reassessment of the Terminal Middle Horizon in the Moquegua Valley, Peru. *Lat. Am. Antiq.* **2019**, *30*, 529–549. [CrossRef]
58. Thompson, L.G.; Mosley-Thompson, E. One-half Millennium of Tropical Climate Variability as Recorded in the Stratigraphy of the Quelccaya Ice Cap, Peru. In *Aspects of Climate Variability in the Pacific and Western Americas*; Peterson, D., Ed.; American Geophysical Union Monograph 55: Washington, DC, USA, 1989.
59. Tung, T.; Vang, N.; Culleton, B.; Kennett, D. *Dietary Inequality and Indiscriminant Violence: A Social Bioarchaeological Study of Community Health during Times of Climate Change and Wari State Decline*; Institute of Andean Studies: Berkley, CA, USA, 2017.
60. Miller, M.; Kendall, I.; Capriles, J.; Bruno, M.; Evershed, R.; Hastorf, C. The Trouble with Maize and Fish: Refining our understanding of the Diets of Lake Titicaca's Inhabitants using Multiple Stable Isotope Methods (1500 BC–AD 1100). In Proceedings of the 59th Institute of Andean Studies Conference, Berkeley, CA, USA, 8–9 January 2014.
61. Clark, N.; Ammann, H. *Damp Indoor Spaces and Health*; Ch.5: Human Health Effects in Damp Indoor Environment; National Academies Press: Washington, DC, USA, 2015.

Publisher's Note: MDPI stays neutral with regard to jurisdictional claims in published maps and institutional affiliations.

© 2020 by the author. Licensee MDPI, Basel, Switzerland. This article is an open access article distributed under the terms and conditions of the Creative Commons Attribution (CC BY) license (http://creativecommons.org/licenses/by/4.0/).

Review

Caral, South America's Oldest City (2600–1600 BC): ENSO Environmental Changes Influencing the Late Archaic Period Site on the North Central Coast of Peru

Charles R. Ortloff [1,2]

[1] CFD Consultants International, Ltd., 18310 Southview Avenue, Los Gatos, CA 95033, USA; ortloff5@aol.com
[2] Research Associate in Anthropology, Anthropology Department, University of Chicago, 5801 Ellis Avenue, Chicago, IL 60637, USA

Abstract: The Late Archaic Period (2600–1600 BC) site of Caral, located ~20 km inland from the Pacific Ocean coastline in the Supe Valley of the north central coast of Peru, is subject to CFD analysis to determine the effects of ENSO (El Niño Southern Oscillation) events (mainly, El Niño flooding and drought events) on its agricultural and marine resource base that threatened societal continuity. The first step is to examine relics of major flood events that produced coastal beach ridges composed of deposited flood slurries—the C14 dating of material within beach ridges determines the approximate dates of major flood events. Of interest is the interaction of flood slurry with oceanic currents that produce a linear beach ridge as these events are controlled by fluid mechanics principles. CFD analysis provides the basis for beach ridge geometric linear shape. Concurrent with beach ridge formation from major flood events are landscape changes that affect the agricultural field system and marine resource food supply base of Caral and its satellite sites- here a large beach ridge can block river drainage, raise the groundwater level and, together with aeolian sand transfer from exposed beach flats, convert previously productive agricultural lands into swamps and marshes. One major flood event in ~1600 BC rendered coastal agricultural zones ineffective due to landscape erosion/deposition events together with altering the marine resource base from flood deposition over shellfish gathering and sardine and anchovy netting areas, the net result being that prior agricultural areas shifted to limited-size, inner valley bottomland areas. Agriculture, then supplied by highland sierra *amuna* reservoir water, led to a high water table supplemented by Supe River water to support agriculture. Later ENSO floods conveyed thin saturated bottomland soils and slurries to coastal areas to further reduce the agricultural base of Supe Valley sites. With the reduction in the inner valley agricultural area from continued flood events, agriculture, on a limited basis, shifted to the plateau area upon which urban Caral and the satellite sites were located. The narrative that follows then provides the basis for the abandonment of Caral and its satellite Supe Valley sites due to the vulnerability of the limited food-supply base subject to major ENSO events.

Keywords: Peru; Archaic period; Caral; CFD models; beach ridges; ENSO events; landscape change; site termination

Citation: Ortloff, C.R. Caral, South America's Oldest City (2600–1600 BC): ENSO Environmental Changes Influencing the Late Archaic Period Site on the North Central Coast of Peru. *Water* **2022**, *14*, 1403. https://doi.org/10.3390/w14091403

Academic Editor: Helena M. Ramos

Received: 3 March 2022
Accepted: 22 April 2022
Published: 27 April 2022

Publisher's Note: MDPI stays neutral with regard to jurisdictional claims in published maps and institutional affiliations.

Copyright: © 2022 by the author. Licensee MDPI, Basel, Switzerland. This article is an open access article distributed under the terms and conditions of the Creative Commons Attribution (CC BY) license (https://creativecommons.org/licenses/by/4.0/).

1. Introduction

The presence of ENSO (El Niño Southern Oscillation) climate variations in the form of long-term drought, flooding, aeolian sand transfer, and sediment deposition/erosion transfer events, and their effect upon the agricultural and marine resource-base sustainability of Peruvian coastal societies, is of importance to understand the influences that affected Andean historical development. While the timing and intensity of El Niño flood landscape deposition and erosion events is manifest from the geophysical analysis of the observed deposition sand and flood slurry layers and erosion/deposition profiles, the soil transfer and deposition geophysics causing landscape changes affecting the agricultural and marine resource zones as a result of such events remain elusive. To understand the geophysics underlying such events, Computational Fluid Dynamics (CFD) analysis is of use using a landscape model of a portion of the Peruvian coastline subject to an El Niño flood. This event produces runoff consisting of a highly viscous slurry mixture containing silt, gravel, rocks and soil particles that proceeds to further erode the rain-saturated landscape to settle and deposit slurry mixed with the captured landscape soil sediments to form a deposition layer on the pre-existing landscape. In the present study, CFD analysis is performed on a model of the Santa-Viru Peruvian north central coastline; the fluid mechanics CFD analysis then duplicates the acts performed by nature to alter the landscape by erosion and deposition flood events. The CFD investigation results substantiate that slurry deposition deposits create coastal linear beach ridges as a result of the flood transported slurry interaction with ocean currents. While linear beach ridge structures are noted in the literature [1–7], the underlying geophysics of their linear structural formation is a problem in fluid dynamics amenable to solution by use of the CFD methodology- this methodology is used in the subsequent sections to present the evolution of coastal and interior valley landscape changes resulting from the ENSO events. Once the fluid dynamics connection between the El Niño flood events and beach ridge formation is established by CFD analysis, then, together with the fluid mechanics origin and dating of major beach ridge formations on the Peruvian coastline, their effects on the agricultural landscape and marine resource base provides information on the sustainability and continuity of the food-supply base of a society as related to the ENSO flood events. As the change in the landscape brought about by the ENSO flood events alters the agricultural field system base of a society, as well as causing damage to the marine resource base through the disturbance of offshore fisheries and shellfish gathering beds, societal continuity and sustainability can be adversely affected. Such ENSO events are subsequently shown to influence and affect the sustainability of Peruvian north central coast (Norte Chico) societies in the Preceramic, Late Archaic Period (2600–1800 BC) as further analysis reveals. Again, the main purpose of the CFD analysis is to show the fluid mechanics physics behind the landscape alteration that caused ancient coastal societies to ultimately collapse as their agricultural lands and food resource base were compromised by climate-related events.

While field system modifications and defensive technologies against flood events play a vital role in societal sustainability, in a worst case condition, flood damage can be irreversible, and the abandonment of pre-existing agricultural field systems occurs, leading to societal dispersal and termination. Changes derived from the ENSO flood and drought events affecting both the agricultural landscape and the ocean littoral affecting the marine resource base of a society are then key elements to understand and interpret societal structural change events. The chapters that follow provide examples of the use of the CFD methods to provide information as to the modification of the agricultural and marine resource base of the Late Archaic society based at Caral, centered in the Supe Valley of Peru, due to multiple major ENSO events—such events present a case for the ultimate collapse and abandonment of Supe Valley and other Norte Chico sites in the ~1800 BC time frame.

2. Evolution of Late Archaic Sites in the Peruvian Supe Valley

The Late Archaic Period north central coast sites in Peru witnessed increased El Niño ENSO flood events that transferred flood sediments from coastal valleys into ocean currents forming a series of extensive beach ridges. The coastal beach ridges containing C14 datable material are therefore key to date major flood events. Typical interior valley landscape sediment layers and beach ridges resulting from datable multiple deposition/erosion events confirm the timing of major ENSO events, as well as intermediate stable climate periods that permit societal continuity and development between destructive ENSO events. As a result of the formation of multiple barrier beach ridges formed from a sequence of the ENSO flood events, the geophysical history of coastal littoral zones reveals that the river drainage to ocean currents was impeded, resulting in bay infilling and the development of coastal marshes behind the beach ridges. This, together with aeolian sand deposits that infilled agricultural land behind and in front of the beach ridges, compromised the productivity of food supply from agricultural lands. In ancient times (and continuing to present times), aeolian sand-dune incursion into the Supe Valley from constant northwesterly winds from exposed beach flats in valleys south of the Supe Valley compromised Supe Valley agricultural lands. Evidence of this transfer process in Late Archaic times is evident from datable sand layers containing organic material noted in excavation test pits. Thus, beach ridge formations and interior valley sand and flood slurry debris deposition layers from a series of the ENSO events datable to the end of the Late Archaic Period in the Supe Valley provide the basis for agricultural land shrinkage and the changes in the marine resource base that played a major role in the collapse of coastal and valley societies in the Late Archaic Period, as revealed in the subsequent chapters.

Many sites within the Supe Valley, with its ceremonial center at Caral, were based upon the trade of marine resources from coastal sites exchanged with agricultural products from valley interior sites [8–16]. Figure 1 details the location of the major Archaic Period sites; Figure 2 details the existence time of major Supe Valley sites while Figure 3 provides the architectural details the Caral site. A probable, but no longer existing, inner city canal is implied from an excavated canal cross-section profile the existence of this early canal is subject of further research as Caral excavations proceed. ENSO landscape disturbances with no possibility of return to previous norms for the agricultural and marine resource of the Supe Valley society of 18 sites (Figure 4) pose a probable reason for the valley site's abandonment after ~1600 BC and indicate that the dynamic landscape change, as a result of the ENSO events, played a role in the collapse of Late Archaic Period sites in the Norte Chico region of Peru.

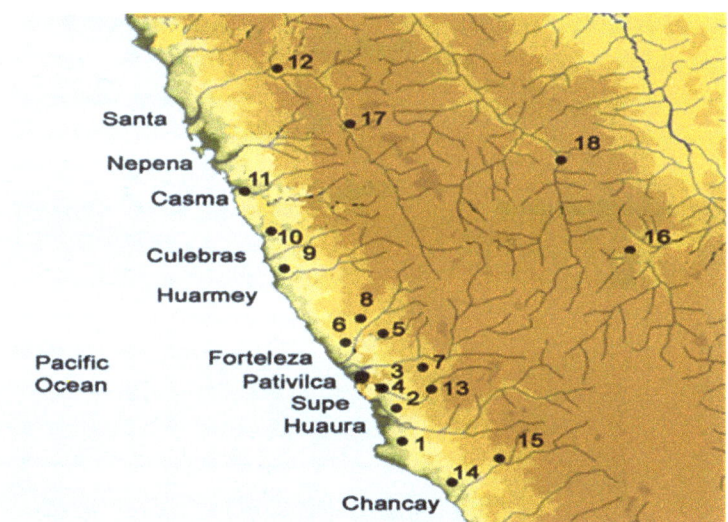

Figure 1. Site map of the coastal Norte Chico river valleys of Peru; the numbers represent major Archaic Period sites. Site 4 is Caral within the Supe Valley. Site 3 is Áspero. The North direction is in the vertical direction. The Chancay to Santa Valley distance is ~400 km.

The Norte Chico region of Peru is characterized by many Preceramic sites (Figure 1) with different existence dates (Figure 2). Within the Supe Valley are many neighboring individual sites to central Caral (Figures 3 and 4), characterized by complex social organization and urban centers with monumental architecture dominated by truncated, stone-faced pyramid structures of which Huaca Major is typical (Figure 5). The T–T and W–W date band notations (Figure 2) refer to a climate anomaly period [17] influencing worldwide oceanic current shifts with probable influence on the frequency and intensity of El Niño events. As these changes occurred towards the end of the Late Archaic Period, some effect on the study areas may be inferred, but further research is needed to track their specificity to Pacific coastal areas.

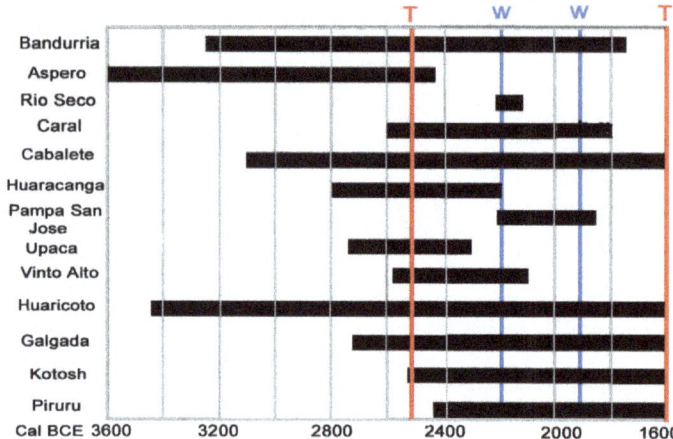

Figure 2. Time duration of the major preceramic sites in the Supe Valley shown in Figure 4 and Appendix A.

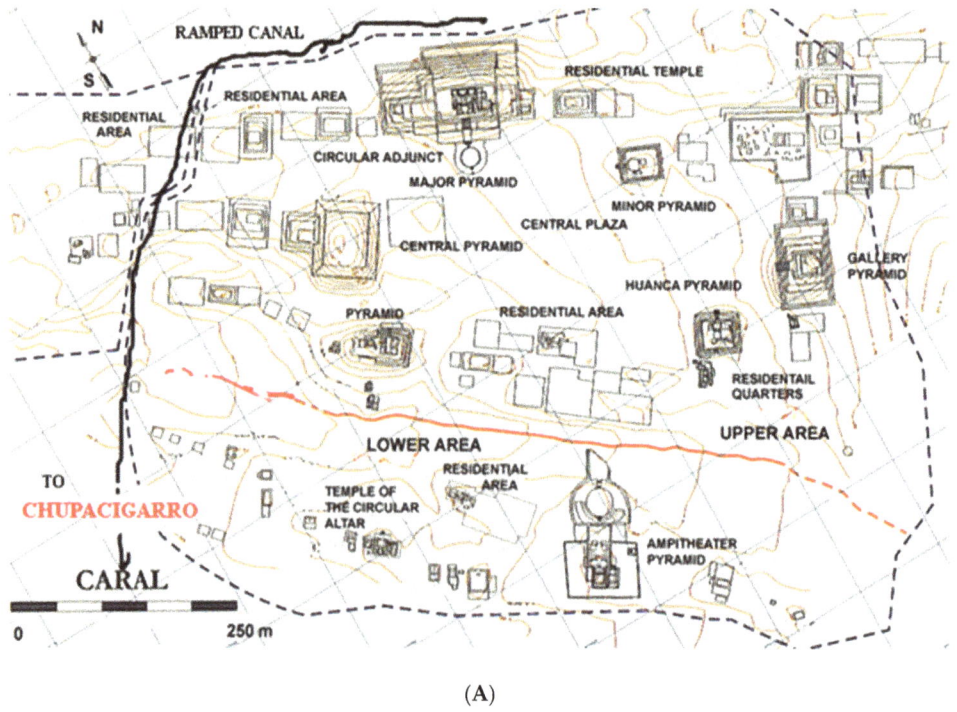

(**A**)

Figure 3. *Cont.*

(B)

Figure 3. (**A**) Caral site map. Note the proposed internal site canal (red line) in the region between the upper north and lower south areas based upon the canal profile data taken next to the Caral excavation house in the Residential Area. (**B**) Continuation Excavated canal cross-section profiles taken at the leftmost red-line canal extension Residential Area shown in Figure 3A thought to exist between the upper and lower regions of Caral.

Figure 4. Site locations in the Supe Valley along the Rio Supe (site names given in Appendix A). The site of Áspero is located on a coastal prominence west of Site 1. Length scale from Site 1 to site 18 is ~80 km.

Figure 5. Caral Major Pyramid (Huaca Mejor)—location presented in Figure 3.

The earliest Preceramic societies from the northcentral coast of Peru developed a cooperative economic model based on agricultural trade, irrigation agriculture, and the exploitation of marine resources to sustain large populations in the Late Archaic Period [8]. The nearby sites associated with Caral in the Supe Valley formed a collective integrated societal complex (Figure 4) participating in this exchange network. Coastal sites exploiting marine resources (fish, shellfish and edible seaweed types) traded with inland sites for agricultural and industrial crops, particularly cotton for fishing nets and lines, as well as gourds for net floats. In the Supe Valley alone, 18 sites (Figure 4) evidenced the success of this economic exchange system [8–16] over long time periods (Figure 2), which experienced, and successfully survived, changes in environmental conditions brought about by Holocene sea level stabilization, Peru Current establishment and the increased frequency of El Niño flood events [17–19]. Again, major flood events were the basis for sedimentary beach ridge deposits inducing river drainage blockage, the creation of coastal marshlands behind drainage barriers, valley water-table height elevation and changes and aggraded sand sea formation behind and in front of beach ridges subject to aeolian sand transport and deposition, all of which influenced the agricultural and marine resource base of valley and coastal sites. Despite these challenges to the food resource base, societal continuity prevailed through relocations of agricultural field systems from coastal to inner valley areas over long time periods; only when ENSO events reached a level of severity without options to continue the food supply base that was sufficient to supply an increasing population did the coastal societies experience a challenge to their continuance.

The beach ridge dates on the north central Peruvian coast and their locations are presented in Figure 6. Note that most of the earliest north central coast beach ridges appear in the ~2000–1600 BC date range, an important period for major site landscape changes, as discussed in the following sections.

Strong coastal winds were the source of aeolian sand dune transfer to interior Supe Valley farming areas from the exposed beach flat areas, as noted by datable sand layers from the excavation pits. The aeolian sand transfer to interior Supe Valley areas originated from the vast sand seas in the Huara Valley south of the Supe Valley. Here, strong northwesterly winds carried sand across the low mountains separating the two valleys, to be deposited on the southern slopes and interior valley margins of the Supe Valley (Figure 7); this effect continues to the present day. As the formation of a major beach ridge from a major ENSO event is noted in the Late Archaic Period record, this initiates a chain of events that threaten both the agricultural and marine resource bases of both coastal and interior valley sites.

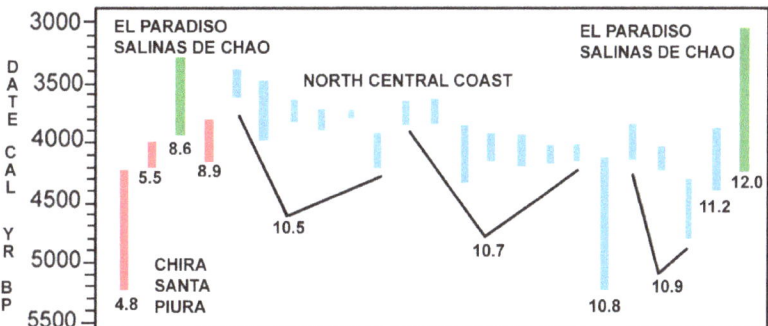

Figure 6. Beach ridge date ranges and locations along the coast of Peru. Band height ranges represent date ±1σ values from the mean value; descending (blue) lines indicate beach ridges in a specific area and their dates. Color bands reflect different information sources.

Figure 7. Aeolian sand deposits on the southern side of the Supe Valley produced from aeolian winds from the southern Huara Valley.

As flood-induced sediment transfer into ocean currents and the formation of beach ridge deposits are fluid mechanics phenomena, recourse to CFD FLOW-3D techniques [20] provide an insight into the sediment formation and deposition processes involved during major ENSO flood events.

To substantiate the CFD details of linear beach ridge formation, recourse to Google Earth satellite photographs of an actual linear beach ridge formation created after a recent major El Niño flood event in the Supe Valley and adjacent valley areas was apparent. The source of this new linear beach ridge on the north Peruvian coast within years after the large 1982 El Niño flood event then verifies that El Niño floods were the origin of a linear beach ridge formation over a relatively short time period. As beach ridge formation dates (Figure 6) are contemporary with large flood sequences occurring in the Late Archaic Period, the CFD analysis provides the rationale that fluid dynamics govern their formation and the geometry behind their linear shape. The fact that later multiple ENSO events occurring after an original beach ridge deposition event may influence dating results through deposition/removal of datable organic material is considered in Figure 6 to provide the mean and one sigma deviation band for an original deposition event. As a major El Niño event leads to a singular deposition ridge, the time sequence of sequential deposition events and the El Niño flood event that caused them can then be determined.

Coastal progradation stemming from sediment accumulation behind the barrier beach ridges is demonstrated by the C14 dating of different mollusk species known to occupy different shallow-water depths; when different mollusk species are found far inland, this indicates the coastline existing at that datable time (Daniel Sandweiss, personal communication). Figure 8 provides an example of shoreline littoral change over time related to the ENSO-induced events, as indicated by the dates associated with waypoint test pits.

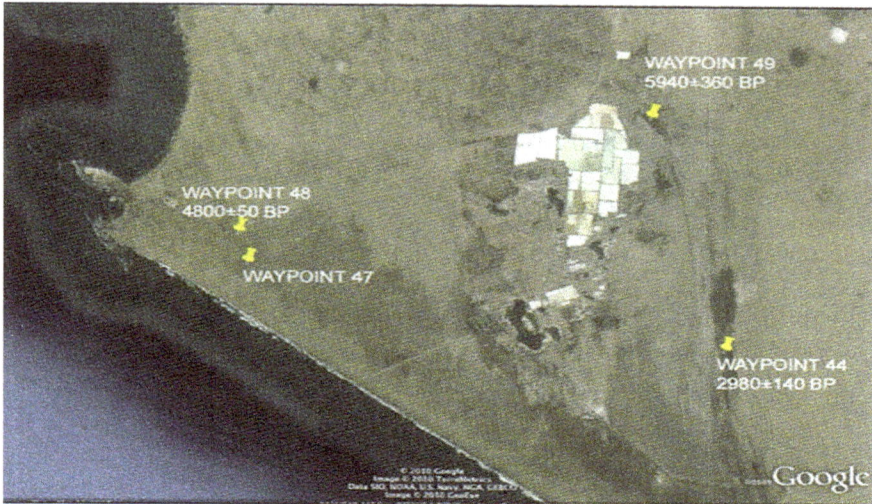

Figure 8. Shoreline changes over time in the area south of the Supe Valley. The dates measured from marine mollusk analysis indicate the variability of the shoreline shape due to ENSO sediment deposition/scouring events over time. (Figure courtesy of Daniel Sandweiss, personal communication).

The geophysical landscape changes affecting the agricultural and marine resource base of Late Archaic societies played a catalytic role in the fate of many Late Archaic Period sites. While research continues into the social, political and economic structure of Late Archaic Period sites and their response to climate change environmental stress to determine the details of societal structure modification [21–28], the present discussion focuses on the underlying fluid dynamics of beach ridge formation and the consequences of induced geophysical changes by flood and aeolian sand transfer events that affected the agricultural and marine resource base of Late Archaic Period sites.

3. The Supe Valley Caral Site

From previous research studies [29–36], it was recognized that complex societies based on irrigation agriculture and marine resource collection arose during the Late Archaic Period on the desert coast of north central Peru (Figures 1 and 4). Labor groups built monumental structures of increasing size and complexity indicating the development of societal structural change with the evolution of a governing managerial class to direct construction projects involving consensual communal labor participation and organization. The earliest Norte Chico region platform constructions contained restricted access rooms [32], indicating some degree of social differentiation [21]. Recent research by Shady and associates [8–15] in the Supe Valley [30–33] demonstrated that this early cultural florescence took place in other north central coast valleys and grew to a size and complexity not previously recognized by earlier researchers. Recent research indicates that Late Archaic Period temples of the north central coast were abandoned by ~1800 cal BC [9,13–15,32–34,36,37], as indicated in Figure 2. Past this period, the Norte Chico region was never again a center for cultural florescence, although a small number of Formative and Initial Period agricultural sites in mid-Peruvian valleys temporarily reoccupied a few of the previously abandoned sites [36]. Several sites at the margins of the northcentral coast originated towards the end of the Late Archaic Period, with sites at El Paraíso in the Chillón Valley to the south and Las Salinas de Chao to the north, and survived hundreds of years well into the Formative/Initial Period as preceramic sites.

For the present analysis, the discussion is focused on Caral in the Supe Valley. The site is located 182 km north of Lima and 24 km inland from the city of Haucho on the central Peruvian coast (Figure 1). Figure 3 indicates major building complexes at Caral. A foundation element for the concentration of Supe Valley sites was an abundance of water for agriculture to support the valley population. Since coastal rainfall is limited to a few centimeters per year, Supe Valley water for agriculture was mainly supplied by springs originating from valley bottomland areas sourced by seepage water transferred from Sierra lakes and man-made reservoirs through valley geologic faults augmented by aquifer seepage from the Supe River; such systems are designated *amunas* (sierra runoff capture-pits and reservoirs to augment the valley groundwater supply). An additional water source amplification of the valley groundwater level originated from canalized lagoons and water reservoirs that formed in low-valley bottomland areas that penetrated the groundwater level (Figures 9–11). These reservoirs, created from penetration of valley depression areas, had irrigation canals to lower valley areas to permit multiple-cropping to sustain valley population increases. As the near-surface water-table surface varies about one meter from the wet to dry seasons, many springs were canalized to irrigate specific bottomland agricultural areas devoted to specialty crops, including varieties of beans, squash and maize types, as well as many fruit varieties and industrial crops, such as cotton and gourds.

Figure 9. Typical interior Supe Valley pool derived from low-valley areas intersecting the high water table.

Figure 10. Typical interior Supe Valley pool derived from low-valley areas intersecting the high water table.

Figure 11. Major reservoir interior to the Supe Valley derived from groundwater penetration into a low-lying area.

As the water-table height was sustained close to a permanent level, valley-bottom field systems (Figure 12) permitted multi-cropping to occur throughout the year. Other archaic Norte Chico valley sites (in the Nepeña Valley in particular) were likewise associated with functional springs and large dams traversing upland valley gullies to trap rainfall runoff water to provide off-season irrigation water for crops supplied only by canals emanating from intermittent river water sources. The typical reservoirs shown in Figures 9–11 are of ancient origin and are still in use today to support extensive valley agriculture.

Figure 12. Supe Valley bottomlands below the southern elevated plateau that situates Caral and other sites shown in Figure 3. The Supe River separates current northern and southern agricultural fields.

4. Early Canal Development in the Supe Valley

Figure 12 indicates the current Supe Valley bottomlands and irrigated areas served by high water-table agriculture and canals originating from springs and reservoirs located on valley bottomlands. The Ramped Canal of the Late Archaic Period age was constructed on the upper plateau sidewall (Figures 3, 12 and 13) that transported water from a Supe River inlet (Figure 14) to the western part of Caral and further on to the inland site of Chupacigarro on a channel built on top of a ramped mounded aqueduct structure. The careful surveying associated with the ramped portion of the canal and its continuance over many kilometers to the Chupacigarro site is notable, given its estimated early provenience at ~2500 BC. This canal system is one of the oldest aqueduct canals yet discovered in Peru and is notable for its length and low channel declination angle. Figure 15 indicates the totality of canal systems in the Caral plateau area.

Figure 15 indicates that the Ramped Canal extension was the water source for the canals on the Caral plateau as well as the water source for the older Chupacigarro site (Figure 4). Figure 16 shows the Ramped Canal path leading to the inland Chupacigarro site now largely buried by drifting sand. The Figure 3 Continuation shows the canal cross-section profiles located near the Supe Valley project excavation house Residential Area; although the source of water for the different canal cross sections remains to be determined by further excavation, it is likely that the water source is from a branch canal from a Ramped Canal extension that supplied an early transverse 'red line' canal running laterally across Caral as indicated in Figure 3. Presently, a large erosion gully divides the north upper and south lower areas of Caral so traces of an earlier transverse canal segment are no longer present to extend the data obtained from the cross-sectional profiles shown in Figure 3 Continuation. Figure 16 shows the remains of the Ramped Canal extension to the Chupacigarro site now buried in sand; traces of its path are evident from a filled earth upstream aqueduct supporting the canal that have been recently excavated. The Supe Valley had (and still has) the advantage of a continuous water supply to source agriculture throughout the year, while adjacent Late Archaic Forteleza, Pativilca and Huara Valley sites only had access to intermittent rainy season runoff for canal irrigation.

Figure 13. The plateau edge above the steep embankment leading from valley bottomlands to the plateau. Along the embankment (now hidden by brush) is the Ramped Canal indicated in Figure 3, leading to the early Chupacigarro interior valley site indicated in Figure 15.

Figure 14. Interior channel from upper reaches of the Supe River leading river water to valley bottom agricultural and reservoir areas.

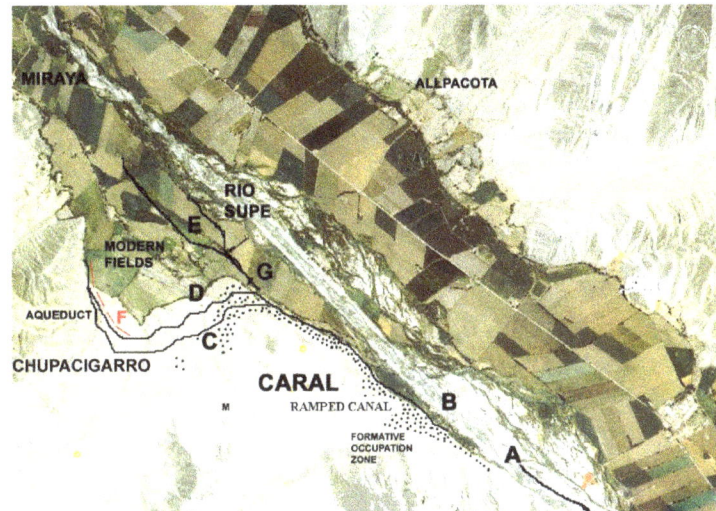

Figure 15. Path of the Ramped Canal to the Chupacigarro site. Current agricultural (brown and black) areas extend to both sides of the Supe River. Figure 14 and the current figure show the canal inlet (A) located in the far upstream reaches of the Supe River. The notation B denotes Early Formative Period sites; C and D canals supplied the early site of Chupacigarro located at (red) F. The E and G locations are modern in-use bottomland fields irrigated by high groundwater levels.

Figure 16. Trace of the canal leading to the Chupacigarro site.

5. Late Archaic Period Climate Change Evolution

Excavation data from Caral [8–15] indicates that marine products transferred from coastal sites were plentiful at interior valley sites, as evidenced by large marine shell and fish bone deposits at locations within Caral. Gourds and cotton grown at interior Supe Valley sites were traded and used for fishing nets and lines at coastal sites indicating that cooperative trade underwrote the valley's economic base. Ancient Supe Valley farming products within coastal valleys included guayaba (*psidium guajava*), pacae (*inga feuillei*), achira (*canna edulis*) as well as avocado, beans, squash, sweet potato, maize varieties and peanuts, attesting to the wide variety of comestibles available for coastal trade. The identification of agricultural products results from current seed-extraction analyses. The key to the importance of Caral and subsidiary Supe Valley sites is that they comprise the earliest New World example of an integrated valley economic unit deriving benefits from valley bottomland spring-sourced canalized irrigation systems, a plateau agricultural ramped canal system (Figure 15) and *amuna* groundwater-level water supply systems for agriculture. This, together with multiple stone-faced pyramid structures within a complex city architectural environment (Figure 3) originating in and past ~2500 BC, designates Caral the place in history as the first city of ancient Peru. The Late Archaic Supe Valley inland and coastal sites initiated and characterized by complex societal development would have developed further complexity, perhaps up to state level for the Norte Chico sites, were it not interrupted by large-scale ENSO environmental-change effects in later phases of its existence.

6. Sea-Level Stabilization, El Niño Floods and Beach Ridges

Sea-level stabilization between 6000 and 7000 cal BP set the stage for Late Archaic Period developments [5,19]. The onset of El Niño rains about 5800 cal BP after a mid-Holocene hiatus had implications for the social processes that found expression in the temple centers of Supe and the surrounding valleys [4,5,18,19]. By ~1500 BC, several preceramic north central coast sites were abandoned (Figure 2), suggesting a common influence of a large-scale environmental change affecting all areas of the Peruvian northcentral coast. Prior to ~1800 BC, the agricultural and marine resource base developed over time in a relatively stable climate period. Accretion processes influencing the geomorphology of the Peruvian north central coast littoral and inland valleys were determined by the deposition of large sediment loads originating from El Niño flood events interacting with oceanic and wave-induced near-shoreline northward flowing currents, together with major aeolian sand incursion events from beach flat areas and wind-borne sand from the southern Huanca Valley; these effects challenged the continuity of the Supe Valley sites. In the later phases of the Late Archaic Period sites, around ~1800 BC, significant changes in the geophysical environment were brought about by a major ENSO flood event and amplified aeolian sand transfer into the interior valley lands, this challenged the continuity of the valley agricultural and marine resource base. Sediment transport and offshore sediment deposition patterns into the Pacific offshore seabed depend upon El Niño flood magnitude, rainfall duration and spatial distribution, valley landscape geometry, landscape soil types, sediment and slurry physical properties, seabed shelf angle, coastal uplift/subsidence and geometric details of river channels and watershed collection areas. Flood sediment load is influenced by earthquake activity that produces large quantities of loose surface material available for runoff transport. Large rainfall events cause changes in the equilibrium profiles of drainages affecting sedimentation and drainage patterns that influence the amount of flood-transported sediment and the formation of offshore sediment deposits in the form of beach ridges. Sediment transfer processes result not only from El Niño flood events, but also from rivers that flood during rainy seasons from high Sierra rainfall runoff and carry sediment into ocean currents and/or deposit sediments behind beach ridges which served as barriers to river drainage into ocean currents. North of 9° S, the continental shelf abruptly widens, and extensive beach ridge plains exist, formed from sediment deposits that trail north from the largest rivers of Peru, which are the Santa, Piura, Chira, and Tumbes rivers. Beach ridges formed from flood sediment slurry deposits are present at Colán, where El Niño rains have eroded an uplifted marine terrace [3,37]. All the northern beach ridge plains originally consisted of eight to nine separate beach ridges, and each plain formed well after late Holocene sea-level stabilization. The beach ridge sequence indicates that the furthest ridge away from the current shoreline formed from the earliest flood event that also altered the coastline shape before the beach ridge. A later flood event acting on the earlier altered coastline littoral, deposited a further beach ridge closer to the shoreline with a further slurry deposit addition ahead of the latest ridge to alter the shoreline littoral. Now there is influence of the later flood event and beach ridge formation on the earlier beach ridge due to erosion and deposition activity to further bury it with later slurry and sand deposits. As later flood events occur, a progression of beach ridges forms together with deposition increments to add to the shoreline. Beach ridge sequences occurring over time continue to modify and add to coastline extension. This process, when subject to CFD analysis in a later chapter, indicates that this sequence of events can be recreated through fluid dynamics analysis.

South of latitude 9° S, where the continental shelf is narrower and the seabed angle steeper [35], Figure 6 shows dated beach ridge formations from a 4.8° to 12.0° south latitude on coastal Peru. The earliest beach ridge dates are from Chira, Piura and Santa coastlines, while the remaining dates are for beach ridges at Salinas de Huacho and El Paraíso, and are shown with their latitude positions in Figure 6. The earliest beach ridges appear at far southern latitudes at 4.8° S, with later ridges occurring to the north. This trend implies that early El Niño activity as a source of deposition material was prevalent after ~5500 BC (but not earlier), consistent with late Holocene sea-level stabilization.

As the shoreline transgressed during post-glacial sea-level rise, prograding beach ridge plains could not form. Subsequent sediment deposits from aeolian sand transfer, intermittent flood and oceanic current sources and silt entrained in farming drainage runoff over millennia promoted bay infilling to constitute the present day shoreline. Only with the relatively stable sea level of the last 6000 years could beach ridge plains form and alter landscape geomorphology and littoral resource suites along the coast. Sea-level stabilization is linked to climate change in the Pacific Basin, with El Niño events starting after a hiatus of about three millennia [5,19], during which northern Peru was characterized by annual warming and a fishery composition absent of small schooling fish, such as anchovy and sardine. With ecological changes accompanying the northward extension of the Peru Current at the start of the Late Archaic Period, small fish species began to dominate the fishery, calling for different capture strategies that required intensive production of cotton nets, fishing lines and gourd floats consistent with cotton dominant among domesticated plant assemblages in all the Late Archaic coastal centers [1,9–12]. The founding dates of Áspero, Vichama and Bandurria on the shoreline of north central coast valleys preceded dates of the inland Late Preceramic temple centers [32], suggesting that access to irrigated agricultural lands was linked to intensified production of cotton and gourds and was key to the integrated economic model existing in the north central coast area.

7. Geophysical Origins of Beach Ridge Formation

Various types of beach ridges observed at different locations along the Peruvian coastline (Figures 17 and 18) result from the complex interaction of river-borne El Niño flood sediments with oceanic and wave action currents to produce beach ridge sequences. Figures 19 and 20 illustrate inland deposition layers originating from the same events.

Post-sea-level stabilization, sediments and aeolian sand transfer began to accumulate west of the Quaternary sea cliff that marks the back of the original Supe Bay and the smaller Albufero and Medio Mundo inlets to the south and the Paraiso Bay, south of Huacho. In time, narrow beach ridges developed from a series of ENSO events inducing sand accumulation behind each ridge; this period was followed by stable progradation that buried each minor ridge by aeolian sand transfer and dune formation. In about ~1800 BC, a major ENSO event created the large Medio Mundo beach ridge along ~114 km of coastline sealing off former fishing and shellfish gathering bays. This event created large scale sand flats that accumulated behind beach ridges and promoted the large-scale aeolian sand dune inundation of coastal plains and inland valley areas compromising a significant part of the agricultural base of the Supe Valley society. The coastline geomorphic change affecting the marine resource base of Norte Chico societies was affected by a combination of flood sediment accumulations amplified by aeolian sand transfer processes that infilled previously established fishing and shellfish gathering areas. Although river-borne sediment constitutes a major source of slurry transport during flood events, additional opportunistic drainage paths originate from areas between river valleys to provide addition sediments to ocean currents. Again, many north central coast preceramic sites terminated occupation by ~1800 BC, suggesting that a major geophysical change over a wide coastal area compromised their economic base, thus, the importance of understanding the geophysics of beach ridge formation and its consequences on the agricultural base of archaic societies. The abandonment of major preceramic sites, with limited Formative Period reoccupation of a limited number of former sites, motivates the discussion to follow as to what these changes were to the economic base of archaic societies and their relation to beach ridge formation.

Figure 17. A linear beach ridge deposit along the Peru coastline resulting from an El Niño flood event. Note the marsh region behind the ridge resulting from the ridge blocking rainfall, river *amuna* discharge. The origin of such deposition events in history (Figure 6) is detailed in the next section.

Figure 18. A further early coastal ancient linear beach ridge, now partially covered by aeolian sand drifts; further stranded beach ridges are to be found closer to the shoreline resulting from later El Niño flood events. Ongoing tectonic uplift helps to strand and separate a sequence of beach ridges.

Figure 19. Deposition silt layers at a coastal Supe Valley site resulting from sequential El Niño flood events.

Figure 20. Interior Supe Valley silt and sand deposition layers from sequential El Niño flood events and aeolian sand transport.

Tectonic coastal uplift rate differences from north to south latitudes influence beach ridge typology. A significant uplift rate may strand and separate sequent ridges from individual flood events, while, in the absence of uplift, single subsea ridges form and sediment deposits accreting landward from the ridge appear to strand a ridge above the land/water interface. Additionally, offshore undersea ridge formation may alter the deposition history of subsequent later ridges by altering seabed shape, river mouth geometry and river positional shifts. Later flood events may also erode and erase previously deposited ridges, and aeolian sand incursion may bury segments of earlier ridges; all these factors, operational over millennia, influence the remains of ridges. The relation between flood events, beach ridge formation and agricultural landscape change, as demonstrated by CFD computer-modeling results, are subsequently detailed to illustrate their connectivity to provide the basis for conclusions supporting the observed geophysical transformations of Supe Valley agricultural and littoral areas in Late Archaic times.

To illustrate the fluid mechanics basis of ridge formation, a three-dimensional computer model of the Peruvian coast from Santa to Viru Valleys was created (Figure 21) using FLOW-3D software [20]. The minor intermediate Chao Valley drainage between the Santa and Viru Valleys was omitted from the CFD model, due to its minor effect on the beach ridge formation compared to the major Santa and Viru river drainages. The intent of the CFD calculations are to show the fluid mechanics transformation of flood slurry movement into ocean currents during an El Niño event that can produce an extensive linear beach ridge; this linear beach ridge formation is unexpected and, intuitively, one may not associate a chaotic flood event and chaotic slurry transport into ocean currents as the source producing a linear beach ridge structure.

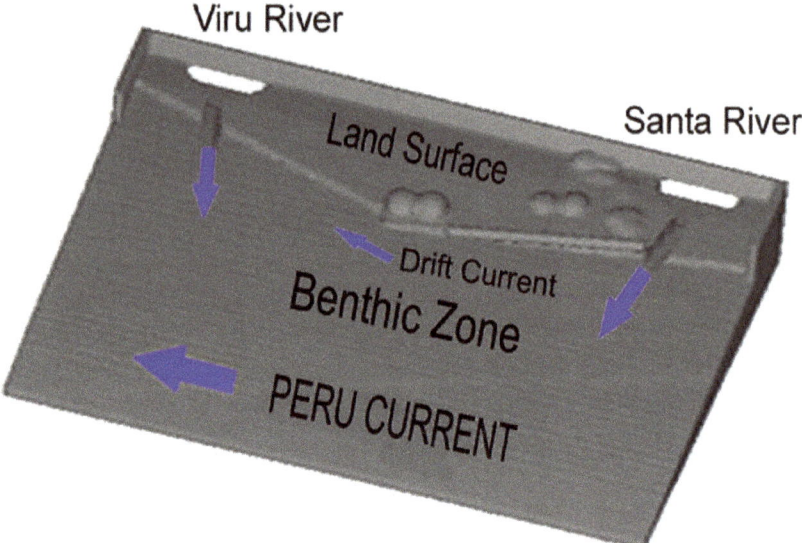

Figure 21. FLOW-3D CFD model of the Peruvian coastline from the Santa to Viru Rivers. The Benthic Zone is the Pacific Ocean; the arrows originating from the Santa and Viru River Valleys represent El Niño flood sediment transfer paths into the Pacific Ocean and offshore arrows denote the direction of the northward Peru Current and the shoreline drift current.

The use of a CFD model is made to demonstrate how fluid dynamics can duplicate observed geophysical events that occur during a major El Niño flood event. Details from the CFD analysis can uncover the geophysics behind field observations and help to understand nature's role in the formation of deposition and erosive landscape structures observed in field studies. The CFD model area in Figure 21 was selected due the availability of data [4,6,7] in the Santa-Viru coastal zone that allows for the qualitative verification of modeling results compared to field observations. It is expected that the CFD results obtained on beach ridge formation in the Santa–Viru coastal area are qualitatively similar to the ridge development in the nearby Supe Valley coastal area, given that the same fluid mechanics-based geophysics applies. Shown in the Figure 21 CFD model are the sloped offshore seabed and coastal land areas, with southern Santa and northern Viru River Valleys providing known data as to beach ridge formation from an El Niño flood event where sediment conduits to the ocean current are the source of beach ridge formation.

During a major flood event, slurry material from the eroding landscape is transported into local rivers and conducted into ocean currents. The interactive geophysics of the slurry interaction with ocean currents and the settling of the slurry to form a beach ridge is then predicted by CFD analysis. Inherent to the CFD analysis are the details of the formation of landscape change from erosion/deposition events as well as beach ridge formation processes; while field observations only record the results of flood events, CFD analysis provides the fluid dynamics mechanisms behind their creation.

The composition of flood-transported slurry material is drawn from the size, gradation, cohesivity and stratification of erodible bank sediments over the coast-to-mountain area watershed and surface material washed into the streambed. Profiling a selection of beach ridge cores provided an indication of the percentage by weight of different-sized sediment particles. High concentrations of large-sized particles (gravel and small boulders) in the wash load damp slurry turbulence, increase the apparent viscosity of the slurry flow and reduce their settling slurry velocity, enabling the early settlement of coarser grains and a larger bed-material load compared to finer grade slurry material. While only river-borne sediment transfer is considered for CFD model purposes, additional opportunistic drainage channels and ephemeral streams develop during flood events, leading to sediment transfer to lower coastal areas that contribute additional sediment to ridge formation/accretion processes and landscape change.

A two-fluid CFD model represents slurry sediment–ocean mixing interaction. Fluid 1 is ocean water characterized by kinematic viscosity v_1 and density ρ_1; Fluid 2 is sediment-laden flood water slurry characterized by high values of kinematic viscosity v_2 and density ρ_2 compared to ocean water. Here, $v = \mu/\rho$, where μ is the absolute viscosity. Depending on the sediment load of the floodwater, v_2/v_1 can range from 1 (no river-borne flood sediment) to 10^3 (heavy river-borne sediment loads with very high absolute viscosity).

For purposes of demonstrating the fluid dynamics phenomena involved, a selection of input properties typical of observed sediment composition is presented. Due to local variations in sediment composition, only a generic, illustrative slurry composition is presented that is typical of the conditions in the area of study. Model river-current velocity is set to a value to induce riverbed erosion and sediment transport mobility. For silt (<0.001 mm diameter, ~15% by weight), sand (0.1 mm diameter, ~30% by weight), gravel (0.1 to 5 mm diameter, ~35% by weight) and an assortment of rock-particle sizes (5 mm to 100 mm diameter, ~20% by weight) composing the sediment solids, estimates of both properties and critical mobility stress levels are provided by [2,38–42] to substantiate the typical v_2/v_1 values used in the CFD simulation. The northward offshore current velocity is set low to represent near cessation during El Niño events. A near-shoreline, northward drift current is induced from the difference between the incoming wave vector angle and a normal vector to the shoreline. The model sea level is set to the stabilized ~5000 BC level. By using model length scales (model area is ~1035 km^2) and slurry velocity ranges approximating actual values, Reynolds and Froude numbers are duplicated, and the computer time is equal to the real beach ridge formation time. The results provide a lower-bound time to determine the flood duration required to deposit beach ridges of known sizes and volumes. Note that the early Quaternary version of the shoreline consisted of deep river-downcut bays consistent with the lower sea level, and that wave-cut bluffs (now inland from the present-day accreted shoreline) resulted from rising sea levels, the subsequent deposition of sediment over millennia is from floods, river, and aeolian sand transport that served to infill this landscape. The present computer model is qualitatively representative of an intermediate stage in this landscape transition process during a flood event. The results of the CFD computations then represent the early stage of beach ridge formation: subsequent millennia of accretion and erosive effects then serve to represent current day observed shoreline patterns.

The results of Santa–Viru coastal zone simulation are subsequently summarized. This zone has a well-documented ridge sequence [4,6,7] and is used to test computer predictions with observed geomorphic features that have survived from early creation stages to present-day stages. Figure 22 shows the initial offshore sediment deposition density distribution from a single, large El Niño flood pulse concentrated in the Santa–Viru Valley area; the density scale ranges from 1.94 slugs/ft^3 (1000 kg/m^3) for seawater to 5.40 slugs/ft^3 (2783 kg/m^3) for heavy flood slurry. For a lower-limit slurry grain size of 0.01 mm, erosion, transport and entrainment are maintained from 0.001 to >10 m/s velocity; for a grain size of 10–100 mm, erosion, transport and entrainment are maintained for >1.0 m/s velocity [2,42]. On this basis, a river near-surface velocity is assumed to be ~1.0 m/s, for purposes of demonstrating the sediment–ocean current mixing and deposition process by CFD simulation. Since slurries of the type encountered in flood debris are highly non-Newtonian power law fluids, shear thickening with an increasing shear rate is expected. Here, a higher apparent viscosity applies, and the kinematic viscosity proposed is used for purposes of the demonstration problem. An observation made of highly viscous slurry motion in river valleys during the catastrophic El Niño flood event in 1982 during my stay on the Peruvian north coast gives credence to the slow slurry velocity used in CFD calculations. Although the velocity profile for Newtonian, viscous channel flow can vary from that determined for non-Newtonian slurries [40], for the present demonstration problem, the slurry is assumed to have a constant absolute viscosity, rather than a shear rate dependent value. The offshore Peru Current velocity is approximated to be ~5 cm/s and the coastal drift current is ~3–5 cm/s, with local drift velocity values computed based upon the geometry of the coastline (Daniel Sandweiss, personal communication). The elapsed time is ~35 hour of continuous El Niño flood activity, with $\nu_2/\nu_1 = 10^3$ indicating a high absolute viscosity, heavy sediment load carried by the flood currents. The offshore average seabed slope is ~0.15° from horizontal for the Santa–Viru north central coast area; the offshore seabed slopes after 3800 BC are estimated from Barrera [35] Figure 11 and Pulgar Vidal [40], as well as the values for oceanic current velocity.

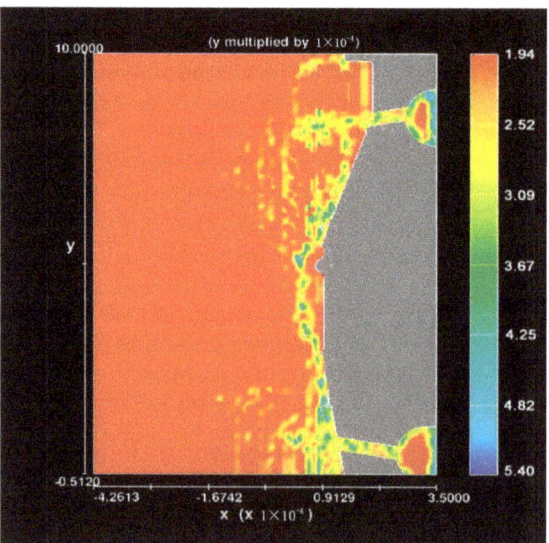

Figure 22. Offshore deposit of sediments. The right hand scale represents sediment density in slugs/ft^3 as described in the text; ocean water has the 1.94 slug/ft^3 scale representation. Intermediate scale values represent the mixing of flood derived slurry with the ocean current.

Figure 22 indicates the formation of a long sediment deposit beach ridge on the coastline; the scale shades represent the mixture density of seawater and the initial flood slurry density emanating from the rivers. Notable is the early deposit of (green) heavy materials and the further offshore deposit of lighter (yellow-orange) slurry component materials. The northward current, together with the near-shore drift current, enhances the northward deposition of the sediment. The predicted high density, larger size sediment compositions appear along the ridge length, as observed from field studies [4], while the transport and sorting of sediment fines continues in time from wave-induced drift currents. For steeper seabed angles, the formation of a close-in ridge deposit to the shoreline occurs; here, a steeper seabed angle results in a greater sea depth closer to the shoreline and, as resistance to a sediment particle's forward motion is related to the hydrostatic pressure encountered on its front projected area (in addition to viscous drag and dynamic pressure effects), sediment deposition occurs more rapidly for steeper benthic seabed angles.

A velocity vector plot (Figure 23) shows that the out-rushing sediment stream creates a flow reversal pattern when encountering ocean and shoreline currents, leading to deposits of lighter sediments back towards the shoreline where low drift velocities prevail. The 'U-turn' of the sediment stream is consistent with the path of least resistance of small and intermediate sizes, low-inertia sediment particles that alter the direction away from the increased hydrostatic pressure resistance encountered further from the shoreline. As offshore Peru and drift current directions are not aligned, the agitation and transport of the deposited fines cause a gradual northward ridge extension over time. Of interest is the disturbance of the ocean current both near and far offshore by the river borne slurry injection as well as currents that show northward flows from the river mouth that influence sediment deposition far north of the river mouth. The CFD calculation results provide a qualitative relation between an El Niño flood event and details of ridge formation: floods with large sediment loads can produce extensive near-shoreline subsea deposits whose size depends upon flood duration and amount/type of sediment available for transport by ocean and drift currents.

For Figure 22, the density of the slurry stream is 5.2 times that of ocean water. Many beach ridges observed in the Santa–Viru sequence are composite, indicating large accumulations from multiple, closely-spaced-in-time flood events; where aggraded sediment material separates ridges, a longer time interval had occurred between major flood events: the use of C14 dating on organic material within the ridges then present the dating of ENSO flood events. Beach ridges contain mollusk shell material indicative of seabed sediments being agitated and entrained during flood events. Once the ridges are stranded on land by surrounding aggraded material, later flood events create new subsea deposits that accrete material to create the ridge sequence noted in the Santa–Viru coastal area, where shallow seabed angles prevail (Figures 23 and 24 illustrates this occurrence). South of this zone, where steeper seabed angles prevail, a composite, a unitary ridge type is predicted. Situations occur in which loose surface sediment is minimal and/or the surface runoff water velocity too low to carry or erode sediment, so that ridge formation is minimal or absent during lower-magnitude flood events.

Figure 24 shows what happens when a later flood event occurs after an earlier beach ridge deposit. Trace sediment deposits occur seaward and inward from the earlier main ridge; this trace ridge alters the river outlet shape and influences river discharge patterns (Figures 22, 24 and 25), as well as altering the local seabed slope from the settled sediment, all of which alter conditions under which subsequent ridges form. In Figure 24, a trace subsea ridge (yellow-brown) was formed close to the shoreline behind a (grey) ridge formed from a prior flood event. The inner sediment deposit is the source of bay infilling and marsh creation from rainfall, river deposits and later flood events. The outer sediment deposit is the source of coastline shape alteration from sediment deposits. A sequence of distinct ridges form from later flood events that gradually accrete sediment between them to both form a ridge sequence typical of that observed in the Santa–Viru area together with alteration of the coastline shape. Based on the observations of the Santa–Viru ridge sequence formed from datable multiple flood events, computer predictions provide the underlying fluid mechanics mechanism to explain ridge formation and their observed orientation, shape, width and composite nature. From Figure 24, previous flood deposition events affect and influence both the inland and offshore deposition history of later flood events that contribute to shoreline growth and shape change as well as river mouth shape. This effect is manifest in Figures 17 and 18 which show inland beach ridges buried by sand aeolian sand deposits and many flood erosion events millennia after their creation.

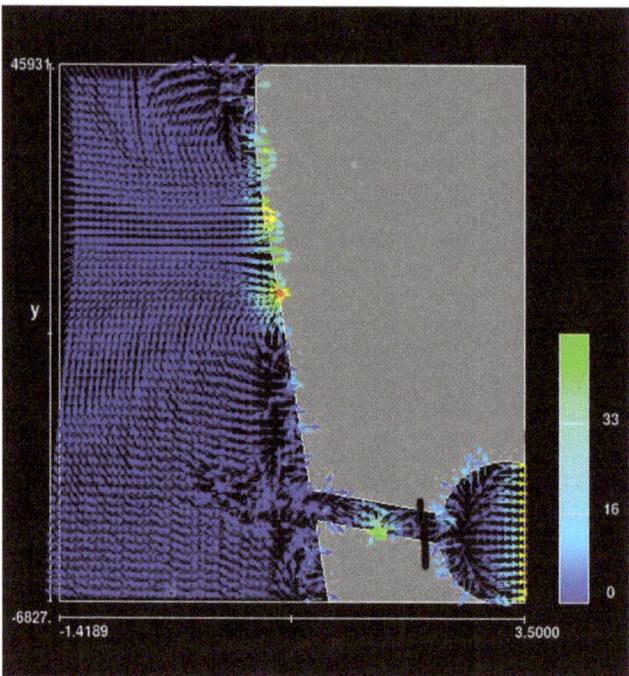

Figure 23. Velocity vector plot indicating turbulent flow at the Supe River outlet indicating the source of differential particle size sediment settling. Velocity scale is in ft/s; length scale is in km.

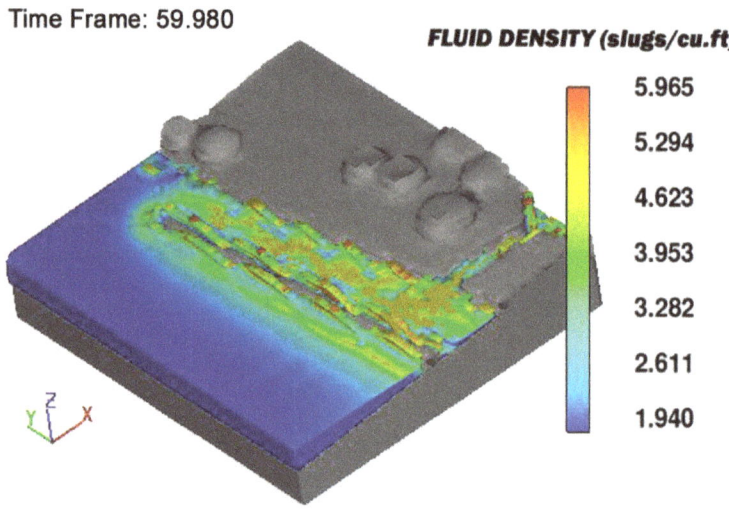

Figure 24. CFD calculated sediment deposition history of a later El Niño flood event interacting with the sediment deposit from an earlier flood deposit event (offshore grey bar ridge area). Note the formation of a water-laden backfilled marsh area between the shoreline and the beach ridge.

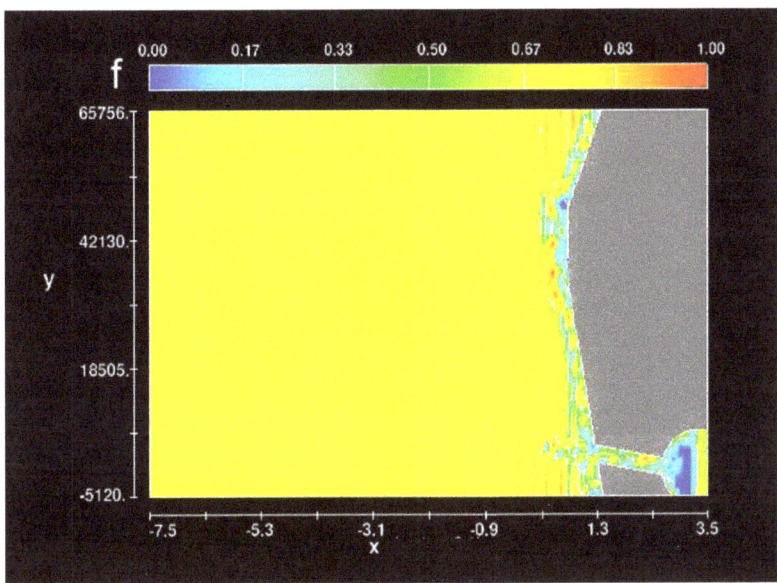

Figure 25. Flood slurry deposition: f = 1 is the ocean water, f = 0 is the flood slurry. Intermediate f values represent the slurry/ocean water mixture. Note the formation of a linear offshore beach ridge from the slurry deposition close to the shoreline as dependent on the ocean bottom slope. Length scale in km.

Figure 25 represents a case in which the seabed slope in the Supe Valley area from the Huara to Forteleza Valleys is steeper than that for the Santa–Viru calculations shown in Figure 22. Figure 26 how a flood deposition event alters the coastline shape deposition event—this is the fluid mechanics version that underlies coastal shape changes shown in Figure 8. The results indicate ridge deposition closer to the coastline, owing to the higher hydrostatic pressure close to the shoreline encountered by low-mass, low-inertia sediment particles; here higher-pressure drag resistance leads to rapid particle settling closer to the shoreline. Subsequent flood events combine to produce a unitary, composite, multi-layered ridge, which is continually reworked and redistributed as currents shift deposits and extend the deposition length.

It is noted that, although many of the above results shown apply to the Santa to Viru Valley coastal area, the results show that large beach ridges well over 50 miles in length can be generated by major El Niño flood events that involve large portions of the Peru coastline over the length scale of the CFD model. The Santa to Viru Valley distance of ~100 km is on the order of the Pativilca–Supe–Huara Valley distance and, typically, major El Niño flood events cover very large portions of the Peru coastline. CFD results applied to similar coastline areas can generate a lengthy beach ridge during a major El Niño flood event, as such events share the same hydrological physics.

Figure 27 is a Google satellite view of the present Supe Valley coastline, indicating a portion of the Medio Mundo beach ridge that created a marsh area from the Supe River blockage. The CFD model counterpart is Figure 24 that indicates a similar marsh area behind an established beach ridge.

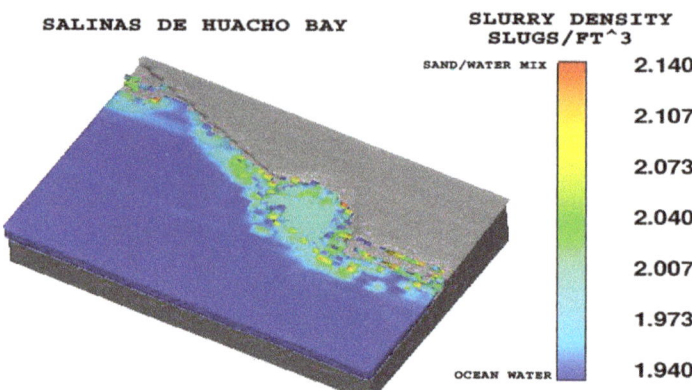

Figure 26. Later El Niño flood events creating a deposition sediment area behind an established beach ridge; the infilling phenomenon infills bays and creates marsh areas behind previously established beach ridges as well as extensions of previous shorelines, as the Figure 26 CFD model demonstrates.

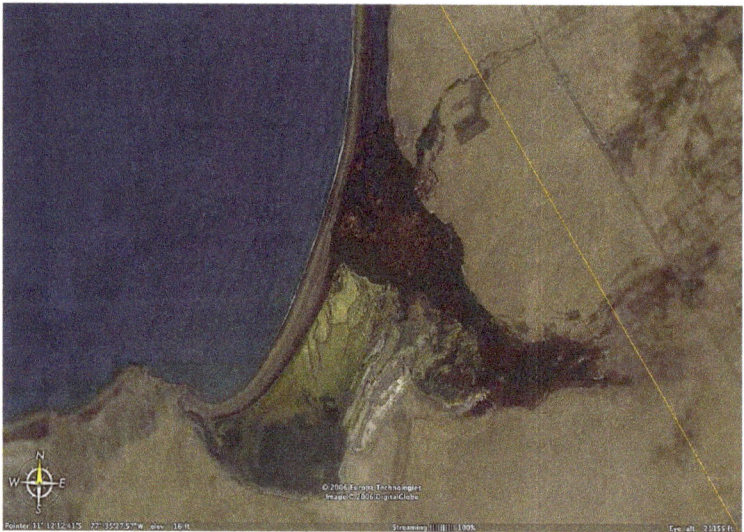

Figure 27. Google Earth satellite photograph of a part of the Medio Mundo beach ridge and infilled marsh area behind the ridge in the Supe Valley coastal area that limits river flows to the ocean. A ridge of this magnitude results from a catastrophic El Niño event (or closely spaced events) and subsequently alters the deposition placement of subsequent El Niño-derived beach ridge placements in different sections of the Peru coastline.

The Figure 26 CFD result represents a case in which an existing offshore submerged sediment ridge allows sediment infilling of a previously existing bay area from a flood event. For this case, the previously existing marine resource base consisting of shellfish and sardine and anchovy net gathering no longer exists, forcing fish gathering at greater offshore distances to sustain the protein food-supply base. As this food-supply change takes time to develop an equal marine base resource to previous values, a major flood event has immediate consequences to sustain the large population of inland sites. Once the beach ridge deposits are in place, the accretion of sediment from later flood, river and aeolian sand transport events leads to the gradual infilling of the shoreline littoral.

That the aeolian sand transfer from the southern Huara Valley was also a continual threat to Caral's city environment was demonstrated by the multiple stone wall capture sand barriers constructed in open inland areas south of Caral (Figures 28–30).

Figure 28. A further stone wall aeolian sand transfer barrier placed at the Ostra base camp to the north of the Santa Valley; such defenses were typical of preceramic societies in the north central coast.

Figure 29. Further stone wall aeolian sand barrier closest to Caral city limits.

CFD-generated Figures 24 and 26 detail the effects of a later flood event that superimposed sediment deposits upon a previously established subsea ridge with results specific to shallow seabed angles. Shown in Figure 25 are the fluid fraction results: f = 1.0 represents ocean water, f = 0.0 represents the sediment slurry stream and intermediate f values indicate slurry–ocean water mixture states. Sediment deposits on top of, and on each side of, the original subsea ridge provide the barrier mechanism for accretion of inland sand and flood deposits. A trace ridge accumulates seaward of the main sediment ridge and affects subsequent ridge development as the sea bottom geometry has been altered together with the river discharge outlet geometry. This result, when repeated for multiple flood events, qualitatively demonstrates how sand and flood sediment accumulates inland and behind (and to a lesser degree, in front of) a beach ridge by multiple flood, river, canal drainage, and aeolian sand/fines transport. Such changes in the seabed geometry from multiple ENSO events alter the composition of fish species available, as well as the availability of shellfish types that can only exist at certain water depths.

Figure 30. Furthest stone wall aeolian sand barrier. The barrier sequence represents efforts to limit sand inundation into urban Caral.

The current research [36] demonstrates the time change in marine-resource dietary patterns derived from beach ridge formations originating from ENSO events occurring over a long time period on the Peruvian shoreline. Here, the cumulative effect of ridge formations proceeding seaward on the seabed from sequential ENSO events gradually infills bays (Figures 24, 26 and 27) and creates a changed marine environment, accommodating different fish species and shellfish types over time. For a major ENSO event, such as that which occurred in the Supe Valley area in the Late Archaic Period, adaptation to a changed marine environment, coupled with induced changes in the valley agricultural environment from coastal zones to limited inland valley bottomlands, eventually led to limitation in the areas in which general agriculture products and certain agricultural varieties could be grown. As each coastal valley had different soil types and landscape geometries, it is expected that the effect of a major ENSO flood event would have different effects on different valleys and affect local sources of food supply.

The composition of beach ridges depends on the soil composition of individual valleys; due to the northward ocean current carrying of dilute slurry materials, beach ridges occasionally contain a mixture of flood sediments from adjacent southern valleys. In certain cases, beach ridges subject to current analysis ~4600 years after their formation may have continuity gaps, due to millennia of landscape erosion and deposition events.

When a mega El Niño flood simultaneously affects multiple river valleys with a steep offshore seabed slope, sediment fields coalesce to form a large unitary ridge spanning the coastline typical of the ~100 km long Medio Mundo ridge observed to span the littoral of five north central coast river valleys. A result of prograding processes behind ridge barriers results in the formation of brackish water lagoons and marshes, such as those observed at the Supe Valley river mouth that represent the physical reality of the CFD prediction. Figure 27 shows a Medio Mundo beach ridge segment and the marsh area that, prior to a major ENSO event, was a large part the agricultural area for the Archaic Period coastal site of Áspero [35], whose existence from ~3600 to 2400 BC is noted in Figure 2. After ~2400 BC, new interior Supe Valley sites proliferated, as additional highland rainfall supplied *amuna* water transfer to the Supe Valley served to elevate the groundwater level and permit extensive interior valley-bottom agriculture [3,35]. As ~1600 BC was the start of major ENSO events [19,37], as noted in Figure 2, it is likely that a major flood event (or sequence of events) negatively influenced nearby coastal agricultural field systems.

To date, the continuous extent of the Medio Mundo beach ridge is not available due to millennia of erosion/deposition events compromising sections of the beach ridge; given its relation to the demise of Preceramic Archaic Period sites in the 1600 BC time period due to a major El Nino event (or events), its approximate formation date is on the order of 1600 BC. However, a more reliable estimate of its formation date can be made by examining the broader history of beach ridges along the northern coast of Peru (Figure 6). The Medio Mundo beach ridge could not have been deposited prior to sea-level stabilization, so it is younger than ~6000 cal BP. Given the ridge-forming processes identified in the region, El Niño floods were active for Medio Mundo to form; this provides a maximum limiting date of 5800 cal BP. Furthermore, rains associated with El Niño events are attenuated to the south. Given these formation date limits, the northernmost dated beach ridge plain (Chira and Colán) began forming ridges earlier than the ridge plains further south. The available dates place the origin these ridges between about 5000 and 5200 cal BP [37,38]. To the south, the earliest Piura date is around 4100 cal BP, and the earliest Santa ridge date is around 4000 cal BP [37,38]. Following this trend, the Medio Mundo ridge should date between ~3900 and 3700 cal BP. This time span overlaps the latest dates for most of the north central coast Late Archaic centers and exists in a time frame necessary to influence the marine resource base of Norte Chico societies.

A further CFD result relates to the Salinas de Huacho area in which vast aeolian sand accumulation infilled bays. During El Niño events occurring in the southern reaches of the north central coast, flood sediment was mainly composed of sand transferred from the Chancay and adjacent southern rivers valleys; a calculation of sand-rich sediment emanating from the Chancay River (slurry density is approximately ~2.34 slugs/ft^3 = 1206 kg/m^3) indicates sand transfer to the Salinas de Huacho area ~25 km north of the Chancay River. Multiple flood events would continue the infilling processes to create the observed vast beach flat area. Again, individual velocity vector patterns, according to the geophysical values (Figure 23), prevail to create different types of near-shore deposits.

The C14 dates of shallow-water mollusks that lived 0.5–1.0 m below shoreline sands indicate the 4000–4500 BC shoreline was about 3 to 4 km from the present-day shoreline (Figure 8), indicating that large scale sediment accretion continuously altered both the marine and inland farming area resource bases through numerous flood and sand transfer processes, similar to what Figure 26 predicts. Similar shoreline and inland infilling processes behind the massive Medio Mundo ridge characterize the north central coastal valleys bounded by that ridge.

A further example illustrates the ridge-formation process in the presence of the irregular coastline of the Sanu Peninsula, which forms the southern boundary of the bay on which Bandurria and Áspero are located (Figure 1). The far offshore Peru Current velocity is ~4 cm/s, while the shoreline drift currents are ~3 to 7 cm/s but vary northward due to coastal geometry effects. Figure 31 indicates that the coastal current caused the shifting and deposit of sediment, creating a curvilinear bay ridge and an inland marsh area, as sediment drainage was blocked by the ridge (Figure 26).

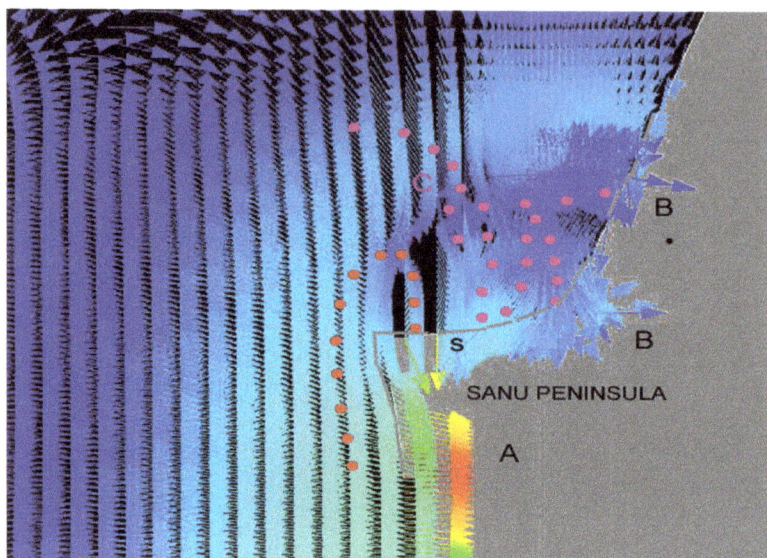

Figure 31. Dot trajectories of the sediment transport paths from the coastal valleys south of the Sanu Peninsula. North of the Sanu peninsula, extreme ocean turbulence and induced rotational flow in the upper concave shoreline region enhance particle settling, resulting in the shoreline shape modification.

Figure 31 shows the dot trajectories of light sediment particles from flood runoff entering the coastal current from southern reaches. Here the shape of a portion of the coastline has influence on the shape of its northward reaches due to the influence on the ocean current velocity distribution. The sediment trail loops around the (gray) Sanu promontory (near site 1 in Figure 1) to deposit sediments along the shoreline, adding to the expansion of the original pre-ENSO event shoreline area. The northward dot sequence shows vortical currents depositing sediment to form a further coastal shoreline extension. The CFD results indicate the qualitative nature of the prograding process; the quantitative determination involves detailed knowledge of the rainfall intensity and duration, geographic extent, event-time duration and the amount/type of surface material available for transport by the eroding action of floods. Sediment transport from the upper-valley areas from flood events, river and aeolian transport, as well as sand transport from southern valleys from ocean currents, accelerated coastal infilling both ahead of and behind the Medio Mundo beach ridge extending from the Huara to the Fortelaza Valley. Based on a survey of the lower Supe Valley, ~3 to 5 km of accreted sand, fines and clays deposited to a depth of 3 to 5 m over the Holocene beach littoral inland of the present shoreline since El Niño floods began at the time of sea-level stabilization. The sand seas in the Huara area, south of the Supe Valley, subject to strong onshore winds, sourced the aeolian transport of vast quantities of sand over the southernmost mountain chain that bounded the Supe Valley to inundate the inland valley farming areas. Sand accumulations appear on the north and south sides of the Supe Valley, as the small, intermittent discharge Supe River presented no barrier to across river valley aeolian sand transport. Buried sand layers covered over by later farming surfaces are present throughout the Supe Valley profiles (Figure 20), indicating continuous aeolian sand and flood sediment transfer events over millennia. As a consequence of the infilled bays and lagoon formation landward of the ridges, coastal lagoons dominated by reeds were created under brackish water conditions; this environment exists in lower valley plains landward of Áspero.

8. Changes in the Agricultural Landscape of the Supe Valley

Figures 23 and 25 show the beach ridge formation from a single large-scale El Niño event to produce the Medio Mundo ridge. This large-scale event sealed off the bay at the Supe Valley and the shallower Albufero and Medio Mundo inlets and the Pariso bay below Huacho and to the South, created the Salinas de Huacho sand flats. To the north, bays and inlets through Bermejo were closed off [1]. With this ridge in place, northward ocean currents narrowed the width of the ridge in time and deposited material westward of the ridge to form new beach areas. In time, another El Niño event followed, with river-borne material cutting through the earlier ridge and depositing new material further westward of the earlier ridge on the beach deposited by the prior El Niño deposit. This process leads to sequentially spaced ridges propagating westward into the ocean; due to northwesterly winds carrying sand, the land between the ridges was gradually buried, leading to the beach ridge areas shown in Figures 17, 18 and 27. The sequences of ridges formed in this manner are clearly observed in the infilled Supe Bay, as well as in the Santa Valley areas among other nearby valleys. With exposed beach flats covered with aeolian sand, plus sand transport over the northern mountains of the Supe Valley, large sand dunes formed in the lower Supe Valley, limiting agricultural land areas. The high water table in the Supe Valley with moist soils provides for increased erosion transport during El Niño flood events, further reducing valley-bottom arable land; this, in combination with sand incursion and a loss of marine resources because of a sequence of major El Niño events, clearly limited Caral's survival. Limited agriculture on the plateau adjacent to Caral (Figure 19) provided the only option left to sustain the small population left in Caral post the ~1600 BC time period.

9. Groundwater Amplification Processes Affecting Supe Valley Agriculture

The limitations to the groundwater drainage due to hydraulic conductivity resistance from accreted sediment and clays behind beach ridges affected areas and led to the gradual elevation of the up-valley groundwater profile, causing numerous springs and water pools to appear in the upper reaches of the Supe Valley (Figures 9–11). Since the lower valley near the coastal delta areas originally comprised most of the agricultural lands, farming land loss and reduced soil fertility due to sand accumulation overlays and deposited flood sediments consisting of eroded sierra gravels and stones, gradually led to agriculture being transferred further inland to narrow bottomland and plateau locations nearer to Caral (Figure 4). This transition was further reinforced by flood events that compromised coastal farming areas. In the near-coastal areas, sediment buildup caused the water table to appear lower with respect to the ground surface limiting spring formation; in the mid-valley locations, the water-table height increased due to near-shoreline clay deposits increasing the aquifer-flow resistance as well as a subterranean geological 'choke' contraction on the Supe River that elevated the local upvalley water table height. To provide surface water to coastal field systems, river or spring flows would have to be channeled from far upriver locations to achieve elevation over coastal plains; no such channels are apparent on valley mountainside margins, indicating the abandonment of coastal agricultural zones. The rough mountain corridor topography incised by erosion gullies and sand deposits covering Supe Valley margins prevented long canals originating from valley neck areas to be extended along valley sidewalls to provide water to lower elevation lands. The reconnaissance of the southern Supe Valley mountain corridor areas yielded no trace of long, high-elevation canals.

As a result of flood sediment accretion over the coastal farming zones and sediment infilling of coastal zones behind the coastal beach ridges forming the Medio Mundo ridge, only narrow, mid-valley bottomland farming areas irrigated by spring-sourced, short canals and *amuna* water supplies remained to replace extensive coastal-zone agricultural areas. As coastal rainfall is on the order of a few centimeters per year, producing an intermittent Supe River flow, springs resulting from inland groundwater elevation and sierra *amuna* sources (lakes and reservoirs) supported valley agriculture throughout the year, albeit in narrow inland valley bottom areas. As testament to the high volume of groundwater underlying the Supe Valley, a current drainage channel adjacent to the access road to Caral from the Pan-American Highway flows continuously throughout the year, with a high velocity drainage flow indicating that water abundance, rather than shortage, to support multi-cropping throughout the year in the present, as in the past. This drainage channel, presumed to have an ancient counterpart, was vital to drain fields of excess irrigation water; this, in turn, limited the salt deposits in agricultural fields that, over time, would limit field system productivity.

A survey of the Supe River choke point revealed Canal A with a river inlet (Figures 1, 14 and 15) to support the ramped canal to Chuapacigarro. Due to the riverbed meander and braiding characteristics of low-slope rivers, rainy season canalized flow to valley bottomland farm areas proved unreliable as the river channel frequently deviated from the established canal inlets. As springs developed in the valley bottomland areas distant from the coastline from *amuna*-based groundwater elevation as shoreline prograding progressed, the changeover to spring-supplied canals provided reliable, year-round irrigation systems that additionally maintained the high-valley groundwater level.

The long, low slope, ramped canal (whose entrance and path is now obscured by dense plant growth (Figures 12 and 13)) supported the spring-sourced Canal A–C (Figure 15) to provide water to the plateau field areas. The embankment ramped canal brought water to Caral and Chupacigarro and was the remedy to add plateau agricultural land areas to supplement the limited narrow inland bottomland areas subject to flood erosion and/or coverage by flood sediment. As an anecdotal note, one local farmer employing valley bottomland for agriculture reported that, as a result of the 1989 El Niño event, 50 hectares of his farmland were washed away; this observation also held true in ancient times, so that devastating floods reduced valley bottomland agriculture irreversibly requiring new lands to be developed on the Caral plateau (Figure 15) to avoid land loss. In the Supe Valley, the excavation house indicates that a canal provided water integral to Caral city precincts, although its path remains unexcavated and is no longer available due to landscape erosion. Canals E, G and C served the site of Chupacigarro, canal D serves modern field systems and an early Canal B–C (Figure 1) provided irrigation water from a spring at the origin point of the ramped embankment canal.

10. Sand Incursions Affecting the Supe Valley Agricultural Base

With the formation of the Medio Mundo beach ridge in the ~1600 BC time frame, sand flats and marsh areas formed near the Supe Valley river mouth near the modern shoreline. Figure 32 illustrates the aeolian sand transport from the southern Huara Valley south of the Supe Valley originating from northwesterly winds. Sand accumulations exist from the archaic times to the present day (Figure 7) in the Supe Valley margins that compromise agricultural soil fertility and, in Late Archaic period, required sand barriers (Figures 27–29) to limit sand incursion into the site of Caral proper. The sand incursion extended to swamp areas behind the Medio Mundo beach ridge and compromised formerly productive coastal agricultural lands in the Late Archaic Period, as well as in the present day.

Figure 32 summarizes the following processes: localized El Niño flood drainage paths (1) combined with river fluvial sediment from Fortelaza, Patavilca, Supe and Huara rivers delivered sediment to coastal areas; (2) flood sediment coalesced into an existing segment of the Medio Mundo beach ridge with ridge geometry determined by sequential sediment transport amounts and the oceanic/drift current magnitude; and (3) sand areas trapped behind the Medio Mundo ridge subject to onshore winds further compromised inland agricultural land areas through inland dune transport. Remnant sand accumulations on the Supe Valley northern side limited river flow, reducing the agricultural potential of the coastal delta area. Increased hydraulic resistance to groundwater drainage from sediment deposits and clay formation in saturated coastal soils backed up the groundwater height and led to increased numbers of springs appearing in the valley bottom areas inland from the coast. This effect was amplified by the northside mountains close to the Supe River exit region that choked the groundwater passage; this subterranean contraction effect required a groundwater height change to provide the hydrostatic pressure necessary to increase the groundwater flow velocity through the choke-point region.

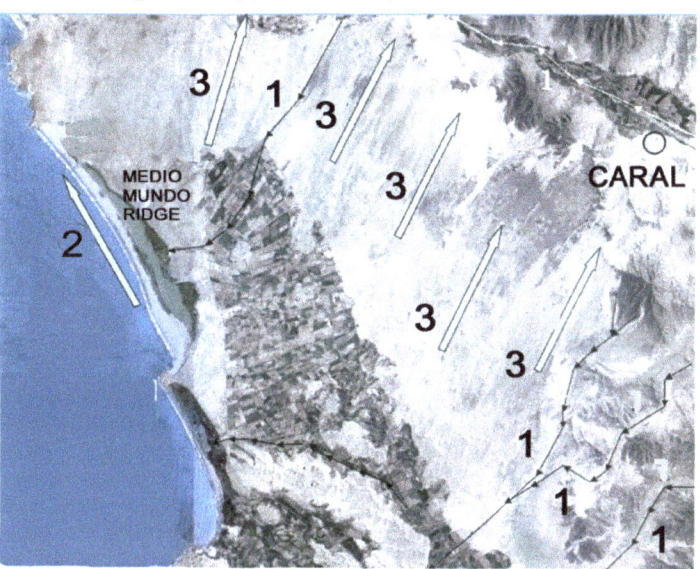

Figure 32. Aeolian sand deposit areas in the Huara valley south of the Supe Valley and urban Caral transferred by constant northeast winds (Google Earth satellite photograph).

Aeolian sand transport over the low mountains between the Huara and Supe Valleys resulted in burying lower valley agricultural fields with overlays of sand, as shown in Figures 7 and 32; additional aeolian sand transport to the urban center of Caral was countered by sand barriers (Figures 28–30) sequentially placed in the southeast canyon open area between the mountain areas. Figure 4 shows the limited extent of the Supe Valley inland bottomlands and the still-existing marsh areas at the Supe Valley mouth bounded by the still-existing Medio Mundo ridge. As sand inundation is constant and compromises soil fertility, modern farmers are forced to use fertilizer additives to sustain maize crops. Figure 15 indicates that a small part of the lower Supe Valley, nowadays has been restored for use; this was achieved as clay deposits near the river mouth region have increased aquifer flow resistance causing a groundwater height amplification that can sustain certain marginal crop types with less water needs not requiring irrigation canal networks in the near coastal bottomlands.

Further east of the valley, *amuna* water supplies and canalized reservoirs (Figure 11) to mid-valley farm areas (Figures 12 and 15) now mainly support maize crops for the very limited population that currently exists in the valley. The aeolian sand-transfer process continues to the present day, and it is surmised that, in the Late Archaic times, sand incursion episodes were major and extended up to the site of Caral, compromising agricultural fields far down the river to the coast. This conclusion is supported by a ~3 cm sand incursion layer deposited atop the final archaic occupational flood sediment deposits in many test pit areas of the site, as illustrated by Figure 20.

In certain Supe Valley areas, the sand layer is stratigraphically overlaid by early ceramic bearing middens dating to the Initial Period (1600–800 BC), indicating some minor reoccupation of the Supe Valley sites, other than Caral proper in the valley in which limited agriculture could exist. Excavation pits reveal that this sand layer is now largely overlain by soil deposits from extensive modern-day farming. Eventually, the Rio Supe carved a stable path to the sea and localized agriculture was returned to near-river margins, in the present day, but both sides of the valley still show remains of the early inundation of archaic sand seas. As sand inundation in the post-Medio Mundo epoch compromised valley agricultural lands and further compromised the marine resource base, an argument for the demise of the Supe Valley society may be proposed at a time close to the creation of the Medio Mundo beach ridge in the Late Archaic Period.

11. Conclusions

Based upon the estimated formation date of the Medio Mundo ridge, a major El Niño event (or sequence of events) started a progression of geophysical landscape changes that compromised both the Supe Valley near coastal agricultural field systems and the coastal marine resource base through flood sediment infilling. Additionally, the flood erosion of thin valley bottomland saturated topsoil led to the reduction in available productive farm areas. ENSO flood and sand incursion into wide expanses of lower-valley bottomlands led to the abandonment of near-coastal agricultural lands and the use of narrow upper-valley land areas that limited agricultural production to lower levels beyond that necessary to maintain high population levels. Further flood events in the ~1800 BC period led to agriculture being moved to small terrace areas (Figure 15) close to urban Caral, which limited the food supply for the growing population of the Supe Valley. This climate crisis also affected other nearby preceramic sites (Figure 2). The effects of the Medio Mundo beach ridge barrier and subsequent closely spaced in time flood events (Figure 6) reduced the agricultural and marine resource base to the extent that large valley populations dependent upon the pre-existing food resource base experienced a collapse of the coastal–inland trade network established during earlier periods in which major flood events were not occurring to any degree. Based upon the decline of food resources related to ENSO events, Supe Valley sites underwent abandonment, as Figure 2 indicates. Figure 6 confirms the frequency of major flood and beach ridge events in the ~1600 BC time period to support the reduction in the agricultural and marine resource base. As further research may show, the large number of sites in the upper Supe Valley region (Figure 4) may have been an attempt to redistribute the valley population around transitory functioning land and water sources, as water and farmland availability for agriculture rapidly decreased in the lower bottomlands part of the valley over a short time period. Although rainfall continued to charge *sierra* basins during El Niño events, the *amuna* source of water to the Supe Valley bottomlands only made matters worse by contributing additional flood water over highly saturated farmland soils. Estimates of the maximum Caral population size [8,10,13] are on the order of several thousand, based upon the number and the extent of excavated housing areas; the reduction in food supplies from increasingly smaller agricultural areas, together with a reduction in the marine resource base, then made life untenable past ~1600 BC.

Ancient (and modern) civilizations of Peru experienced recurrent ENSO episodic climate change patterns inducing floods, drought and landscape change through inflation and deflation cycles that affected their cities and agricultural base. Despite these challenges, several of these societies demonstrated continuity throughout time by relocating their population to areas with more land and water resources and/or instituting large-scale inter- and intravalley water-transfer projects [43,44]. While such changes provided a form of societal continuity for several societies, other societies vanished from the archaeological record when their agricultural systems did not respond or permit modification to use alternate water sources for their agricultural fields. While some societies managed to overcome environmental challenges by technological innovations applied to modify agricultural landscapes to maintain food productivity, other Andean societies unable to implement successful modifications, due to the irreversible damage to agricultural and marine resource areas, and unable to return to their current resource base, terminated their existence from the archaeological record [22,27,43–45]. For Late Archaic Period Norte Chico societies, landscape changes induced by the establishment of large beach ridges from a major ENSO flood event severely altered the agricultural and marine resource base, to the extent that the intra-valley trade network no longer functioned. As food resources diminished from reduced farming areas and decreased marine resource availability, changes in social structure to accommodate a population out of balance with the available food supply was a likely source of population decline or resettlement to other life-sustaining areas, although further details of this transformation of Caral society structure is now only in its early stages of research.

El Niño flood deposition events formed subsea ridges that initiated progradation processes infilling coastal zones trapped by the ridges. A comparison of the duration dates of preceramic coastal societies (Figure 1) to beach ridge dates (Figure 2) indicates an overlap period accompanied by intense El Niño activity. By ~1800 BC, most local sites were depopulated (Figure 2), indicating a common cause for the abandonment of the central north coast area. The effects of the gradual loss of the marine and agricultural base of north central coast societies were likely contributing factors to the abandonment of these major sites (Figure 2). Evidence of flood events from depositional silt layers and later marsh formation in the regions west of Áspero [34] verify major flood-event consequences in the Late Archaic Period. Within the Supe Valley, excavation profiles reveal sedimentary layers dispersed with sand layers indicative of major erosion and deposition events from ENSO events. Recent research [36] in the Norte Chico region related to subsistence changes in the Preceramic and Initial Periods indicates that the presence of littoral changes brought about by ENSO events caused a shift in the dietary composition of site inhabitants. Bay infilling in the Huaca Negra area (close to the present-day town of Barranca) was apparently slower, due to different landscape and valley geophysical conditions than in bay areas to the north, permitting longer-term shellfish gathering in shallow-bay areas, as well as a shift in netting small schooling fish to catching fish species found in deeper offshore waters. The gradual changes in the landscape and offshore bed geometry particular to different coastal valleys from sequential ENSO events permit, in certain cases, continued, but limited, availability of a modified food resource base sufficient to reinstitute previous food supply norms. Only when such transformations are possible, can societal continuance occur, but, in a limited condition, compared to previous norms.

Since the inland valley bottomlands and sierra foothill areas at the western edge of the Cordillera Blanca Andes were the source of most springs and water basins that penetrated the groundwater level (Figures 9–11), agriculture was limited to up-valley narrow bottomlands and limited ramped canal plateau areas (Figure 19) as a result of the geophysical landscape changes and aeolian sand incursion from exposed beach flats. As marine resource extraction was the purview of coastal communities and inland sites that supported farming, reciprocal product trade diminished between inland and coastal communities as a result of a major ENSO event (or series of events) that altered previous trade-basis norms.

To sustain large-scale agriculture in the gradually infilling coastal environment, river or spring water would have to be channeled onto land surfaces lower than the riverbed choke point; this would require canal inlets originating far upriver to achieve elevation over the near-coastal land surface and canal construction on the steep and erosion-incised mountainside corridor topography on the upper reaches of the Supe Valley to revitalize lower-valley agricultural lands.

Sand accumulation on mountain slopes limited ambitions for canal extension to lower-valley areas. Extensive surveys of the southern mountain corridor flanks of the Supe Valley revealed no high-level canal construction. Thus, the coastal area was progressively removed from agricultural exploitation and could not be irrigated by canalized river sources. Since coastal areas decreased in agricultural productivity over time from erosion of topsoil, overlays of flood sediments and aeolian sand deposition, transfer of agriculture to narrow valley-bottom farming areas in inland valley locations could not support a large population. The disruption of the marine resource base from bay infilling and sediment deposits over mollusk shell beds accompanied the loss of farmlands and the viability of the economic model upon which Supe Valley society was based. A later Formative/Initial

Period occupation occurred at some valley sites with limited construction overlay over earlier temple sites; the sand layers between the construction phases attest to large sand incursions during the hiatus period.

The results presented relate to the investigations conducted in the Supe Valley and reconnaissance of the coastal areas of the adjacent Fortaleza, Patavilca and Huara valleys. These valleys contain Late Archaic Period sites and yield terminal C14 dates for these sites consistent with those sites in the Supe Valley [30,31,41]. The present analysis extends investigations [34] detailing reasons for the collapse of the agricultural and marine resource base of the Supe Valley society in the Late Archaic Period from the geomorphic changes described to date and are causative elements contributing to societal disruption from previously established societal norms established over long time periods with stable environmental conditions.

Several climate-driven events that altered the ecological conditions beyond recovery have influenced Andean prehistory. Notable is the collapse/transformation of the southern Moche V society in the 6–7th Century AD by cycles of high rainfall, severe drought and sand incursion into their Moche–Chicama Valley homeland [1]; the collapse of the altiplano Tiwanaku society in the 12th Century AD due to extended drought [37,43,44]; the collapse of the Lambeyeque Valley Sican and Wari societies in the 12th Century AD due to extended drought; the collapse of the Chimú intravalley (Moche Valley) canal systems in the 11th Century AD [43] and El Niño flood catastrophes experienced by the Chirabaya [45] in far-south Peru. To this list, Caral is a further example based on the rapid decline in the agricultural and marine resource base, which exerted a profound influence on the continuance of the economic model of Supe Valley sites. Given the abandonment of major Late Archaic sites in ~1800 BC in nearby valleys, the environmental change based upon the formation of the Medio Mundo beach ridge was likely a key event for the similar fate experienced by the Supe Valley societies.

12. Further Text Notes

1. Evidence of the ENSO events: Refs. [5–7,37]; Figure 6 shows the sequence of events in the Medio Mundo date range over the extensive length of the Peruvian coastline.
2. Bay infilling results: Figure 27 shows a satellite view of the infilled bay at the mouth of Supe Valley due to the beach ridge blockage of river drainage paths creating a marsh area.
3. Farming of the Supe Valley bottom areas: as later phases of coastal Áspero were contemporary with early phases of Caral origination in 4600 BC, early agriculture was located close to the Supe River coastline, close to Áspero, with its high water table and flood transferred fertile *sierra* soils. The land area was extensive, prior to a Medio Mundo ENSO flood event that transformed the land area into the later marsh area location.

Mid-valley Caral amuna water systems for agriculture then provided a safer mid-valley location for Caral and subsidiary sites, given its *amuna* water supply that could accommodate substantial population increases. As later ENSO events created sediment blockage of previous coastal farming areas and marsh creation, mid-valley agriculture with mainly *amuna* water supplies and reservoir-based (Figures 9–11) canal systems permitted agricultural expansion to support the population increase.

1. Contemporary ENSO events and unsuitability for agriculture: Santa-Viru coastal area flood events were contemporary with Fortaleza–Huara coastal area flood events caused by a major Medio Mundo event (or closely timed series of events); the destruction of mid-valley agricultural soils from flood-amplified, valley bottomland erosion further amplified by *amuna* supplied saturated soli thin farming layer erosion, together with the fact that the Supe River runs over relatively flat land causing the Supe River to meander under a flood event and overspill its banks and erode riverside farming areas, indicates that Caral food supply sustainability was continually challenged by flood events in early phases of its development.

2. Drainage channels: these are seen in modern times alongside the entry road to Caral; due to the *amuna* supplied high water-table amplification, drainage channels were the only means to regulate local water-table height for specific crop types, both in ancient and present times. Since ancient mid-valley agricultural systems were likely in form similar to what exists today, it is likely that a similar drainage system was in use in ancient times.

3. Level of groundwater post-flood and duration: Supe Valley groundwater remains within a meter of the farming surface due to the continuous *amuna* water supply that continues from ancient times to the present day. The Supe Valley is unique compared to most other Peruvian valleys as it has too much water and needs a drainage network to control the water-table height for agriculture. Later ENSO events threatened agricultural sustainability as flood water easily washed away thin saturated farming topsoil, given the current-event example described in the manuscript.

4. Abandonment of the valley bottom: Figure 15 indicates the valley-bottom story; initially, agriculture in the early Late Archaic Period was close to the Supe Valley coastline area in which sediments over large river-mouth areas had fertile farm soils washed down from mountain areas. Later, as flood events became more common, flood events and beach ridges created marsh areas in which farmland once existed; as mid-valley near-river farm areas were the next alternative after coast area abandonment, these areas were, in time, also compromised as saturated *amuna* thin soil layers were easily washed away from floods. What remained, as Figure 15 shows, was that agriculture was moved to the upper plateau area on which Caral was located, along with far up-valley sites where Supe River water could be used for local, but smaller, agricultural fields. There were long stretches of time between destructive events on agricultural systems, where mid-valley agriculture could be successively used between major El Niño flood events. Presently, Figure 12 (photo taken about 30 years ago) shows the current status of mid-valley agriculture, where *amuna* water still supplies the high groundwater now mainly used for maize crops; the earlier 1959 El Niño event compromised existing near-river lands and, after several years, with a new soil layer deposited from mountain soil deposition and commercially added fertilizers, lands were used, once again, to continue agriculture.

5. Productivity of areas: Supe valley crop types used in Late Archaic times are described in the text; the variety of inner-valley crops (plus extensive cotton-planting areas) provided reciprocal export trade comestibles to coastal areas for their fish and shellfish exports. The productivity and fertility of crop types was likely enhanced by plowing under leaf material for their mineral content; this decomposition fertility increase was aided by the high moisture level of mid-valley farming soils. There were valley areas of different heights that provided different groundwater moisture levels for different crop types; for example, cotton growth requires constant water availability, so cotton plants would be situated on lower-height land areas where the groundwater height was closer to the land surface. Other crops would accordingly be assigned to land areas in which the moisture level was appropriate for their growth.

6. Population supportable with agriculture before and after Enso events: initially, at the early phases of the Late Archaic Period with the limited Supe Valley population, Caral and near-coastal sites were in the preliminary stages of development using near-coastal valley delta agricultural lands. Population levels were likely on the order of hundreds and likely drawn from local valley tribal groups realizing the agricultural potential of the Supe Valley from its water abundance. Later, as ENSO events originated and intensified, coastal lands, once fertile sources of agriculture and marine resources, diminished in productivity leading to mid-valley agricultural land development; this proved a positive move, as population growth could then be continually supported from the high water-table *amuna* water supply for multi-cropping of a wide variety of crops. Population levels were in the low thousands after this relocation which likely promoted additional valley sites (Figure 4) and provided

the labor source for new pyramid constructions (Figure 5). As ENSO events progressed and intensified, agricultural lands and marine resources were compromised in later time stages leading to site abandonment of most Supe Valley sites as Figure 2 indicates. This led to the collapse of inland to coastal trade, as food resources diminished to non-survivorship levels.

7. From field reconnaissance visits starting from approximately 20 plus years ago to a recent revisit in 2010, notable landscape features were apparent: linear beach ridge formations and coastal landscape changes from recent ENSO activity (Figures 17 and 18). In order to determine their origins and given that their features derived from the fluid mechanics of El Niño flood events, recourse to CFD methods were initiated to show how these feature could originate. The manuscript's many computer results (Figures 21–26) now demonstrate the feature origins and help remove conjecture and speculation as to their origins. To date, the CFD results prove how beach ridges form linear shapes and how landscape change originated from El Niño flood events conclusively.

Funding: All finding provided by author.

Institutional Review Board Statement: Not applicable.

Informed Consent Statement: Informed consent was obtained from all subjects involved in this study given in the Acknowledgement section to follow.

Data Availability Statement: All data provided by author.

Acknowledgments: The author wishes to acknowledge the contributions of his colleagues Michael Moseley, Daniel Sandweiss, Donald Keefer and especially Ruth Shady-Solis, the official director of Caral excavations during five years of fieldwork at Caral. All the above-individuals cited contributed their specialist knowledge and research findings to this manuscript.

Conflicts of Interest: The authors declare no conflict of interest.

Appendix A Supe Valley Site Names

(1) Bandurria; (2) Vichama; (3) Aspero; (4) Upaca; (5) Pampa San Jose; (6) Caballete; (7) Vinto Alto; (8) Haricanga; (9) Galivantes; (10) Culebras; (11) Las Aldas; (12) La Galgada; (13) Caral; (14) Rio Seco; (15) Las Shicras; (16) Kotosh; (17) Huarico; and (18) Piruru. The Allpacota site (Figure 15) lies between sites (8) and (9).

References

1. Moseley, M.E.; Wagner, D.; Richardson, J.B., III. Space Shuttle Imagery of a Recent Catastrophic Change along the Arid Andean Coast. In *Paleoshorelines and Prehistory: An Investigation of Method*; Johnson, L., Stright, M., Eds.; CRC Press: Boca Raton, FL, USA, 1992.
2. Knighton, D. *Fluvial Forms and Processes: A New Perspective*; Hodder Education, Part of Hachette Livre: London, UK, 1998.
3. Rogers, S.; Sandweiss, D.; Maasch, K.; Belknap, D.F.; Agouris, P. Coastal Change and Beach Ridges along the Northwest Coast of Peru: Image and GIS Analysis of the Chira, Piura and Colán Beach Ridge Plains. *J. Coast. Res.* **2004**, *20*, 1102–1125. [CrossRef]
4. Sandweiss, D. The Beach Ridges at Santa, Peru: El Niño, Uplift and Prehistory. *Geoarchaeology* **1986**, *1*, 17–28. [CrossRef]
5. Sandweiss, D.; Richardson, J.; Reitz, E.; Rollin, H.; Maasch, K. Geoarchaeological Evidence from Peru for a 5000 BP Onset of El Niño. *Science* **1996**, *273*, 1531–1533. [CrossRef]
6. Wells, L. Holocene Landscape Change on the Santa Delta, Peru: Impact on Archaeological Site Distributions. *Holocene* **1992**, *2*, 193–204. [CrossRef]
7. Wells, L. The Santa Beach Ridge Complex: Sea Level and Progradational History of an Open Gravel Coast in Central Peru. *J. Coast. Res.* **1996**, *12*, 1–17.
8. Shady, R. Sustento Socioeconómico del Estado Pristino de Supe-Peru: Las Evidencia de Caral-Supe. *Arqueol. Y Soc.* **2000**, *13*, 49–66. [CrossRef]
9. Shady, R.; Haas, J.; Creamer, W. Dating Caral, a Preceramic Site in the Supe Valley on the Central Coast of Peru. *Science* **2002**, *292*, 723–726.
10. Shady, R.; Leyva, C. *La Ciudad Sagrada de Caral- los Origines de la Civilización Andina y la Formación del Estado Pristina*; Proyecto Especial Arqueológico Caral-Supe; Instituto National de Cultura Publicación: Lima, Peru, 2003.
11. Shady, R.; Creamer, W.; Ruiz, A. Dating the Late Archaic Occupation of the Norte Chico Region of Peru. *Nature* **2004**, *432*, 1020–1023.

12. Shady, R.; Quispe, E.; Novoa, P.; Machacuay, M. *Vichama, Civilización Agropesquerade Végueta, Huara: La Ideologia de Nuestros Ancestros, 3800 Años de Arte Mural*; Servicios Graphicos JMD S.R.L.: Lima, Peru, 2014.
13. Shady, R. *La Ciudad Sagrada de Caral-Origines de la Civilización Andina*; Museo de Arqueología; Universidad Nacional de San Marcos Publicaciónes: Lima, Peru, 2001.
14. Shady, R. *The Social and Cultural Values of Caral-Supe, the Oldest Civilization in Peru and America and Its Role in Integral and Sustainable Development*; Proyecto Especial Arqueologico Caral-Supe, Instituto Nacional de Cultura Publicación: Lima, Peru, 2007; Volume 4, pp. 1–69.
15. Shady, R. Caral-Supe: Y Su Entorno Natural y Social en Los Origenes de la Civilización. In *Andean Civilization*; Marcus, J., Williams, P., Eds.; Cotsen Institute of Archaeology 63: Los Angeles, CA, USA, 2009; pp. 99–117.
16. Shady, R. *Caral- La Ciudad del Fuego Sagrado*; Interbank Publishers: Lima, Peru, 2004.
17. Weiss, H. Beyond the Younger-Dryas: Collapse as an Adaptation of Abrupt Climate Change in Ancient West Asia and the Eastern Mediterranean. In *Environmental Disaster and the Archaeology of Human Response*; Bawden, G., Reycraft, R., Eds.; Maxwell Museum of Anthropology Paper No.7: Albuquerque, NM, USA, 2000; pp. 75–98.
18. Sandweiss, D.; Quilter, J. *El Ñino, Catastrophism and Cultur, 17e Change in Ancient America*; Dumbarton Oaks Research Library and Collection: Washington, DC, USA, 2009.
19. Sandweiss, D.; Maasch, K.; Burger, R.; Richardson, J.; Rollins, H.; Clement, A. Variation in Holocene El Niño Frequencies: Climate Records and Cultural Consequences in Ancient Peru. *Geology* **2001**, *29*, 603–606. [CrossRef]
20. *Flow Science FLOW-3D User Manual, V. 9.3*; Flow Science, Inc.: Santa Fe, NM, USA, 2020.
21. Moseley, M.E.; Willey, G.R. Áspero, Peru: A Reexamination of the Site and its Implications. *Am. Antiq.* **1973**, *38*, 452–468. [CrossRef]
22. Renfrew, C. System Collapse as Social Transformation: Catastrophe and Anastrophe in Early State Societies. In *Transformations-Mathematical Approaches to Culture Change*; Renfrew, C., Cooke, K., Eds.; Academic Press: New York, NY, USA, 1979.
23. Paulson, A. Environment and Empire: Climatic Factors in Prehistoric Andean Culture Change. *World Archaeol.* **1996**, *8*, 121–132. [CrossRef]
24. Moore, J. *Cultural Landscapes in the Ancient Andes*; University Press of Florida: Gainesville, FL, USA, 2005.
25. Manners, R.; Migilligan, F.; Goldstein, P. Floodplain Development, El Niño, and Cultural Consequences in a Hyperarid Andean Environment. *Ann. Assoc. Am. Geogr.* **2003**, *97*, 229–249. [CrossRef]
26. Contreras, D. Landscape and Environment: Insights from the Prehispanic Central Andes. *J. Archaeol. Res.* **2000**, *18*, 241–288. [CrossRef]
27. Bawden, G.; Reycraft, R. Exploration of Punctuated Equilibrium and Culture Change in the Archaeology of Andean Ethnogenesis (Chapter 12). In *Andean Civilization*; Marcus, J., Williams, P.R., Eds.; UCLA Cotsen Institute of Archaeology Press: Los Angeles, CA, USA, 2009; pp. 195–210.
28. Bawden, G.; Reycraft, R.M. *Environmental Disaster and the Archaeology of Human Response*; Maxwell Museum of Anthropology Papers No.7, University of New Mexico Press: Albuquerque, NM, USA, 2000.
29. Lanning, E. *Peru before the Inkas*; Prentice Hall Publishers: Englewood Cliffs, NJ, USA, 1967.
30. Haas, J.; Creamer, W. Cultural Transformations in the Central Andes in the Late Archaic. In *Andean Archaeology*; Silverman, H., Ed.; Blackwell Publishers: London, UK, 2004.
31. Shady, R.; Kleihege, C. *Caral, La Primera Civilización de América*; Universidad de San Martin de Porres Publicaciónes: Lima, Peru, 2008; pp. 112–167.
32. Feldman, R. Preceramic Corporate Architecture: Evidence for Development of Non-Egalitarian Social Systems in Peru. In *Early Ceremonial in the Andes*; Donnan, C., Ed.; Dumbarton Oaks Publications: Washington, DC, USA, 1985; pp. 71–92.
33. Creamer, W.; Ruiz, A. *Archaeological Investigation of Late Archaic Sites (3000–1800 BC) in the Pativilca Valley, Peru*; Fieldiana Anthropology 40; Field Museum of Chicago Publication: Chicago, IL, USA, 2007.
34. Sandweiss, D.; Shady-Solis, R.; Moseley, M.E.; Keefer, D.; Ortloff, C.R. Environmental Change and Economic Development in Coastal Peru between 5,800 and 3,600 Years Ago. *Proc. Natl. Acad. Sci. USA* **2009**, *106*, 1359–1363. [CrossRef] [PubMed]
35. Barrera, A. *Bandurria*; Servicios Gráficos Jackeline: Huara, Peru, 2008.
36. Chen, P. Marine-Based Subsistence and its Social Implications in the Late Preceramic Initial Period: A Different Pattern from Huaca Negra, North Coast of Peru. In Proceedings of the 58th Institute of Andean Studies Conference, Berkeley, CA, USA, 5–6 January 2018.
37. Ortleib, L.; Fournier, M.; Macharé, J. Beach Ridge Series in Northern Peru: Chronology, Correlation and Relationship with Major Holocene El Niño Events. *Bull. Inst. Fr. Etudes Andin.* **1993**, *22*, 121–212.
38. Richardson, J. The Chira Beach Ridges, Sea Level Change, and the Origins of Maritime Economies on the Peruvian Coast. *Ann. Carnegie Mus.* **1983**, *52*, 265–275.
39. Bain, A.; Bonnington, S. *The Hydraulic Transport of Solids by Pipeline*; Pergammon Press: New York, NY, USA, 1970.
40. Pulgar Vidal, J. *Geografía del Peru, Novena Edición*; Editorial PEISA: Lima, Peru, 2012.
41. Haas, J.; Creamer, W. Crucible of Andean Civilization: The Peruvian Coast from 3000 to 1800 BC. *Curr. Anthropol.* **2006**, *47*, 745–775. [CrossRef]
42. Abulnagy, B. *Slurry Systems Handbook 3(3)*; McGraw-Hill Publishers: New York, NY, USA, 2002; pp. 183–210.
43. Ortloff, C.R.; Kolata, A.L. Climate and Collapse: Agro-ecological Perspectives on the Decline of the Tiwanaku State. *J. Archaeol. Sci.* **1993**, *16*, 513–535. [CrossRef]

44. Ortloff, C.R. *Water Engineering in the Ancient World: Archaeological and Climate Perspectives on Societies of Ancient South America, the Middle East and South-East Asia*; Oxford University Press: Oxford, UK, 2010.
45. Reycraft, R. Long-term Human Response to El Niño in South Coastal Peru circa A.D. 1400. In *Environmental Disaster and the Archaeology of Human Response*; Bawden, G., Reycraft, R., Eds.; Maxwell Museum of Anthropology and Archaeological Papers: Albuquerque, NM, USA, 2000; Volume 7, pp. 99–120.

Case Report

Inka Hydraulic Engineering at the Tipon Royal Compound (Peru)

Charles R. Ortloff [1,2]

1. CFD Consultants International, Ltd., 18310 Southview Avenue, Los Gatos, CA 95033, USA; ortloff5@aol.com
2. Research Associate in Anthropology, University of Chicago, 5801 S. Ellis Avenue, Chicago, IL 60637, USA

Abstract: The Inka site of Tipon had many unique hydraulic engineering features that have modern hydraulic theory counterparts. For example, the Tipon channel system providing water to the Principal Fountain had a channel contraction inducing critical flow as determined by CFD analysis- this feature designed to induce flow stability and preserve the aesthetic display of the downstream Waterfall. The Main Aqueduct channel sourced by the Pukara River had a given flow rate to limit channel overbank spillage induced by a hydraulic jump at the steep-mild slope transition channel location as determined by use of modern CFD methods- this flow rate corresponds to the duplication of the actual flow rate used in the modern restoration using flow blockage plates placed in the channel to limit over-bank spillage. Additional hydraulic features governing the water supply to agricultural terraces for specialty crops constitute further sophisticated water management control systems discussed in detail in the text.

Keywords: Inka; Tipon; precolumbian; CFD; water systems; flow rates; aqueduct; fountain; critical flow

Citation: Ortloff, C.R. Inka Hydraulic Engineering at the Tipon Royal Compound (Peru). *Water* 2022, 14, 102. https://doi.org/10.3390/w14010102

Academic Editor: Goen Ho

Received: 13 October 2021
Accepted: 9 December 2021
Published: 4 January 2022

Publisher's Note: MDPI stays neutral with regard to jurisdictional claims in published maps and institutional affiliations.

Copyright: © 2022 by the author. Licensee MDPI, Basel, Switzerland. This article is an open access article distributed under the terms and conditions of the Creative Commons Attribution (CC BY) license (https:// creativecommons.org/licenses/by/ 4.0/).

1. Introduction

The site of Tipon, located in the proximity of Cuzco, provides an example of Inka hydraulic engineering knowledge and civil engineering practice as demonstrated by the design and operation of the site's complex water system. The water engineering knowledge base is revealed by analysis of the Tipon's use of river and spring-sourced surface and subterranean channels that transport, distribute, and drain water to/from multiple agricultural platforms, reservoirs, and urban occupation and ceremonial centers. Complex intersecting surface and subterranean channel systems that regulate water flows from diverse sources provide water to Tipon's thirteen agricultural platforms to maintain different ground moisture levels to sustain specialty crops. Additionally, within the site are fountains and multiple water display features that combine sophisticated hydraulic engineering with aesthetic presentation. To understand the water technology used by the Inka to design the site's water system, use of modern hydraulic theory is employed to examine key elements of the Principal Fountain and the Main Aqueduct to determine the design intent and civil engineering knowledge base used by Inka engineers. Results of the analysis show an Inka hydraulic technology utilizing complex engineering principles similar to those used in modern civil engineering practice centuries ahead of their formal discovery in western hydraulic science.

2. Site Description

The site of Tipon, located in Peru approximately 17 km east of Cuzco along the Huatanay River at south latitude 13°34′ and longitude 71°47′ at 3700–4000 masl is known for its many unique hydraulic features coordinated in a practical and aesthetic manner to demonstrate Inka knowledge of water control principles. The site indicates an early Middle Horizon (600–1000 AD) presence evidenced by an encircling 6.4 km long outer wall (Figure 1) attributed to Wari control of the enclosed area [1]. The ~2 km² interior site area was under Inka control past ~1200 AD and was later converted into the royal estate of Inka Wiracocha in the early 15th century [2–4] as evidenced by the royal residence and ceremonial compounds of Sinkunakancha and Patallaqta shown in Figure 1. The site was composed of thirteen major agricultural platforms [5] as shown in Figures 1 and 2; the lowermost platforms are associated with nearby ceremonial centers (Figures 1 and 3). The agricultural platforms were irrigated from water supplied from a branch of the Main Aqueduct sourced from the Rio Pukara (Figure 1); further branches of the Main Aqueduct provided water to the lowermost ceremonial areas (Figure 1). Each agricultural platform had a drainage channel at its base that collected post-saturation aquifer groundwater seepage as well as rainfall runoff; excess water was led to side channels (Figures 4–8) directed to lower site occupation and ceremonial use areas. A fraction of the collected water was used to irrigate the next lower agricultural platform while a further portion of a platform's drainage channel water was directed to an easternmost collection channel to provide water to lower-level ceremonial and domestic occupation buildings and site drainage. A further portion of water from agricultural platform seepage collection channels passed through a series of interconnected surface and subsurface channels to provide water to special water display areas (the Principal Fountain in particular) and then on to domestic and elite residential and ceremonial areas (Figure 1) at lower site areas.

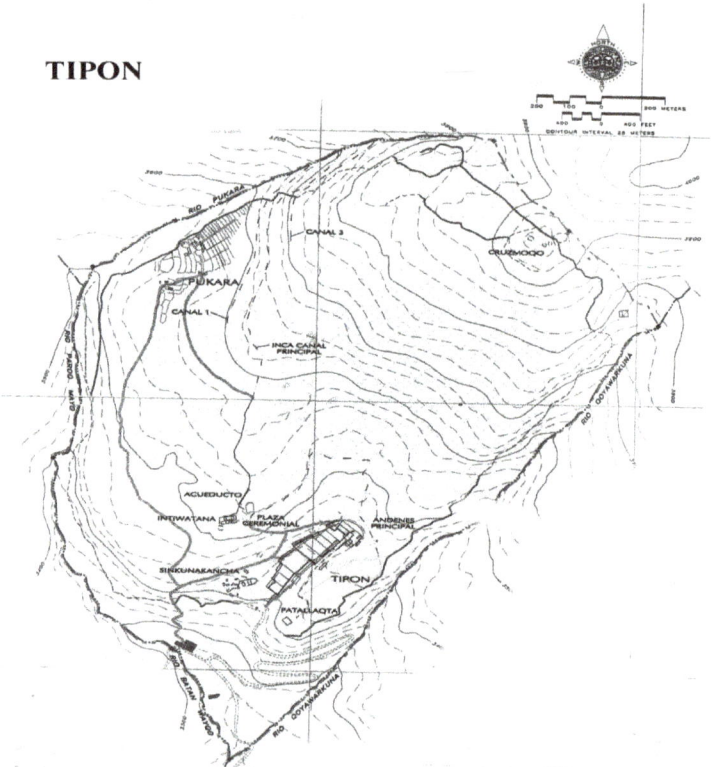

Figure 1. The site of Tipon close to the Inka capital of Cuzco. Note multiple site water channels derived from the northern Rio Pukara through terrace systems at Pukura that supplied Main Aqueduct water directed to the thirteen agricultural terraces. Note the dashed-line easternmost drainage channel east of the agricultural platforms shown in Figure 2 directing water to the Patallaqta area. (Map courtesy of Ken Wright, Wright Water Engineers, Boulder, CO).

Historically, Inka rulers had the state privilege to build a private retreat away from the central administrative center in Cuzco, to balance their civil duties with an environment that can facilitate the intimacies of family life with friends and associates and where they could enjoy a contemplative life with attendants to serve all needs. These centers had the highest technical level of architecture and water systems that Inka engineers could produce consistent with the elevated status of Inka royalty. Notable in this regard was the site of Machu Picchu serving the Inka Pachacuiti, the site of Choquequirao serving Topa Inka, the site of Chinchero serving Inka Tupac Yupanqui and the site of Quispiquanca serving Inka Huayna Capac among other royal sites. The Tipon site, as further discussion reveals, is a further example of the best water engineering known to Inka water engineers.

Figure 2. View of the thirteen agricultural terraces at Tipon. The terraces are located to the right of the Plaza Ceremonial shown in Figure 1. (Photo by C. Ortloff).

Figure 3. View of the walled Sinkunakancha ceremonial plaza location shown in Figure 1 with water supply from a lower branch channel from the Main Canal and a leftmost channel from the northern Pukara area from Rio Pukara. (Photo by C. Ortloff).

The multiplicity of channels and their water sources serving different agricultural platforms and display fountains are described in [4,5] to show the complexity of Tipon's water supply and distribution systems. Figures 4, 5 and 7 illustrate several of the overfall channels between agricultural platforms sourced by sequential seepage collection channels. The origin water source to the uppermost agricultural platform is from the Pukara River supplied Main Aqueduct (Figure 1) and a local high level spring source; all interconnected channel structures are indicative of a complex water control design distributing water from higher to lower elevation agricultural platforms. Figures 4–8 illustrate the continuance of overflow channel water from elevated platforms shown to continuation channels leading water to lower elevation destinations. Figure 8 shows a channel intersection from a base aquifer seepage channel, to supply water, to the Waterfall display structure shown in Figures 9 and 10. The totality of water supplies to the Waterfall display structure is given in [5] (p. 47); all the water supply systems had flow rate supply and drainage controls to ensure a constant water supply to the Waterfall display fountain during seasonal water availability changes, preserving the aesthetic display of the Waterfall display structure throughout the year. As the Rio Pucara flow rate supply to the main Aqueduct was seasonally variable, use of blockage elements and additional water supply control features in adjoining water distribution and drainage channels was employed to maintain a given flow rate to the Waterfall display structure through seasonal water supply changes—this is a further indication of the water engineering technical base available to Inka engineers. Note that the water intake from the Rio Pukara to the site has a flow rate drainage control to the Rio Qoyawarkuna (Canals 1 and 3, Figure 1) to limit site water supply during the rainy season. Figure 9 shows an additional water display feature in the easternmost water collection channel (Figure 7). A spillage basin is supplied by an overfall stream originating from the east drainage collection channel water to direct water to lower site areas (Figure 8); this feature is a further example of an aesthetic display expected from a site occupied by Inka royalty where the best available engineering practice was in use for both practical and display purposes.

Figure 4. Drop structure conveying water from a higher to lower agricultural platform. Typically, a portion of the received water into the lower platform channel was distributed laterally into a channel network to provide irrigation water for the next lower platform; excess water was channeled to an easternmost drainage channel. This drop structure is associated with the base of one of the lower platforms shown in Figure 1. (Photo by C. Ortloff).

Each of the higher elevation agricultural platforms contained a base open channel collecting aquifer drainage and rainfall runoff, a portion of which was subsequently passed on to lower platforms as irrigation water or drainage water led to other site areas (Figure 5).

Measured amounts of passage water from one platform to a lower platform was necessary to provide the correct moisture level to grow different crops on different individual platforms—this consideration explains the multiplicity of water portioning structures derived from multiple drainage paths and the inherent complexity of the site water control system.

Figure 5. Successive Water drop structures collecting excess agricultural platform drainage water beyond that necessary for agricultural purposes from upper platforms and channeling water to lower site ceremonial, reservoir and occupation areas. This terrace drainage structure is to the right of the Plaza Ceremonial shown in Figure 1. (Photo by C. Ortloff).

Figure 6. Typical water collection channel at the base of a lower platform shown in Figures 1 and 2. A fraction of the aquifer seepage collection water was directed further to a lower agricultural platform while remaining water was directed through a series of channels to the easternmost water collection channel and/or to lower occupation, ceremonial and reservoir areas. The easternmost collection channel is located to the right of the terraces shown in Figure 2. (Photo by C. Ortloff).

Figure 7. Multilevel side channel collecting individual platform aquifer seepage, excess rainfall runoff and excess water not used for specific purposes directed to lower site levels and the easternmost drainage channel. The figure is associated with the three lowermost platforms shown in Figures 1 and 2. (Photo by C. Ortloff).

Figure 8. Typical seepage collection channel at a platform base collecting water to distribute to a lower terrace and/or a channel to direct water to various areas of the site. This typical structure is to be found at the base of platforms shown in Figure 2. (Photo by C. Ortloff).

Figure 9. Spillage basin supplied by the easternmost drainage channel directing water to lower site occupation and reservoir areas. The easternmost channel is located to the right of all platforms shown in Figure 2 and is to the right of the Plaza Ceremonial shown in Figure 1. (Photo by C. Ortloff).

Figure 10. Waterfall display sourced by upper platform seepage water, several auxiliary channel water supplies and a spring source [5]. This structure is associated with platform 10 as described in the text and located in the terrace system shown in Figure 1. (Photo by C. Ortloff).

Agricultural platform numbering convention [5] (p.34–35) starts with Platform 1 at the lowermost south altitude and sequentially proceeds upward to Platform 13 at the highest altitude (Figures 1 and 2). In one case, water was led from Main Aqueduct sources to combine with a natural spring source on agricultural platform 11 to provide water to the Principal Waterfall Fountain (Figure 10) designed for aesthetic as well as for drainage control and water distribution purposes. The Rio Pukara (Figure 1) was the river water source [5] (p. 39) for the uppermost altitude platforms 11-NW, 12 and 13 through the Main Aqueduct (Figure 1 and shown later in Figures 16 and 17) while the spring on platform 11 was the water source for platforms 1 to 10 and part of 11 as well as several side platforms (Figure 1). Among the more prominent hydraulic features of Tipon, the Principal Fountain on platform 11 had an elaborate spring supplied, multichannel branched water delivery system leading to the Waterfall (Figure 10) with its four independent waterfall streams. One challenge to Inka water engineers was the design and flow rate regulation of the water distribution system upstream of the Waterfall to promote equal flows in each of the four water distribution channels (as indicated in Figure 10); this required use of a settling basin area upstream from the four overflow streams so that each overflow stream had the same water height and velocity to produce similar overfall stream geometry in each of the four channels with no basin currents to distort the equal flow conditions to each overflow channel. How this was done is described in detail in the subsequent Principal Fountain section by use of a unique water control system upstream of the Waterfall Fountain shown in various viewpoints in Figures 11–13.

Figure 11. Channel cross-section geometry change made to induce critical flow in the contracted channel section to downstream channels supplying the Figure 9 Waterfall. This structure is associated with the upstream water supply to the Waterfall. (Photo by C. Ortloff).

Figure 12. Transverse surface wave structure in the downstream contracted channel section indicative of induced critical flow conditions. This is an alternate view of the water flow pattern shown in Figure 11. (Photo by C. Ortloff).

Figure 13. Alternate view of the channel cross section change supplying water to the Waterfall. Transverse surface wave pattern typical of critical flow is apparent in the contracted channel section. This is an alternate view of the water flow pattern channel shown in Figure 11. (Photo by C. Ortloff).

Yet a further hydraulic feature is the Main Aqueduct (Figure 1, Figures 15 and 16) vital to supply water from the Rio Pukara source to central Tipon through different branch channels. Inherent to the water regulation system is the main drainage channel [5] (p. 34) shown in Figure 12 located at the eastern edge of several of the platforms to convey excess water past that required for different crop types on individual platforms to a lower drainage area. Figure 8 shows a water redistribution basin derived from the easternmost drainage channel (Figure 14) with a water overfall directed to provide water to a lower elevation destination. Encoded within the complex multi-channel water supply and distribution system of Tipon are examples and demonstrations of the hydraulic science base available to Inka hydraulic engineers. The following discussion explores, by use of modern hydraulic engineering methods, the hydraulic knowledge base available to Inka engineers and used to construct the elaborate water system supplying the multi-channeled Waterfall (Figure 10). The analysis presented serves to yield added information about Inka water technology not previously reported in the literature and provide refined estimates of important flow parameters in two main hydraulic structures at Tipon.

Figure 14. View of a portion of the easternmost drainage channel directing terrace seepage and rainfall runoff to lower occupation and ceremonial areas as well as for excessive water site drainage. The easternmost channel is located to the right of the Plaza Ceremonial terraces shown in Figures 1 and 2. (Photo by C. Ortloff).

3. The Principal Fountain

Channel measurement data [5] (p. 42) is used for the Principal Fountain analysis and calculations related to its water supply; this data is for channel sections that precede later century repair and reconstruction modifications. In conformance with modern civil engineering practice, English unit convention is used for hydraulic engineering calculations.

A first example used to determine the scope of Inka water technologies used at Tipon derives from examination of the channel system shown in Figures 11–13 supporting water flow to the Principal Fountain Waterfall. A channel contraction occurs from a 0.9 m width, ~2.5 m long channel to a 0.4 m width, ~10.5 m long channel upstream of the Principal Fountain area on platform 11 as illustrated in these figures. Both channel sections have a rectangular cross section and have the same mild (low) slope [5] (p. 47). The water source to the upstream wider channel section derives from eight separate water supply conduits [5] (p. 47) together with a major spring indicative of the totality of water control systems used to maintain a constant flow rate to the Waterfall during seasonal changes in water supply. Measured flow rates into the (reactivated) wide channel from two different tests yielded 0.68 ft^3/s and 0.58 ft^3/s leading to an average 0.63 ft^3/s (0.02 m^3/s) flow rate [5] (p. 47). The question arises as to the water engineering design intent of the abrupt width change of the channel section shown in Figures 11–13.

To understand Inka hydraulic engineering in terms of modern hydraulics technology, use of the Froude number (Fr) is convenient to explain water behavior [6–10]. For shallow depth D (ft) flows, the Froude number definition is given as Fr = V/(g D)$^{1/2}$ where V is the water velocity (ft/s), and g is the gravitational constant (32.2 ft/s^2). Physically, Fr is the ratio of water velocity V to the gravitational wave velocity (g D)$^{1/2}$—when Fr > 1, water velocity exceeds the signaling gravitational wave velocity so that water has no advance warning of a downstream obstacle—this leads to the creation of a sudden hydraulic jump at an obstacle as illustrated from a flume test shown in Figure 19. In this figure, a shallow, high velocity water flow (Fr >> 1) encounters a plate obstacle at the leftmost exit region of a hydraulic flume causing an elevated, highly turbulent elevated hydraulic jump. In physical terms, for Fr >1, there is no upstream awareness of the presence of an obstacle until the obstacle is encountered by the water flow as the gravitational wave signaling velocity that informs the flow that obstacle exists (g D)$^{1/2}$ is much less than the V flow velocity. For Fr < 1, the gravitational wave signaling velocity travels upstream of the obstacle faster than the water velocity V to inform the incoming water flow that an obstacle lies ahead. This causes the water flow to adjust in height and velocity far upstream of an obstacle to produce a smooth flow over the obstacle. Calculation of the upstream water height readjustment for such Fr < 1 flows is given in [6–10]. Here the 'obstacle' for the present application is the large contraction in channel width shown in Figures 11–13. In the proceeding discussion to follow, Fr > 1 flows are denoted as supercritical, Fr < 1 flows are denoted as subcritical and Fr =1 flows are denoted as critical. While the presence and characteristics of subcritical, critical, and supercritical flows can be calculated from modern hydraulic theory, a simpler method exists to determine flow types that were used by Inka engineers. Insertion of a thin rod into a flow that produces a downstream surface V-wave pattern is indicative of supercritical flow; if the surface pattern shows upstream influence from the rod, then subcritical flow is indicated. If only a local surface disturbance around the rod is noted, then critical flow is indicated. This simple test can determine different flow regime types to promote the various usages associated with the different Fr flow regimes. This knowledge base was known to Inka water engineers and recorded in some form yet unknown to researchers; as the Inka had no known writing system, details of their water technology only remain from modern analysis methods applied to their water systems.

Figure 15 is derived from the Euler fluid mechanics continuity and momentum equations and is useful to describe the flow transition from sub- to supercritical flow in the Principal Fountain supply channel shown in Figures 11–13 where water frictional effects are minor consideration effects on flow patterns. As all water motion is governed by the mass and momentum conservation equations, Figure 15 indicates flow transitions based on Froude number change due to channel geometry changes. The x-axis represents (1/2) Fr_1^2 incoming flow conditions into a channel; the (1/2) Fr_2^2 conditions represent outgoing flow conditions in a resized channel section. The W2/W1 curves represent width ratios of the incoming to the downstream channel shape change. W2/W1 > 1 represents channel expansion and W2/W1 < 1 represents channel contraction such as shown in Figures 11–13 that illustrate different views of the same channel contraction section. While modern conservation differential equations are a method of summarizing governing water flow principles, comparable results were obtainable by codified and recorded observation methods developed by Inka water engineers albeit in notations yet to be discovered. Note that Figure 15 assumes no base height change along the Figure 11 channel.

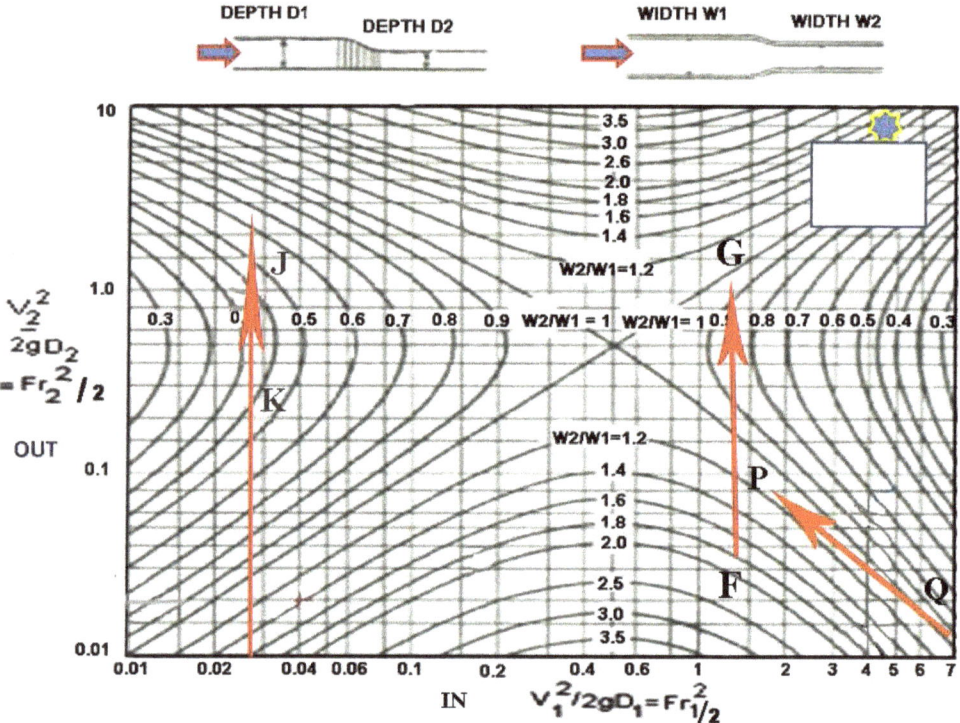

Figure 15. Flow plot describing flow transitions in channels. The x-axis describes Froude number channel entry (IN) conditions; the y-axis describes Froude number channel exit conditions (OUT). The W2/W1 curves describe the contraction (W2/W1 < 1) or expansion ratio (W2/W1 > 1) of a channel sequence. Figure derived from [10] (Bakhmeteff 1932:254, his figure 193b). Here the high Froude number range is applicable to the Main Aqueduct segment in the hydraulic jump area as later detailed. (Figure additions by C. Ortloff).

Using the subscript notation (1) for flow conditions in the wide channel section and (2) for contracted width conditions, the Figure 10 width contraction ratio is W2/W1 = 0.44. The (1) flow entry value using the average flow rate is based on a ~0.3 ft water depth for which $V_1 = 0.72$ ft/s and Froude number is $Fr_1 = V/(g D)^{1/2} = 0.23$, $(Fr_1^2/2 = 0.03$ in Figure 15). The contracted channel (2) Froude number is $Fr_2 \approx 1.14$, $(Fr_2^2/2) = 0.65$ in Figure 15 based on ~0.3 ft depth. The channel contraction shown in Figures 11–13 takes water flow from subcritical flow (Fr < 1) in the wide channel to a near critical (Fr ≈ 1) flow in the narrowed channel. From [5] (p. 34–35), the flow rate per unit width in the contracted rectangular cross section channel is $q_2 = 0.68/W2 = 0.52$ ft^3/s ft and the critical water depth is given by $y_c = (q_2^2/g)^{1/3} = 0.2$ ft in agreement with the previous critical water depth value calculated for near critical Fr $\approx$ 1 flow in the contracted width (2) channel. The special case for which Fr = 1 is accompanied by flow surface wave instability derived from the presence of translating large scale vortex motion within the body of the water. Figures 12 and 13 illustrates a surface ripple pattern consistent with near critical Fr $\approx$ 1 flow in the contracted channel region as well as a slight decrease in water depth resulting from the (1) to (2) width change transition.

The parallel ripple wave structure normal to the flow direction is consistent with $\sin \theta = 1/\text{Fr}$, where θ is the half angle of the surface wave so that when Fr $\approx$ 1, $\theta \approx 90°$ verifying the transverse surface ripple wave structure shown in Figures 12 and 13. Surface wave instabilities associated with Fr = 1 flow are to be avoided in the supply channel as a uniform, stable flow is required to produce the proper aesthetic display in the downstream fountain waterfall channels.

Figure 15 illustrates the flow Froude number transition arrow (K to J) resulting from the wide (1) to narrow channel (2) shape change that incorporates transition from sub- to near critical flow. This indicates that Inka engineers designed the (2) contracted channel section to support near-critical Fr $\approx$ 1 flow, but not exact critical Fr = 1 flow. In modern channel design practice, ~0.8 < Fr < ~1.2 is the prescribed Fr range to obtain highest flow rates that accompany Fr $\approx$ 1 flows to avoid surface wave instabilities associated with translating large scale vortex motion below the water surface associated with exact Fr = 1 critical flow. The Fr$_2$ Froude number range in the contracted (2) channel lies between ~0.8 < Fr < ~1.2 and the compute Fr$_2$ ~ 1.14 value is consistent with near stable flow according to modern hydraulic engineering standards. The width reduction construction shown in Figures 11–13 thus yields the narrowest supply channel (2) at the maximum flow rate per unit channel width without significant internal surface wave structures causing transient flow instabilities to be passed on further downstream to the Waterfall area. Observing the narrow channel leading to the stilling reservoir ahead of the waterfall, it appears that Inka hydraulic engineers wanted to limit disturbances and currents within the reservoir upstream of the overflow streams to help promote equal, stable flows into the four Waterfall channels. The contracted channel (2) then serves this purpose as no upstream influence exists from flow disturbances in the downstream direction for Fr > 1 flows. The contracted width channel led to transecting channels which when blocked, served as a reservoir with a further channel to the platform immediately ahead of the four-channel waterfall (Figure 10). This channel design promoted near symmetrical flow conditions in the reservoir so that waterfall channels on each side of the inlet channel had symmetrical input flows from the reservoir. As flow into the channel is directed into a smaller channel directly across from the supply channel (Figures 11–13), there is an increase in the Froude number—but this value is still close to the critical Froude number as shown in the Figure 15 process path connecting F to G. With Fr > 1 flow in contracted channels ahead of the waterfall, no upstream disturbances can affect the stability and aesthetics of the Waterfall.

Further, near critical flow is associated with the maximum flow rate per unit channel width that a channel can support; this is closely achieved for Fr$_2$ = 1.14 conditions. Flow stability is achieved when Fr is either slightly less than or more than the critical Fr = 1 condition—knowledge and use of this hydraulic engineering practice is evident in the Principal Fountain design and is vital to produce a constant, stable water delivery flow to the Waterfall area. The presence of near critical flow in both channel (2) and its continuance channel had the advantage of preventing upstream influence of any downstream channel flow resistance element (channel bends, wall roughness effects, non-symmetric flow into reservoirs creating surface waves and vortices disturbances) from creating flow instabilities in the subcritical (1) wide channel that would translate instabilities into downstream flow patterns. For Fr > 1 flows in all downstream channels, uniform flow to the waterfall is guaranteed to preserve its aesthetic display.

The contracted channel flow design is important as: (1), the flow from the spring source channeled into the wide channel (1) to a downstream narrow width channel (2) is associated with the maximum flow rate per unit channel width; (2) the near critical flow values in the two channels downstream of the wide section channel (1) eliminate flow disturbances propagating upstream derived from flow into the reservoir connecting channel and other downstream resistance sources that would alter the stability of flow to the Waterfall; (3) the channel leading to the reservoir immediately upstream of the Waterfall, when partially blocked to induce slow water entry into the reservoir, serves to create a stilling basin to help eliminate any reservoir water motion that would challenge symmetrical flow delivery to the fountain area and; (4), any disturbances from flow into the downstream stilling basin are not propagated upstream into the wide (1) near critical flow channel to destabilize its flow to downstream channels. Why is limiting upstream influence important? If Inka engineers chose a channel design that had a wider width throughout than those shown in Figures 11–13, then the value of Fr < 1 would exist in all channels so that unstable disturbances from flow width transitions to the Waterfall display area would propagate in both up- and down-stream directions. An example disturbance associated with subcritical flows into sequentially wider channels would be a sudden water velocity decrease that would be felt as an "obstacle" and have an upstream influence on the subcritical flow in the upstream wider channel section. The low velocity, unstable upstream water height changes would then interact with the incoming spring flow producing transient surface wave instabilities that propagate downstream in all wider channels and would lead to an erratic, non-aesthetic waterfall display. This negative design was anticipated by Inka engineers and avoided by their design shown in Figures 11–13 with its many benefits.

An overview of the Inka technology used for the Principal Fountain reflects modern hydraulic design principles to preserve fountain aesthetics during seasonal water supply changes. Knowledge of channel width change effects on flow regime change from sub- to supercritical lows to achieve stable flow at the maximum flow rate to the Principal Fountain is apparent in the channel and is consistent with modern practice.

As the Late Horizon Inka society (1400–1532 AD) was familiar with water systems of conquered and occupied territories, the Inka had access to the hydraulic knowledge base of contemporary and earlier societies appropriate for use in their royal estates, cities, urban areas, and agricultural systems. As surveys of several contemporary and earlier Andean societies' water system technologies are available in the literature [3,11–39] as well as descriptive expositions of water systems at other Inka and Wari sites it is instructive to determine if water technologies used by the Inka, particularly at Tipon, had borrowings from earlier predecessors. Given that application of different water technologies from different predecessor and contemporary societies were specific to their environmental and water resource conditions, only a limited number of these technologies would be applicable given the mountainous terrain and water resource constraints in the main Inka occupied Cuzco area compared to those conditions for Peruvian coastal societies.

A further question, beyond observation and analysis of Tipon water systems and the water technologies used in the design of these systems, relates to indigenous innovation not previously noted from hydraulic engineering precedents derived from Inka and from other ancient Andean societies. Given the multiple sources of water engineering knowledge available to the Inka, a further question arises as to how close to modern hydraulic engineering practice Inka water systems were. As codified observations of hydraulic phenomena provide a common basis for hydraulic engineering construction in both ancient and modern practice, analysis of the Inka hydraulic constructions at Tipon reveal an early application of hydraulic principles predating their later discovery in western science many centuries later as subsequent discussion reveals.

While appropriated usage of hydraulic engineering knowledge was available to Inka engineers from Inka conquered domains, native innovation and inventiveness would have played a role given the mountainous terrain of the Inka homeland that would require special technologies to irrigate and productively farm. As many different land area types were available for farming by *mit'a* labor extracted from Inka conquered populations transferred to different geographical regions together with the reciprocity and gift giving strategy of the Inka to assimilate different Andean societies into their multiethnic state, transferred water control and distribution technologies were an important part of the Inka strategy to expand and control their state structure. Additionally, control of water systems for urban and agricultural use demonstrates aspects of political power exercised by Inka elites over states conquered by the Inka military as well as symbolic manipulation of sacred water symbols [3,25–35] important to show Inka connections to deities controlling rainstorms and droughts. The importance of understanding Inka water control technologies lies in its importance to maximize agricultural production through elaborate irrigation systems both within the Cuzco area and Inka conquered territories; here surplus production was vital for Inka storage facilities [36] that served as a defensive measure against extended drought periods. As Inka royalty controlled portions of agricultural lands for state governance functions involving ritual bonding ceremonies that included social participation of all classes of Inka society, agricultural success through knowledge of water irrigation technology was vital to demonstrate the management intelligence and reliance on the Inka elite class to the civilian population. While agricultural success based upon water control technology was a key concern of the Inka state, the provision of potable water to Cuzco inhabitants (and at other Inka sites) through display fountains and water basins involved aqueduct design technologies and pressurized pipeline systems to provide water from distant spring and river sources. Investigation and analysis of water technologies used for the Principal Fountain and the Main Aqueduct are therefore vital to expose aspects of Inka water technology.

As for Andean precedents for channel width change to regulate Froude number, examples from the Late Intermediate Period (~900–1300 AD) Chimu Intervalley Canal show a similar hydraulic technology designed to regulate Froude number within the ~0.8 < Fr < ~1.2 range to ensure stable flow conditions at a prescribed maximum flow rate of 4.6 ft^3/s to reactivate drought desiccated irrigation canals in the Moche Valley. This was achieved by changing the channel cross section geometry consistent with channel slope and wall roughness variations to make the required flow rate close to maximum flow rate the channel can transport with Fr $\approx$ 1 flows. In this sense, the Inka use of channel width changes to regulate Froude number may derive from the Chimu precedent as both technologies involve creation of stable, near critical flows maintained by channel cross sectional shape changes. Unstable, pulsing flow in the Chimu Intervalley Canal would have the effect of accelerating erosion of unlined canal banks as well as causing oscillatory forces on canal stone-lined banks that would cause leakage into unconsolidated foundation soils. The Inka use of a subsidiary channel to maintain the design flow rate to the fountain during seasonal changes in spring flow rate is further acknowledgment that flow addition/subtraction water control engineering was necessary to preserve the fountain's aesthetics. It is noted that the Inka transferred experts from conquered societies to their Cuzco homeland to support advances in ceramic, textile, and metal-working technologies—likely such transfers also included water engineering experts to support Inka dominance and occupation of their conquered territories.

Noting that the Figure 10 Waterfall streams are close to the back wall, the x distance of impact of the overfall stream from the vertical wall on to the flat bottom base platform is $x = V_0 (2z/g)^{1/2}$ where V_0 is the water velocity just before the fountain overfall edge and z is the height difference between the overfall edge and the flat bottom base here given as ~4.0 ft. As x is approximately ~0.25 ft, then V_0 ~ 0.5 ft/s. As the input flow rate (assuming canal extraction/supplements are not present) is ~0.63 ft^3/s and, as four channels supply the fountain, then the water height in each of the four channels is ~0.3 ft which is consistent with the four-waterfall input channel depth measurements. The point made here is that to achieve an aesthetic fountain display, a precise and equal water velocity into all four channels was required. This was achieved by the channel geometry change (Figures 11–13) to transition Fr < 1 flow in the wide channel to a near critical Fr > 1 flow in subsequent contracted channels to eliminate downstream flow resistance effects on upstream flow conditions. Additionally, as seasonal rainfall varies, affecting supply/drainage canal flow rates and the platform 11 spring flow rate, the supply channel's upstream flow additions/subtractions from aqueduct and spring were necessary to maintain a constant flow rate to the fountain.

The realization that the channel contraction effect (W2/W1 in Figure 15) leads to an Fr ≈ 1 near maximum flow rate in the contracted (2) channel was a necessary first step in the Inka design. This feature was necessary to deliver a stable water flow to the downstream delivery system to eliminate instabilities that would compromise equal water delivery to all four Waterfall channels. The attention paid to flow stability and elimination of upstream influence effects that would compromise downstream flow delivery stability demonstrates Inka water management techniques to achieve waterfall aesthetics. In modern hydraulic engineering terms, the Froude number of the incoming water stream determines the Froude number in an expansion W2/W1 > 1 or contraction W2/W1 < 1 stream (Figure 15). Here Inka water engineers demonstrated knowledge of this concept to design channel expansion/contraction control of water flows as being Froude number dependent. Together these design features indicate a sophisticated hydraulic technology in use to preserve fountain characteristics during water seasonal supply changes.

4. The Main Tipon Aqueduct

The Main Canal's aqueduct [5] (p. 60, PLATE) upstream of the site of Intiwatana (Figure 1) demonstrates several additional facets of Inka hydraulic engineering. A 40 by 25 m reservoir above the Intiwatana sector conducted water from the Main Aqueduct supplied from the Pukara River (Figure 1) to supply fountains in the adjoining main plaza (Figure 1) as well as branch canals to the Main Fountain. Section 4 of the Main Aqueduct included a long, steep channel of ~16 degrees measured declination slope followed downstream by a ~1.7-degree mild slope (Figures 16 and 17). Channel width dimensions range from ~0.65 to ~0.8 ft while channel depths range from ~0.8 to ~1.0 ft indicating near-constant dimensions of the rectangular cross section aqueduct channel [5] (pp. 53–63). Upstream of the aqueduct is a built-in water diversion channel [5] (pp. 53–63) activated by a movable sluice plate to control the water delivery flow rate to the aqueduct—excess flow was diverted to a drainage channel. As flow in the steep section is supercritical (Fr > 1) and approaches normal depth asymptotically on a steep slope [5] (pp. 6–9, p. 47, p. 35), when the flow approaches the downstream mild slope, horizontal channel section, a hydraulic jump occurs early in the flat channel section [5] (p. 60) as no channel width and shape change is present to lower or cancel the hydraulic jump. Figure 18 schematically shows the formation of the hydraulic jump (HJ) at the lower slope segment of the aqueduct.

This hydraulic structure occurs at the slope transition A-B coordinate location shown in Figure 17. A typical hydraulic jump shown in Figure 19 is representative actual jump at the A-B location. (Drawing by C. Ortloff).

In terms of the Froude number, supercritical Fr > 1 flow on the steep section of the aqueduct is converted to subcritical Fr < 1 flow in the aqueduct flat section through creation of a hydraulic jump. For flow downstream of the hydraulic jump on the near horizontal mild slope (Fr < 1), the aqueduct slope continues downhill on a mild slope to the final destination in the city center. In this part of the aqueduct, although the water velocity increases, the post-hydraulic jump flow has no significant upstream effect on the location and height change of the hydraulic jump. The hydraulic jump feature provides a means to calculate the flow rate in the aqueduct channel given that the hydraulic jump flow height produced by the supercritical to subcritical flow transition is to be contained within the channel wall height to eliminate spillage from the channel. As water spillage represents a waste of precious water intended for agricultural and urban use purposes, the canal design reflects Inka engineer's concern to eliminate water wastage. To determine the correct canal flow rate to eliminate spillage, modern hydraulic engineering analysis methods are employed. Here trial water depths and the associated flow rate calculated from the

Manning equation [5] (p. 39) were used to determine if the hydraulic jump height produced at the base of the steep aqueduct section is contained within the channel wall height dimensions. As the correct flow rate in the Main Aqueduct is as yet unknown to eliminate spillage, trial water flow heights (h_1) and associated trial flow rates for different steep slopes [5] (p. 60) together with the water velocity consistent with trial heights and slopes, permits trial Fr_1 values to be calculated at the base of the steep slope. From the trial Fr_1 and h_1 values, the hydraulic jump height relation shown as Equation (1) below [5] (p. 47) is used to determine the hydraulic jump height h_2; the post hydraulic jump Froude number Fr_2 on the mild aqueduct slope is determined [9] by Equation (2). Based on calculations of the trial hydraulic jump water height change from Equation (1), the correct hydraulic jump water height and the correct aqueduct flow rate is determined when the h_2 water height is fully contained within the channel height to eliminate channel spillage. The height change h_2 from the original trial water height h_1 due to a hydraulic jump is given as:

$$h_2/h_1 = (1/2) [(1 + 8\, Fr_1^2)^{1/2} - 1] \tag{1}$$

and the post hydraulic jump Froude number Fr_2 is from [7–10]:

$$Fr_2 = \{(1/8) [1 + 2\, (h_1/h_2)]^2 - 1/8\}^2 \tag{2}$$

Figure 16. Main Aqueduct Canal Section 4 showing the steep sloped channel section and adjacent stairways. Hydraulic jump occurs at midpoint in figure where channel angle changes from steep to a mild, near horizontal slope. The channel section shown is part of the 'aqueducto' (Figure 1) that supplies water to the terraces (Figure 2) and is located near to the Plaza Ceremonial (Figure 1). Coordinate locations, vertical from A, horizontal from B, give the location of the hydraulic jump. (Photo by C. Ortloff).

Figure 17. Alternate view of the Main Aqueduct Section 4 canal steep slope section shown in Figure 17. The hydraulic jump location occurs where the channel slope angle changes from steep to near horizontal shown on left side of figure. The hydraulic jump location on the aqueduct is located at the junction at the A and B coordinates. (Photo by C. Ortloff).

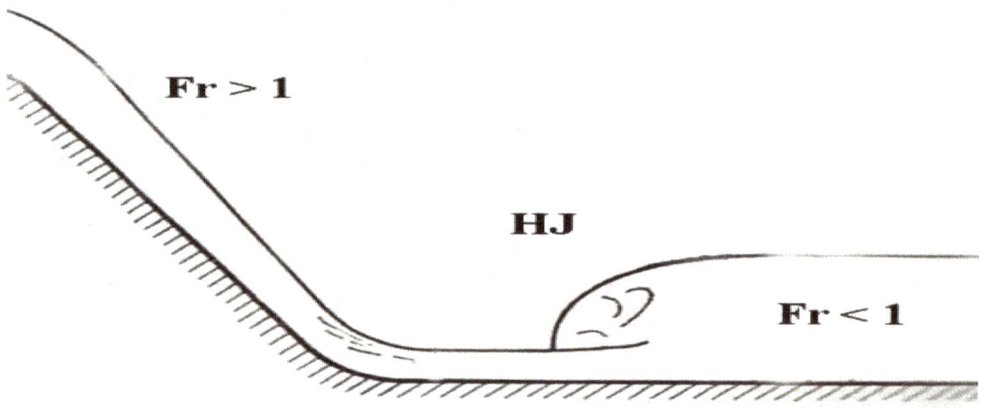

Figure 18. Schematic representation of the formation of a hydraulic jump (HJ) due to the aqueduct slope change from a steep to a mild near-horizontal slope.

Figure 19. Actual hydraulic jump caused by a high velocity (right-to-left) Froude number water flow (Fr >> 1) encountering a submerged plate obstacle at the leftmost exit of a hydraulic flume. (Photo by C. Ortloff).

Using elements of [5] (p. 60) Table for different trial water depths at different trial slopes where the volumetric flow rate is in ft^3/s, H is the post hydraulic jump total water height in feet, Fr_1 the supercritical Froude number characterizing flow at the base of the steep slope channel, Fr_2 is the subcritical Froude number characterizing the post hydraulic jump flow, h_2/h_1 is the hydraulic jump height change ratio, the h_2 water height can be determined. D_L is the trial lower height normal depth at the base of the sloped section for trial flow rates. From Equations (1) and (2), Table 1 results follow.

Table 1. Flow Rate Results for the Major Aqueduct.

D_L (in)	Slope (Degrees)	Velocity (ft/s)	Flow Rate (ft^3/s)	Fr_1	h_2/h_1	H (ft)	Fr_2
2	8.53	6.6	0.80	2.85	3.65	0.61	0.42
2	11.3	7.6	0.92	3.29	4.18	0.70	0.24
2	16.7	9.3	1.12	4.01	5.19	0.87	0.34
4	8.53	8.6	2.08	2.63	3.25	1.08	0.45
4	11.3	9.9	2.40	3.03	3.81	1.27	0.41

Based on Table 1 results, the maximum flow rate (assuming the aqueduct flow diversion channel plate is completely raised and flow in the aqueduct is from the Rio Pukara River source only), is on the order of ~1.12 ft^3/s (~0.03 m^3/s) and is sufficient for water to be safely contained in the channel without spillage given the ~1.0 ft depth and ~1.0 ft width of the channel [5] (p.62). Note here that the flow rate obtained in [5] by a different analysis method is of the same order of magnitude as the presently determined value. For this case, the 16.7° slope case closely matches the measured aqueduct steep slope. For higher aqueduct flow rates on the same steep slope, Table 1 indicates a large degree of spillage as the hydraulic jump water height exceeds the channel wall height dimension. An upstream

Section 1 part of the channel has a water diversion structure [5] (p.56) used in conjunction with an adjustable height sluice plate used to divert a fraction of the Main Aqueduct's water into a side turnout irrigation canal. This diversion structure was used in times of heavy rainfall periods to divert water from the aqueduct channel to prevent spillage in downstream portions of the Main Aqueduct. Some evidence of a slight widening of the channel is apparent from Figure 16 in the near-post hydraulic jump region. This effect, if deliberate and more substantial in width, would reduce spillage and permit a higher flow rate on the order of ~2.0 ft^3/s. This feature, however, is not evident in the aqueduct design. The Figure 15 Q-P arrow indicates the Froude number transition from supercritical Fr_1 to subcritical Fr_2 due to the hydraulic jump located at the steep-mild slope transition point for contained flow within the channel; results shown indicate consistency with the governing equations of fluid motion. For a flow rate higher than ~1.12 ft^3/s, only the lower amount would be deliverable to destination sites and the rest subject to aqueduct channel bank overflow. This consideration defines the intent of Inka water engineers to design the aqueduct for the maximum flow rate fully contained within the aqueduct channel. Based upon Figure 1, the upstream channel ahead of the steep slope section appears to have a mild slope supporting subcritical flow; therefore, only the steep to mild slope segment of the aqueduct shown in Figures 15 and 16 prove useful in determining the aqueduct maximum flow rate. Since the Main Aqueduct was to supplement site water supplies during the drier parts of the year when the Rio Pukara source was at a low flow rate, the water contribution to the urban Intiwatana area (Figure 1) and its storage reservoir would be vital to maintain its potable water source for site inhabitants as well as to supply adjacent canal systems for agricultural use throughout seasonal changes in the river water supply. Again, the sluice plate placed far upstream in the aqueduct channel together with supply and drainage offshoot channels served the purpose to regulate the channel flow rate to a design value to eliminate flow spillage during seasonal changes on water supply.

The question arises as to why the Main Aqueduct flow rate is important to determine. As the aqueduct was the water source for the ceremonial plaza fountains and reservoir and had connection to several of the agricultural platforms (Figure 1), knowledge of its maximum flow rate determined the geometry of several downstream channels designed to accommodate its water flow without overflows or spillage. Additionally, water delivery amounts (ft^3/s-unit land area) for specialty crops on different agricultural platforms with different soil compositions were important to provide the correct moisture levels for plant growth and determine water supply and reservoir structure dimensions within the plaza to maintain their functionality and aesthetic presentation. Thus, Inka designers had the means to determine channel width, depth, channel slope, and wall roughness to tailor flow characteristics dependent upon Froude number to support different delivery flow rates to different agricultural platforms, the aesthetic Waterfall display, and ceremonial plaza usages much in the same way that modern hydraulic engineering practice dictates.

As for hydraulic engineering precedents observed from earlier Andean societies by Inka water engineers, particular attention was paid to limiting canal flow rate beyond design values to prevent canal over-bank spillage. In a (possibly late Moche or Chimu) Jequetepeque Valley Canal, a choke consisting of opposed stones a given distance apart (throat) was installed in a water conveyance channel to limit canal flow rate to a design value. Here the dual stone separation distance determines critical flow at the throat limiting the flow rate. Excessive runoff from El Niño rains flowing into the canal causing a flow rate excess over the design flow rate (as determined by the choke geometry) causes water to be backed up ahead of the choke; the excessive water height was then shunted into an elevated side weir upstream of the choke that emptied water into a lower drainage area thus limiting damaging erosive overbank flow spillage. Elsewhere along the ~50 mile long Chimu Chicama-Moche Valley Intervalley Canal built in the 900–1000 AD time period between the Chicama and Moche valleys of north coast Peru, drainage chutes were placed high up along canal walls to convey excessive water from El Niño flood events away from the main canal [22,24]. For this canal, for excessive flow rates over the design flow rate

(4.6 ft^3/s), sophisticated channel shaping was used to create vortex regions in concave channel side pockets that effectively narrowed the streamline path of the channel flow to convert a sub- to super-critical flow. The amplified water height from the hydraulic jump was then diverted into an elevated side weir and then conducted to a diversion channel that led to a lower farming area. From known cases observed from Chimu (and possibly late Moche) hydraulic engineers' work, special attention was paid to preserving water transport canals from erosive damage due to water spillage over canal banks. As the Inka incorporated the Chimu Empire by conquest, water technology experts were exported back to the Cuzco Inka capital to serve in advisory roles for hydraulic engineering projects. The preoccupation of Chimu (and hydraulic engineers from different societies) to preserve their agricultural systems by innovative adaptive strategies under climate change duress (principally flood and drought) is documented in [21]. Given the same preoccupation of the Inka society to preserve their vital agricultural and urban water supply systems, importation of relevant hydraulic technology was a vital concern. As a further example of technology available to Inka engineers, the complex channel shaping noted at the exit of a steep sloped channel at Tiwanaku's Lukurmata area created a hydraulic jump of zero height necessary to limit erosion damage of an unlined canal; this was accomplished by channel geometry change to induce critical flow at the exit of the steep sloped channel. The technology to raise the channel's flow rate by reducing or eliminating the hydraulic jump height by widening the canal width and changing the local slope to limit the hydraulic jump height, although available to Inka engineers, was not evident to limit over bank spillage as Figure 16 indicates no substantial channel width or slope change in the approximate flat slope part of the canal. In modern hydraulic practice, a hydraulic jump can be eliminated when the aqueduct mild slope section is made equal to the critical slope together with channel widening to produce a neutralizing reach condition.

A further precedent involving control of groundwater for agricultural purposes involves Tiwanaku raised field agricultural systems. Water supplied to raised field swales (located between elevated crop-planted berms) through a channel was used to regulate groundwater/swale water height for different crop growth requirements This provides, in theory, an analog to Tipon's controlled moisture content for different crops on their agricultural platforms. An elevated channel side weir cut into in the Tiwanaku Pampa Koani water supply channel directed excess water above the weir bottom height to an adjacent channel that led drainage water directly to Lake Titicaca. Remaining water was then transported to raised field system swales to set the local groundwater/swale level. The elevated weir flow control regulated the flow rate from the main channel to supply the correct amount of water to the field system swales for different crop water requirements. This control system was vital to maintain the productivity of the raised field systems bordering Lake Titicaca during high seasonal rainfall periods as the supply channel side weir could be blocked thus transporting all water directly to Lake Titicaca. This water control system limited field system saturation destructive to agriculture to maintain the required groundwater and swale water height to support crops on raised field berms. Given differences in distance between Bolivian altiplano Tiwanaku and Peru's Cuzco, and centuries between the Tiwanaku Middle and Inka Late Horizon times, necessary agricultural water supply technologies were vastly different, yet it remains questionable if the Inka understood and utilized aspects of the Tiwanaku technology to control berm moisture levels for different crop types. As all Andean societies were aware of water control technologies to maintain the correct moisture level in field systems (or platforms) for different crops, this consideration was independently developed by societies inhabiting different ecological niches to provide the agricultural support base of their society. Inka water engineers were aware of different water supply systems developed by different societies in different ecological zones as Inka conquest of all lands and societies necessitated continued use and further development of these lands for their occupation.

5. The Site of Moray

Given the complex water supply system of Cuzco that included multiple spring-supplied channels as well as pressurized pipe systems supplying potable fountain water to city inhabitants, Inka water technology was in an advanced state of knowledge. It may be speculated that the complex terrace system at Moray (Figure 20) near Cuzco played a role as an experimental facility to help Inka engineers design the many water transport facilities used in agriculture and urban use in the Inka realm. The site has temperature and moisture level variations on terraces that vary with altitude from the lowest level base surface that may have played a role in determining the best conditions for different crop types. The terrace moisture level changes derive from their height intersection with the local aquifer water content which varies with depth from the ground surface. Given that the site is located in a high rainfall zone in the Andean foothills, the uppermost terraces experience a high aquifer water content given their closeness to the ground surface; lower terraces at greater depths from the ground surface experience low aquifer water content.

Figure 20. The Inka agricultural platform system at the site of Moray located some ~15 km southwest of the Cuzco area. (Photo by C. Ortloff).

Given that the individual platforms at Tipon (Figure 2) have regulated water content through a complex water distribution network to provide the optimum aquifer moisture conditions for given specialty crops, it may be surmised that Moray served as an Inka agricultural test center to determine best water supply conditions for specialty crops in advance of the Tipon design. The knowledge gained from the Moray site would then be of use to design Tipon's individual terrace water control systems in advance. While this interpretation of the function of the Moray site is controversial with some researchers appropriating a 'ceremonial' or 'religious' function of the site given the intelligence of the Inka to design complex water systems using their advanced knowledge of hydraulic principles as demonstrated at Machu Picchu and other Inka royal estates [37–39], nothing is left to chance in the design of water systems that have a practical use. As previously discussed, the Inka had access to water engineering specialists brought to Cuzco from conquered territories to develop their water engineering base—this is reflected in the many Inka sites with elaborate water engineering structures.

Of interest are the many words in the Quechua language used by the Inka related to water. Many words relate to the hydraulic technology that was prevalent in Late Horizon Inka times such as *pincha*, a water pipe, *rarca*, an irrigation ditch, *patqui*, a channel, and *chakan*, a water tank, while other words describe water motion such as *pakcha*, water falling into a basin, and *huncolpi*, a water jet. While these words were commonplace among the

Inka public given their dependence upon water for agricultural and urban use, there must have been a more complex vocabulary to describe more complex hydraulic phenomena used by Inka hydraulic engineers involved in the design Tipon, Machu Picchu and other water systems given the complex water technology used in those systems.

6. Conclusions

The utilization of multiple spring water sources together with a river sourced aqueduct used to supply surface and subsurface channel networks controlling the water supply and drainage systems of Tipon was part of an intricate Inka design to control the moisture content of the agricultural platforms for specialty crops suitable for the royal occupiers of the site. The use of subsidiary channels with independent water sources intersecting main channels was part of a complex water control system designed to supplement (or drain) water to achieve design flow rates to key site areas. For the Principal Fountain, the input water flow rate was carefully controlled by subsidiary canals, a spring source and aqueduct flow to control the flow rate to ensure fountain aesthetics during seasonal changes in spring water supply. For the Main Aqueduct, the flow rate into the aqueduct from the Rio Pukara source was regulated by a movable sluice gate to ensure wasteful aqueduct spillage was eliminated. The maximum aqueduct delivery flow rate (for the sluice gate in full-open position) was determined by the maximum contained water height in the hydraulic jump region of the aqueduct to eliminate overbank spillage. The realization that a control sluice gate to divert excess flow from the aqueduct beyond its design flow rate was a necessary part of the aqueduct system design to eliminate spillage was acknowledgement of the thought process behind Inka hydraulic engineering practice. In summary, the presence of intersecting canals as part of the design for both the Principal Fountain and the Main Aqueduct indicate supplemental (or drainage) water controls designed to achieve precise flow rates necessary for the intended destination purpose of these hydraulic features. As revealed by modern hydraulic theory illustrated by Figure 13, the relation between input channel Froude number (Fr_1) and its relation to the Froude number (Fr_2) in an expanded (or contracted) channel is a complex hydrodynamic procedure. As observed from the analysis of the Principal Fountain and the Main Aqueduct, Inka engineers understood, likely by trial-and-error observations and recordings, that the channel contraction geometry they chose for the water supply to the Principal Fountain converted the low speed subcritical flow in the wide channel to high speed, near critical flow, in the contracted channel section to give the Principal Fountain its proper function and aesthetics. The use of near critical flow in the supply channels to the Waterfall Fountain eliminated upstream resistance influence that would influence flow stability and waterfall aesthetics. As with all hydraulic engineering projects, both ancient and modern, application of an engineering knowledge base underlies the design and function of complex water supply and delivery systems to ensure successful operation. In modern hydraulic engineering practice, test work involving models placed in a hydraulic flume provide visual confirmation of theoretical flow predictions and, for many cases, empirical correlation equations derived from test observations to describe flow phenomena. It may be assumed that a similar form of observational data was available to Inka engineers to know in advance how to design channel geometry changes to achieve a desired effect. Inherent to the design of the Tipon water network is as yet unknown Inka format for analyzing and recording observations of water engineering experiments and tests necessary to achieve the Tipon system's success. This process would involve pre-scientific notations for water velocity, flow rate, water height change, flow stability and farming aquifer moisture levels, among other parameters, and their mutual interaction to predict water flow patterns and efficient farming productivity results.

The engineering base used to design complex water facilities would involve use of *yupanas* (and *quipus*) for analysis and recording of flow phenomena much in the same way that a canal surveying problem can be represented by *yupana* calculations used by Chimu water engineers. The use of multiple canal water supply and drainage systems for agricultural, royal, and commoner living compounds, ceremonial center functions, experimental agricultural research as well as for the aesthetic fountain displays at Tipon and within Cuzco, are prime examples of Inka knowledge of hydraulic engineering principles and constitute a notable contribution to the history of hydraulic engineering. The water system designs exhibited at Tipon are notable in their reliance on a modern water engineering practice and, as such, anticipate discovery of modern hydraulic principles by western science by many centuries.

It is of interest to compare the state of hydraulic knowledge in ancient Andean societies with that of continental Europe in similar AD centuries. The water engineering technology exhibited by 300–1100 AD Tiwanaku, 900–1480 AD Chimu and the 1480–1532 AD Inka societies and further Andean societies indicate a reservoir of advanced hydraulic technologies comparable in many ways to modern technology. Roman water technology exhibits comparable levels of advanced hydraulic knowledge. For both ancient Old and New World societies thus far investigated, accumulated hydraulic knowledge relied upon test, experiment, and nature observations put into codified formats and recording procedures that are, unfortunately, lost due to lack of surviving written records or other forms of recording hydraulic knowledge.

To recover ancient societies' versions of hydraulic science, use of modern hydraulic engineering methodologies involving computer simulations, theoretically derived equations, and laboratory test procedures applied to analyze ancient water structures provide one way to uncover the lost knowledge used in their design and operation to bring forward the design intent of ancient engineers—although their techniques of obtaining and applying this knowledge and the format and data recording and data storage methods used to produce sophisticated water system designs is as yet unknown. Only in the 17th and 18th centuries in Europe did the invention of mathematical descriptions of physical phenomena using calculus methodologies together with concepts of mass, momentum, and energy conservation equations lead to basic calculation methods applied to predict fluid motion. In this regard Bernoulli's 1738 publication *Hydrodynamica* initiated and advanced hydraulic calculation methodology to levels that continued in complexity to the present day. That ancient societies of both Old and New Worlds utilized advanced hydraulic technologies remains a subject from which future research can uncover further details—most fascinating is that ancient hydraulic science demonstrates alternate ways to describe and utilize the hydraulic knowledge base only discovered in later centuries.

Funding: Individually funded, no external funding.

Data Availability Statement: All data by provided by author.

Acknowledgments: The author wishes to thank Ken Wright, director of the Wright Water Engineers, for use of his Figure 1 site map and site analysis presented in his book Tipon: Water Engineering Masterpiece of the Inca Empire that inspired a visit to Tipon to examine the site's many unique hydraulic engineering features. Ken's pioneering work at Tipon and many other ancient Peruvian sites has brought forward new aspects of water engineering accomplishments of pre-Columbian societies previously unknown in the archaeological literature and added a new dimension to the history of water science from ancient Andean societies.

Conflicts of Interest: The present author declares no conflict of interest.

References

1. McEwan, G.; LeVine, T. Investigations of the Pikillacta Site: A Provincial Huari Administrative Structure in the Valley of Cuzco. In *Huari Administrative Structure Prehistoric Monumental Architecture and State Government*; Isbell, W., McEwan, G., Eds.; Dumbarton Oaks Research Library and Collection: Washington, DC, USA, 1991; pp. 93–119.
2. Bray, T. *The Archaeology of Wak'as: Explorations of the Sacred in the Pre-Columbian Andes*; University Press of Colorado: Boulder, CO, USA, 2016; p. 171.
3. Bauer, B.; Covey, A. The Development of the Inka State (AD 1000–1400). In *Ancient Cuzco: Heartland of the Inka*; Bauer, B., Ed.; University of Texas Press: Austin, TX, USA, 2004.
4. Mithin, S. *Thirst: Water and Power in the Ancient World*; Harvard University Press: Cambridge, MA, USA, 2012; pp. 274–277.

5. Wright, K.; McEwan, G.; Wright, R. *Tipon: Water Engineering Masterpiece of the Inca Empire*; American Society of Civil Engineers Press: Reston, VA, USA, 2006; ISBN 0-7844-0851-3.
6. Woodward, S.; Posey, C. *Hydraulics of Steady Flow in Open Channels*; John Wiley and Sons: London, UK, 1941; p. 140.
7. Morris, H.J. *Wiggert Open Channel Hydraulics*; The Ronald Press: New York, NY, USA, 1972; p. 190.
8. Chow, V.T. *Open-Channel Hydraulics*; McGraw-Hill Book Company: New York, NY, USA, 1959; p. 451.
9. Henderson, F. *Open Channel Flow*; New York, the Macmillan Company: New York, NY, USA, 1966.
10. Bakhmeteff, B. *Hydraulics of Open Channels*; McGraw Hill Book Company: New York, NY, USA, 1932; p. 254.
11. Ortloff, C.R. Engineering Aspects of Tiwanaku Groundwater Controlled Agriculture. In *Tiwanaku and its Hinterland: Archaeology and Paleoecology of an Andean Civilization*; Kolata, A.L., Ed.; Smithsonian Institution Press: Washington, DC, USA, 1996; Volume 1, pp. 153–168.
12. Kendall, A. *Aspects of Inka Architecture Description, Function and Chronology*; British Archaeological Reports, International Series 242; Oxford University Press: Oxford, UK, 1985; Volume 1.
13. Brundage, B. *Lords of Cuzco*; University of Oklahoma Press: Norman, OK, USA, 1967; pp. 379–380.
14. D'Altroy, T. *The Inkas*; The Blackwell Publishing Company: Oxford, UK, 2003.
15. Moseley, M.E. *The Inkas and their Ancestors*; Thames & Hudson: New York, NY, USA, 2001.
16. Patterson, T. *The Inka Empire: The Formation and Disintegration of Pre-Capitalist State*; Berg Publishers: New York, NY, USA, 1991.
17. Earle, T.; D'Altroy, T. The Political Economy of the Inka Empire: The Archaeology of Power and Finance. In *Archaeological Thought in America*; Lamberg-Karlovsky, C., Ed.; Cambridge University Press: Cambridge, UK, 1989; pp. 183–204.
18. Rowe, J. Inca Culture at the Time of the Spanish Conquest. In *Handbook of South American Indians 2: The Andean Civilization*; Steward, J., Ed.; Bureau of American Ethnology, Government Printing Office: Washington, DC, USA, 1946; pp. 183–330.
19. Morris, C. Inka Strategies of Incorporation and Governance. In *Archaic States*; Fenman, G., Marcus, J., Eds.; School of American Research Press: Santa Fe, NM, USA, 1998; pp. 293–309.
20. Wright, K.; Wright, R.; Zegarra, A.; McEwan, G. *Moray: Inka Engineering Mystery*; American Society of Civil Engineers Publication: Reston, VA, USA, 2011; ISBN 978-0-7844-1079-0.
21. Ortloff, C.R. *Water Engineering in the Ancient World: Archaeological and Climate Perspectives on Societies of Ancient South America, the Middle East and South-East Asia*; Oxford University Press: Oxford, UK, 2010.
22. Ortloff, C.R. *The Hydraulic State: Science and Society in the Ancient World*; Routledge Press: New York, NY, USA, 2020.
23. Ortloff, C.R. *Hydraulic Engineering in Ancient Peru and Bolivia*; Encyclopaedia of the History of Science, Technology and Medicine in Non-Western Cultures; Springer Publications: Heidelberg, Germany, 2014.
24. Ortloff, C.R.; Moseley, M.E.; Feldman, R. Hydraulic Engineering Aspects of the Chimu Chicama-Moche Intervalley Canal. *Am. Antiq.* **1982**, *47*, 572–595. [CrossRef]
25. Zuidema, T. *Inka Civilization at Cuzco*; University of Texas Press: Austin, TX, USA, 1990.
26. Quilter, J. *The Ancient Central Andes*; Routledge Press: New York, NY, USA, 2014; pp. 274–275.
27. Urton, G. *Inka Myths*; University of Texas Press: Austin, TX, USA, 1999.
28. Rowe, J. An Account of the Shrines of Ancient Cuzco. *J. Inst. Andean Stud.* **1979**, *17*, 2–80. [CrossRef]
29. Moore, J. *Cultural Landscapes in the Ancient Andes*; University Press of Florida: Gainesville, FL, USA, 2005.
30. Moseley, M.E.; Tapia, J.; Satterlee, D.; Richardson, J. Flood Events, El Niño Events, and Tectonic Events. In *Paleo-Enso Records*; International Symposium Extended Abstracts; Ortlieb, L., Machare, J., Eds.; OSTROM: Lima, Peru, 1992; pp. 207–212.
31. LeVine, T. *Inka Storage Systems*; University of Oklahoma Press: Norman, OK, USA, 1992.
32. Murra, J. Rite and Crop in the Inka State. In *Culture in History*; Diamond, S., Ed.; Columbia University Press: New York, NY, USA, 1960; pp. 393–407.
33. Hyslop, J. *Inka Settlement Planning*; University of Texas Press: Austin, TX, USA, 1990; p. 137.
34. Rouse, H. *Elementary Mechanics of Fluids*; Dover Publications: Mineola, NY, USA, 1978; p. 145.
35. MacLean, G. Sacred Land, Sacred Water: Inka Landscape Planning in the Cuzco Area. Ph.D. Thesis, University of California, Berkley, CA, USA, 1986.
36. Niles, S. Style and Function of Inka Agricultural Works near Cuzco. *J. Andean Stud.* **1982**, *20*, 163–182.
37. Wright, K.; Zegarra, V. *Machu Picchu: A Civil Engineering Marvel*; ASCE Press: Reston, VA, USA, 2000.
38. Wright, K.; Zegarra, V. Ancient Machu Picchu Drainage Engineering. *J. Irrig. Drain.* **1999**, *125*, 360–369. [CrossRef]
39. Wright, K.; Kelly, J.; Zegarra, V. Machu Picchu: Ancient Hydraulic Engineering. *J. Hydraul. Eng.* **1997**, *123*, 838–843. [CrossRef]

Article

The Masterful Water Engineers of Machu Picchu

Kenneth R. Wright

Wright Water Engineers, Inc., 2490 W. 26th Ave., Ste. 100A, Denver, CO 80211, USA; krw@wrightwater.com

Abstract: The water engineering achievements of the Inca at Machu Picchu, when defined in technical terms common to modern engineers, demonstrate that the Inca were masterful planners, designers, and constructors. They demonstrated their technical skills through the planning, design, and construction of water supply, fountains, terraces, foundations, walls, and trails. The site of Machu Picchu was a difficult place to build, with high precipitation, steep terrain, and challenging access. Nonetheless, the Inca had the uncanny ability to plan public works and infrastructure in a manner that fit this problematic site and lasted for centuries.

Keywords: Machu Picchu; Inca; ancient water engineering; hydraulics

1. Introduction

It was in the 1913 issue of National Geographic magazine that Hiram Bingham first announced to the world that the "Inca were good engineers" [1]. History Professor Bingham came to this conclusion after his 1912 clearing and mapping of Machu Picchu.

Started in A.D. 1430–1450 as a royal estate of the emperor Pachacuti, the Machu Picchu site was an unlikely place to build what would become, 550 years later, South America's most well-known archaeological site (Figure 1).

Figure 1. The Machu Picchu site is abutted by the mountains of Machu Picchu and Huayna Picchu.

When we think of Machu Picchu, we picture its perfectly battered walls and intricately carved temples, huacas, and niches. While these features, alone, are a testament to the engineering skills of the Inca craftspeople, this article explores the water-related infrastructure of Machu Picchu and the remarkable ingenuity and foresight that it demonstrates.

Machu Picchu is an unlikely place to construct a royal estate due to its remoteness, geologic faults, water availability, and landslide potential. Built on a ridge between the two mountains of Machu Picchu and Huayna Picchu to house a resident population of 300 and a peak population of 1000, the Inca engineers worked with nature to create a community with a reliable water supply, good storm drainage, flat areas for agriculture, and building foundations that would meet the challenges of steep, unstable slopes and high rainfall.

They constructed a remarkable road system, which would connect Machu Picchu with the outside world, not only via Cusco, but also downstream to the Vilcabamba region.

An analysis of the Inca civil engineering achievements at Machu Picchu, ranging from their use of hydrologic and hydraulic principles to their erosion control and soil stewardship, makes it clear that they adopted technologies from other peoples and the earlier empires of Wari and Tiwanaku [2]. Then, as now in the twenty-first century, technology transfer was a key component to successful engineering achievements.

While Machu Picchu is judged to have been mostly abandoned by A.D. 1540, final abandonment did not happen until A.D. 1572 [3]. The site sat alone and isolated, except for a few Quechua Indians, until Bingham made his discovery in 1911 [4].

2. Materials and Methods

Andean archaeologist Alfredo Valencia Zegarra and I led a team of water engineers and hydrologists from Wright Paleohydrological Institute, who conducted approximately eight site visits to Machu Picchu over a six-year period. Our goal was to use our professional expertise as a basis for examining the remains of this ancient site to (1) document the site from an engineering, hydrological, and hydraulic perspective, and (2) develop a theory of what the designers and builders of the site knew about engineering, hydrology, and hydraulics. The team used photographic documentation, measurements, notes, excavations, interviews with experts, laboratory testing, ice core data, and outside research to develop a picture of what the Inca knew. Some of this was accomplished by what we call "reverse engineering," the process of evaluating existing systems to determine the principles that were used in their design. A series of papers on these topics was developed, which served as the basis for detailed mapping and a book, Machu Picchu: A Civil Engineering Marvel [5]. This article provides an overview of the team's findings.

3. Results

3.1. The Physical Setting

The ridge-top site of Machu Picchu, at an elevation of 2430 m, lies between two regional geologic faults (Figure 2), and is a steep 450 m above the Urubamba River. These faults formed a graben upon which Machu Picchu was built, and which relates to the up thrusted twin peaks of Huayna Picchu and Machu Picchu to the north and south at 13°9′ south of the Equator in the headwaters of the Amazon River basin. Machu Picchu experienced moderate temperature variations and high annual rainfall of approximately 2000 mm, as shown in Table 1 [5].

Table 1. Estimated annual precipitation at Machu Picchu by decade based on comparison of modern climatological data at Machu Picchu to data from the Quelccaya ice cap [6], located approximately 250 km southeast of Machu Picchu. Average annual precipitation at Machu Picchu was estimated at 1960 mm. (Reproduced from Machu Picchu: A Civil Engineering Marvel [5], with permission from the American Society of Civil Engineers [ASCE], the publisher.)

Decade	Equivalent Annual Precipitation (mm/yr)
1450–1459	1770
1460–1469	1900
1470–1479	1830
1480–1489	1770
1490–1499	1860
1500–1509	2020
1510–1519	2150
1520–1529	1980
1530–1539	2220
1964–1977	1960

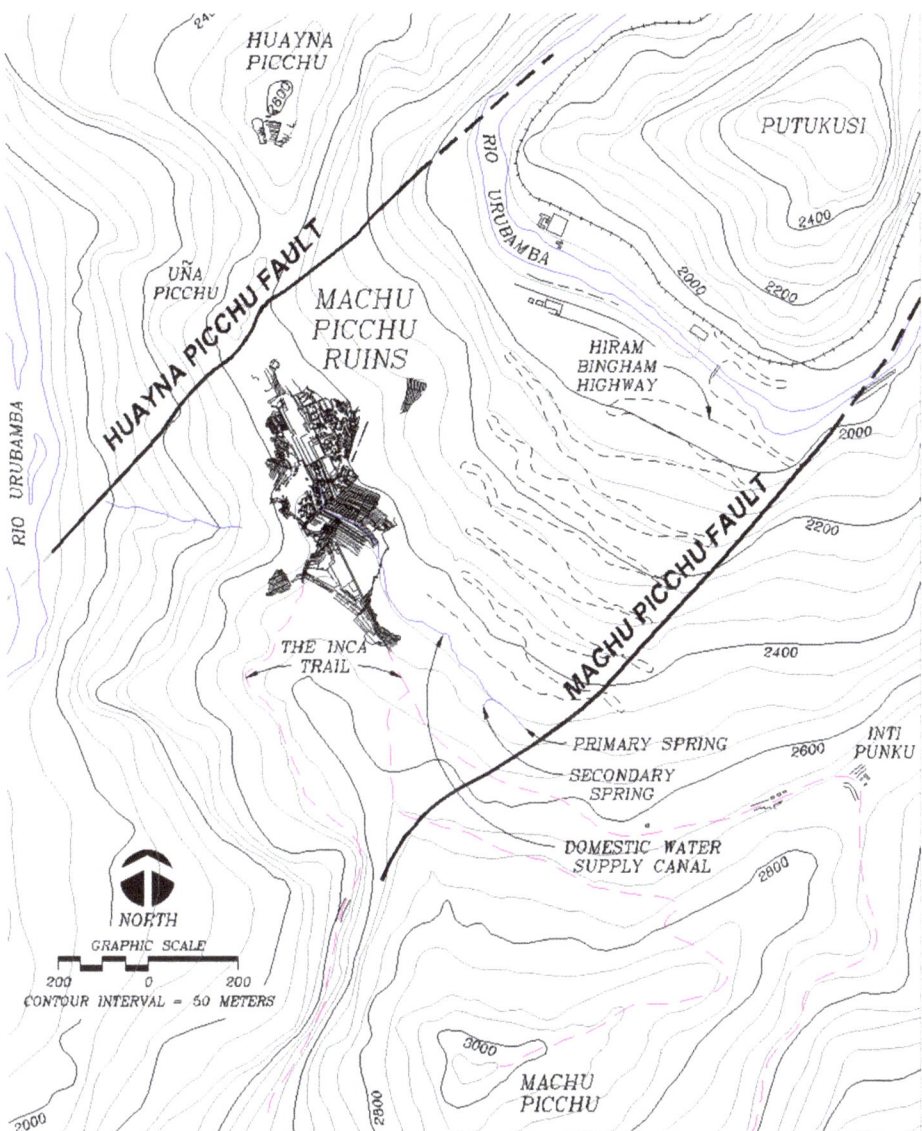

Figure 2. Machu Picchu lies between two faults. (Reproduced from Machu Picchu: A Civil Engineering Marvel [5], with permission from ASCE, the publisher.)

The rainfall, typical ground slopes of 50 percent, local faulting, and soils that are prone to landslides presented the Inca engineers with many challenges to overcome. Overcome them they did, for Bingham's 1912 photographs and descriptions show Machu Picchu to have been nearly free of damaging landslides and failed foundations after 340 years. It is known, however, that the Inca did experience some failures such as a large earth slide near the Main Drain (Figure 3 shows the layout of Machu Picchu and key features such as the Main Drain).

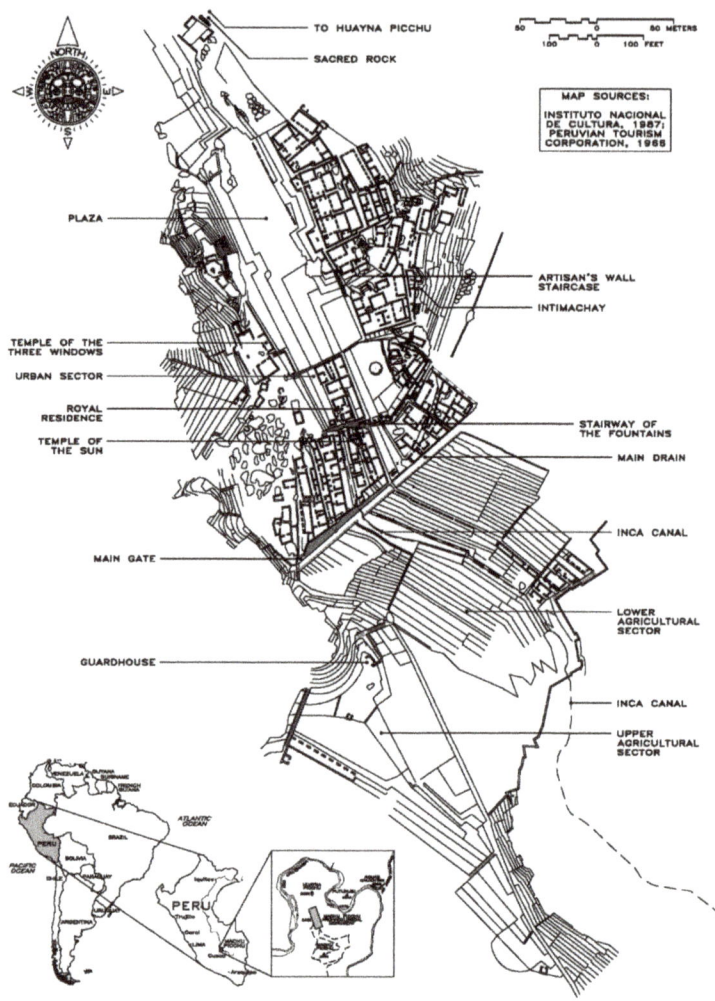

Figure 3. Site map showing layout and key features of Machu Picchu site.

3.2. Water Supply Canal

The Machu Picchu geologic fault caused fracturing and crushing of the adjacent hard granite rock, which in turn resulted in a natural spring water source on the north slope of Machu Picchu Mountain. [5]. The Inca, using a clever groundwater interception structure (Figure 4), carefully developed this spring, which then fed a long domestic water supply canal (Figure 5). The canal traversed the steep mountainside on a narrow terrace formed by a sturdy and well-founded wall—in some places, 4 m in height. The canal, 749 m in length, was built at an engineered slope with a typical hydraulic cross-section of 13 cm × 12 cm, making it capable of handling up to 300 L per minute. The measured spring yield is 23 to 125 L per minute, as shown in Figure 6.

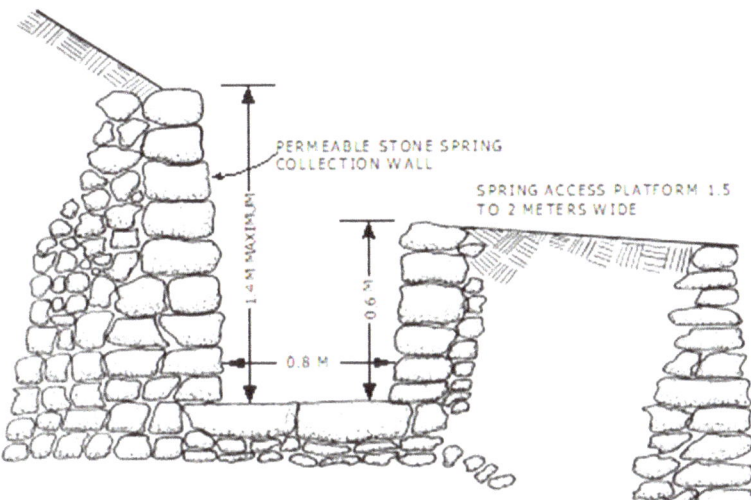

Figure 4. Cross-section of the Inca Spring collection system, which enhanced the domestic water supply for Machu Picchu. The yield of the spring was maximized by a permeable stone wall set into the steep hillside and a collection trench, which in turn fed the canal shown below (Reproduced from Machu Picchu: A Civil Engineering Marvel [5], with permission from ASCE, the publisher.)

Figure 5. The canal at Machu Picchu runs past grain storehouses on a finely constructed terrace.

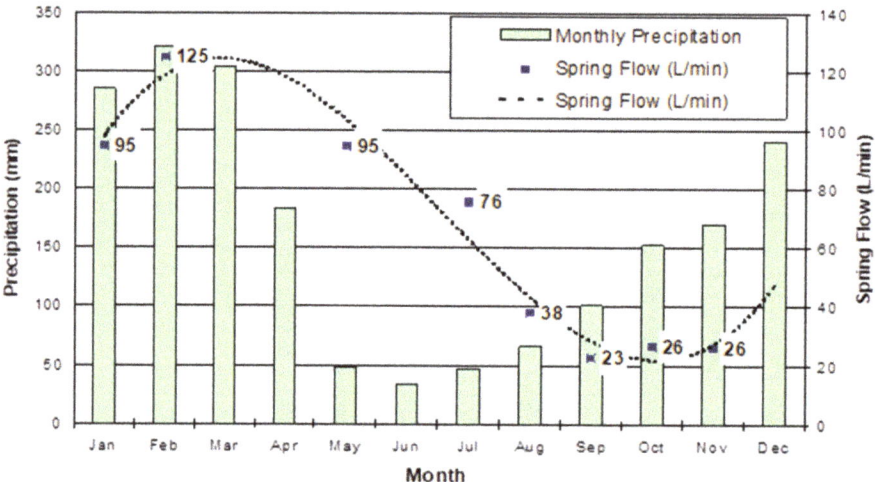

Figure 6. Variations of Inca Spring flow by season. The green bars indicate average precipitation while the squares represent average seasonal measurements of spring flow. The dotted line shows interpolated spring flow, representing a lag time of several months. Based on interpretations of Quelccaya ice cap data [6], modern precipitation data correspond closely with estimated precipitation during the period of Machu Picchu occupation. (Reproduced from Machu Picchu: A Civil Engineering Marvel [5], with permission from ASCE, the publisher.).

The Inca engineers were in the process of constructing a branch canal off the main canal at the time of abandonment, which would have taken advantage of the excess capacity of the main canal.

3.3. Fountains

The Inca canal delivered its water supply by gravity to Fountain #1 at an elevation of 2437 m, which then defined the location of the entrance to the Royal Residence, allowing the emperor to have first use of the pure spring water. Here, the Temple of the Sun, the Sacred Fountain, and the Wayrona were also built. Additionally, incorporated into the special complex were Fountains #2 through #6 (Figure 7).

Downhill, another ten fountains were constructed in series for domestic water supply purposes with the last fountain being a private water supply accessible only from the Temple of the Condor. The remaining water was then carried in a buried conduit and then an open channel for discharge as waste into the Main Drain (a.k.a. Dry Moat) (Figure 8).

The fountains of Machu Picchu are well engineered for the volume and flow of water provided by the Inca Canal. The approach channels were carefully carved to create a jet that would allow for the filling of Inca water vessels called aryballos (Figure 9). The hydraulic design of the fountains allowed for reasonable operation for a flow of between 10 and 100 L per minute. At flows of more than 100 L per minute, a control orifice in Fountain #4 would reject excess water for overflow down the granite stairway of the fountains [5].

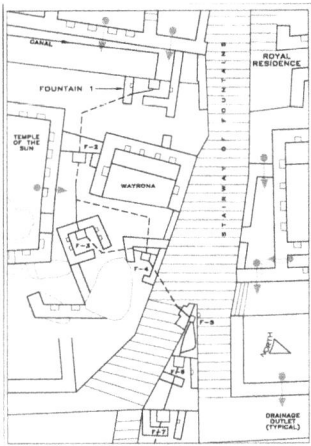

Figure 7. Map of canal and fountains showing relationship to Royal Residence. Fountain #3 (F-3) is the Sacred Fountain. (Reproduced from Machu Picchu: A Civil Engineering Marvel [5], with permission from ASCE, the publisher.)

Figure 8. The main drain or "dry moat" was the route for wastewater to exit the Machu Picchu site. The wastewater flowed to the rainforest below, where lush vegetation minimized the erosive potential of the drainage.

(a) (b)

Figure 9. Inca engineers designed water jets that were sized perfectly for the range of flows (**a**) and which allowed for clay Inca water jugs known as aryballos (**b**) to easily be filled.

3.4. Agricultural Terraces

The numerous agricultural terraces at Machu Picchu total 4.9 ha in area and were built to withstand the ravages of gravity and time. This is made clear by photographs Hiram Bingham took of Machu Picchu in 1912 that showed intact walls. The longevity of the agricultural terraces was due in part to advanced drainage design that incorporated adequately sloped surfaces, good subsurface drainage, and a remarkable network of surface drains. Another aspect of the terrace construction that assured stability was the carefully-thought-out wall foundations that resisted both settlement and sliding.

The terraces had an annual nutrient-producing capability of some 172,000,000 kJ of nutrients—enough to furnish annual sustenance to approximately 55 people. Therefore, it is assumed that food was imported from the floodplain of the Urubamba River below and perhaps from outlying terraced areas such as Intipata [5].

3.5. Drainage, Erosion Control, and Soil Stewardship

Nowhere are the fruits of good drainage, erosion control, and soil stewardship more evident than at Machu Picchu. The terraces were found by Bingham to be free from erosion and sedimentation even after nearly 400 years of lying unattended because the Inca engineers were as good at engineering substructure as they were at engineering the beautiful walls for which they are known. Narrow, steep terraces climb up every possible slope, including the Intiwatana pyramid and the crown of Huayna Picchu, to keep soil in its place and erosion at bay. Figure 10 shows the typical subsurface drainage employed in the fill between the terrace walls.

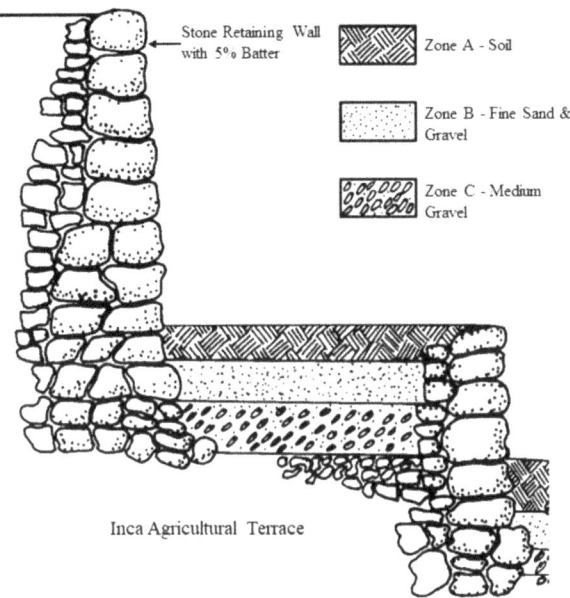

Figure 10. Good subsurface drainage in the agricultural terraces ensured that much of the rainfall percolated down into the permeable zone. (Reproduced from Machu Picchu: A Civil Engineering Marvel [5], with permission from ASCE, the publisher.)

With nearly 2000 mm in annual precipitation, it was imperative that extensive storm drainage engineering be incorporated not only into the agricultural terraces, but also that urban drainage be planned and engineered throughout the ridge-top royal estate. To this effect, there were a total of 129 formal drainage outlets incorporated into the urban sector wall construction. These are not ordinary wall holes, but carefully planned and adequately sized structural openings established at just the right elevation to drain interior floor surfaces.

Runoff from thatched roofs and compacted urban surfaces was high, and for that reason a relatively safe set of drainage criteria was used that accounted for the high rainfall amounts. By examining the size and pattern of drainage outlets at Machu Picchu, it was possible to define the empirical hydraulic criteria used by Inca engineers, as shown in Table 2 [5].

Table 2. Urban surface runoff criteria [1] for wall drainage outlets. (Reproduced from Machu Picchu: A Civil Engineering Marvel [5], with permission from ASCE, the publisher.)

Primary	Magnitude
Tributary area per drainage outlet [2]	200 m^2
Drainage outlet size, typical	10 cm by 13 cm
Drainage outlet capacity, maximum	650 L/min
Design rainfall intensity	200 mm/h
Rational formula runoff "C"	0.8
Design flow per drainage outlet	500 L/min

[1] The author does not assume that Inca had formalized criteria. Parameters shown represent approximate empirical equivalents. [2] The Temple of the Condor is an exception with only one drainage outlet for approximately 0.045 ha; however, subterranean caverns under the Temple of the Condor drain most of the surface runoff.

The masterful engineering of Machu Picchu is not only what is visible to the eye, but also the 60 percent of the building effort that lies underground in the form of foundations and subsurface drainage.

3.6. Inca Trail

In an empire that stretched nearly 4000 km from north to south, and crossed mountain ranges higher than the alps, the Inca built a road system that was a wonder of the ancient world. They had the design skills, labor forces, and organizational capacity to build a trail system that represented not just routes for travel from one place to the next, but a stunning achievement in road development.

The Inca Trail is a well-constructed road with carefully placed stones. A roadway network as grand as the Inca Trail system needed supporting infrastructure in the form of terraces and culverts to manage unstable hillsides and provide drainage. The outside walls of these terraces, even the highest ones, are distinctive in that they are often vertical, without the sloping pitch of typical Inca architecture. Stations to house military guards to control trail use were placed at regular intervals.

My team and I studied the Inca Trail system extending from the hub of Machu Picchu, including the well-known trail hiked by many tourists from the Kilometer 88 Railroad Stop to Machu Picchu, and the long-buried East Flank trail from Machu Picchu to the left bank of the Urubamba River. There, the trail linked with the trail that took ancient runners, military personnel, and Inca travelers down into the Amazon basin [7].

The Inca Trails features many of the same hallmarks of modern road systems: tambos (overnight rest stops), fountains (cafes), and grade-controlled paths that allowed for good drainage and comfortable travel. Terraces are integrated for erosion control and to form flat land surfaces for farming. Finely built granite stairways are provided where the terrain is steep. Many sets of convenient "flying stairways" allowed the field workers ready movement from one terrace level to another without having to return to the access staircases.

During our East Flank excavations, we unearthed two ceremonial fountains, which sprang into life and flowed again after the debris was removed (Figure 11). To assure that the two ceremonial fountains would not overflow during periods of excessive water yield, a hydraulic bifurcation was built upstream of the upper fountain so that excess water would be diverted into a stone-formed conduit for discharge to an adjacent drainage channel. This is evidence that the Inca engineers understood the fundamental mechanics of water flow, and that they had the ability to enhance the hydraulic operation of the fountains to account for periods of low flow.

Figure 11. This long-buried fountain along the Inca Trail still operated after centuries of neglect. This photograph was taken a year after it was uncovered during East Flank trail clearing.

4. Discussion

Inca design standards were relatively uniform. The building construction details at Machu Picchu demonstrate consistent engineering that followed Imperial Inca standards. Building groups, individual buildings, entrances, and niches at Machu Picchu help tell a story of centralized authority. Building designs that are thousands of miles apart have been well documented by Inca scholars as to shape, style, and spacing. The uniform batter of Inca walls helped to ensure stability. The placement of stone in a course to avoid planes of weakness also contributed to soundness and longevity, but the Inca investment in geotechnical site preparation and foundations was the most important reason why the public works construction has endured.

The Inca transportation system included paved trails, stone stairways, tunnels, retaining walls, and hanging bridges over rivers and gulches for which abutments still exist. Then, high above the trails, the Inca built storehouses for foodstuffs, so that a traveling military legion or the local people would did not go hungry.

The ultimate testaments to the sound engineering of Machu Picchu are that the site has withstood the moist elements of Peru for 550 years and that many of its features—canals, fountains, walls, terraces, and drainage systems—still function as designed. The Inca did not have a written language and they left behind no designs or plans to demonstrate their engineering skills. The remaining structures at Machu Picchu and the insight that not all things that are constructed do last, inform my judgement that the Inca knew what they were doing.

5. Conclusions

The water engineering work at Machu Picchu, ranging from the Inca Spring and canal, to the fountains and drainage, to the engineering of the trail system, represents extensive Inca knowledge of the fundamentals of hydrology and hydraulics. The design of the spring collection works and canals shows that the Inca knew how to efficiently maximize their water supply. They also understood the volume and variability of the water supply, in that the Inca canal provided the proper grade and capacity for the volume of water. The Inca created fountains that were not only beautiful, but provided jetted water streams for filling water vessels and an overflow system to address higher volumes of flow. The Inca also understood good drainage and erosion control practices, as demonstrated by the longevity of Machu Picchu despite annual average precipitation of approximately 2000 mm. The Inca trails further demonstrated their builders' understanding of drainage, erosion control, and water supply. All of this was accomplished in a challenging location with steep, unstable slopes and high rainfall. In short, the Inca engineers knew how to build public works and infrastructure in a manner that fit this difficult site and lasted for centuries.

Funding: This research received no external funding.

Conflicts of Interest: The author declares no conflict of interest.

References

1. Bingham, H. In the Wonderland of Peru. *Natl. Geogr. Mag.* **1913**, *23*, 387–574.
2. McEwan, G.F. *The Incas: New Perspectives*; Norton & Company, Inc.: New York, NY, USA, 2008.
3. Rowe, J.H.; UC Berkeley, Berkeley, CA, USA. Personal communication, 1997.
4. Bingham, H. *Machu Picchu A Citadel of the Incas*; Yale University Press: New Haven, CT, USA, 1930.
5. Wright, K.R.; Valencia Zegarra, A. *Machu Picchu: A Civil Engineering Marvel*; ASCE Press: Reston, VA, USA, 2000.
6. Thompson, L.G.; Moseley-Thompson, E.; Morales, B.M. One half millennia of tropical climate variability as recorded in the stratigraphy of the Quelccaya ice cap, Peru. *Geophys. Monogr.* **1989**, *55*, 15–31.
7. Wright, K.R.; Wright, R.M. *The Inka Trails near Machu Picchu. The Great Inka Road: Engineering an Empire*; Matos Mendieta, R., Barreiro, J., Eds.; Smithsonian Books: Washington, DC, USA; New York, NY, USA, 2015.

Article

Engineering Resilience to Water Stress in the Late Prehispanic North-Central Andean Highlands (~600–1200 BP)

Kevin Lane

CONICET-UBA-IDECU (Instituto de las Culturas), Moreno 350, Buenos Aires CABA 1091, Argentina; kevin.lane@cantab.net

Abstract: The Andes are defined by human struggles to provide for, and control, water. Nowhere is this challenge more apparent than in the unglaciated western mountain range Cordillera Negra of the Andes where rain runoff provides the only natural source of water for herding and farming economies. Based on over 20 years of systematic field surveys and taking a political ecology and resilience theory focus, this article evaluates how the Prehispanic North-Central highlands Huaylas ethnic group transformed the landscape of the Andes through the largescale construction of complex hydraulic engineering works in the Cordillera Negra of the Ancash Province, North-Central Peru. It is likely that construction of these engineered landscapes commenced during the Middle Horizon (AD 600–1000), reaching their apogee under the Late Intermediate Period (Huaylas group, AD 1000–1450) and Inca (AD 1450–1532) period, before falling into disuse during the early Spanish colony (AD 1532–1615) through a combination of disease, depopulation, and disruption. Persistent water stress in the western Pacific-facing Andean cordillera was ameliorated through the construction of interlinked dams and reservoirs controlling the water, soil, and wetlands. The modern study of these systems provides useful case-studies for infrastructure rehabilitation potentially providing low-cost, though technologically complex, solutions to modern water security.

Keywords: central-Andes; engineered landscapes; political ecology; Prehispanic; resilience; water security; wetland management

Citation: Lane, K. Engineering Resilience to Water Stress in the Late Prehispanic North-Central Andean Highlands (~600–1200 BP). *Water* 2021, *13*, 3544. https://doi.org/10.3390/w13243544

Academic Editors: Ognjen Bonacci and Jihn-Sung Lai

Received: 30 September 2021
Accepted: 24 November 2021
Published: 11 December 2021

Publisher's Note: MDPI stays neutral with regard to jurisdictional claims in published maps and institutional affiliations.

Copyright: © 2021 by the author. Licensee MDPI, Basel, Switzerland. This article is an open access article distributed under the terms and conditions of the Creative Commons Attribution (CC BY) license (https://creativecommons.org/licenses/by/4.0/).

1. Introduction

Water is a critical human resource, and cultural transformations in the Prehispanic South American Andes have been defined by how they have managed and harnessed it [1–3]. Nowhere is the need to control this resource more evident than along the drier western Pacific-facing Andean mountain-range [4]. Here, aside from the Cordillera la Viuda in central Peru [5], the whole of the mountain range is bereft of glacial ice cover and therefore relies exclusively on seasonal rainfall and winter sea-fog (*garúa*) for hydrological replenishment. Furthermore, yearly precipitation in the Cordillera Negra is half (*c.* 500 mm) of the average levels for the nearby ice-capped Cordillera Blanca (*c.* 1000 mm) [6]. Therefore, the Cordillera Negra not only does not benefit from glacier runoff but is also considerably drier than the other mountain ranges in the region.

In this article, I study the Ancash highlands of the north-central Andes focusing primarily on the Late Intermediate Period (AD 1000–1400) and the Huaylas ethnic group that inhabited this mountain range. As one of a large number of balkanized ethnic groups following the demise of the Middle Horizon (AD 600–1000) Wari in the central Andean highlands [7], the Huaylas constituted a very loosely bonded cultural unit with a common language, economy, set of beliefs (ancestor worship and especially water cult), and material culture.

In the latter case, the Huaylas shared a crude ceramic style, known generically as Aquilpo [8], with other neighboring ethnic groups to the north and south of the Ancash region. Practicing a highly specialized form of late Prehispanic agropastoralism [9], the

Huaylas settled the area roughly bounded by the Cordillera Negra to the west, Cordillera Blanca to the east, the Central Andean Huayhuash Cordillera to the South and the Cañon del Pato and Santa Valley to the North [10], during the period immediately preceding the Inca expansion into the area.

Against a backdrop of a long highland drought [2] followed by fluctuating climate and water availability [11] for the 500 years following *c.* AD 1000 in the Andes, and taking an explicitly political ecology approach [12,13], this article analyses how the Huaylas generated societal resilience through large-scale landscape transformation. This involved the wholesale construction of dams, reservoirs, artificially irrigated wetlands, terraces, canals, and other hydraulic infrastructure, which, in turn, converted this otherwise marginal landscape into a highly productive zone able to sustain substantial human populations.

Landscape transformation and water availability were generally generated through the construction of reservoirs and dams towards water storage both physically and geologically [14]. Although scholars [15,16] have previously argued that the available evidence indicates that dams and reservoirs (*represas, reservorios,* and *estanques*) were not a major hydrological feature of either Prehispanic water storage or water flow regulation in the Andes, this has recently been revised given the overwhelming evidence for these features across large swathes of the central Andes, e.g., [17–21], with their use extending into Northwest Argentina [22].

That this use of technology provided the wherewithal for large local populations is attested by the large number of Late Intermediate Period settlements in the area. Perhaps surprisingly, this intricately constructed landscape was instigated and maintained by the Huaylas at a community and village level (respectively the *ayllu* and *llacta*) without obvious elite or state interference *contrario sensu stricto* [23], see also [24].

Incorporation into the Inca Empire brought with it state-sanctioned modifications to the hydraulic system without overtly affecting the underlying economic lifestyle and societal rationale of the Huaylas, beyond a shift to greater internal social hierarchy, including the possible Inca appointment of hereditary *curaca* or chiefs, such as Huacachillac Apu, Lord of the Huaylas [25]. Nevertheless, given the mountainous and often inaccessible nature of the central Andean highlands, small-holder farming has persisted and is still a way of life throughout the region [26–28]. In such cases, water management, as in the past, tends to still be organized at the community level [29].

Here, I evaluate the Huaylas landscape and hydraulic transformation in the Cordillera Negra as an important proxy for how community-based societies during the Late Intermediate Period in the Andes actively modified their environment providing the means for successful exploitation of the local ecology, especially water, under the aegis of a vibrant agropastoralist political economy [30], thus, leading to increased water security for these Prehispanic populations. In turn, these segmented acephalous ethnic groups were the backbone of Late Intermediate Period highland society and represented the constituent blocks of the later Inca and their eponymous empire.

In turn, I use this Huaylas case-study as a proxy for how technologically savvy groups can harness hydrological resources to provide resilience and water security under climate stress. Taking this comparison further, we advocate that Huaylas, and late Inca, landscape transformation and integrated approach to water management during the Prehispanic Period provides a potential model towards a modern best-practice use of available water resources, especially in a time of renewed hydraulic demand and deteriorating climatic conditions.

1.1. Climate and Community during the LATE Intermediate Period

The Middle Horizon (AD 600–1000), the period immediately prior to the Late Intermediate Period (AD 1000–1400), in the highlands was characterized by a long cycle of higher than average precipitation [11] and with it a concomitant cultural effervescence seen in the emergence of large consolidated cultural entities, such as the Central Andean Tiwanaku [31] and Wari polities [32]. These subsequently collapsed during the eleventh and twelfth century ushering in a period of balkanized highland communities [7].

While interpretations based around deteriorating climatic conditions are fraught with difficulties [33], nevertheless, climate post-AD 1000 took a definite turn for the worst. A recently published article shows that drought conditions in the Southern Andean highlands commenced in the mid-tenth century with arid condition extending into the thirteenth century [34], see also [35,36] for pertinent revisions of Arnold et al.'s article. It is during the Late Intermediate Period that we have the start of the warmer Medieval Climate Anomaly (MCA) dated at between ~1050 and ~1300 AD followed by a highly unstable phase leading to the onset of the Little Ice Age (LIA) ~1400 AD [37,38].

Variations on this theme see the LIA starting somewhat later at ~1500 A.D., although this is based solely on the Quellcaya data whose results are skewed by the Amazon Basin signal [39]. Nevertheless, the various authors coincide that there is a warmer, drier period during the early LIP followed by a long period of unsettled climatic conditions before the advent of colder, wetter conditions towards the end of the LIP [11,40]. These broad conclusions are also lent support by proxy data from sedimentary δ18Ocal levels taken at Laguna Pumacocha located in the eastern Central Andes, which suggests that the period between AD 900–1100, during the MCA, was particularly arid, whilst the late phase between AD 1400–1820, during the LIA, was very wet [41].

Therefore, the paleo-environmental evidence would suggest that, although the agricultural frontier might have benefitted from the warmer conditions prevalent during the eleventh to early fourteenth century, the increased aridity would likely have necessitated a greater investment in water procurement technology. In turn, this might have been a prime motivator behind increased terrace construction during and after the Middle Horizon [42] as well as potentially, investments in wetland management [43].

It is possible that rural communities at this stage gravitated around the use of an Andean highland resource suite involving maize (and potatoes) combined with domesticated camelid exploitation concentrated on the intermediate *kichwa/suni* (2300–4100 m) ecozones [44]. While it is likely that this agropastoralist economic package had a long pedigree [45], the increasingly specialized agropastoralism enhanced through recourse to hydraulic engineering was most likely a late Middle Horizon and Early Intermediate Period innovation.

The unsettled climatic conditions that pervaded throughout most of the Late Intermediate Period might have favored an increase in pastoralism given the mobile nature and environmental threshold tolerance of camelid vis-à-vis individual plant species. Likewise, this possibly led to increasing herder predominance within these bimodal, segmentary agropastoralist communities, e.g., [46,47].

Needless to say, segmentary social organization is very prevalent in herder-dominated societies both in the Andes and elsewhere [48,49], lending support to the idea that Late Intermediate Period segmentary organization may well have had its roots in the increasing importance of herding and herder society during this period [50]. Similarly, scholars have noted the segmentary organization prevalent in community anarchism as a potential viable model for understanding past Andean kin-group and village organization [51]. This type of organization has also been described as 'ordered anarchy' [52] and describes an acephalous political structure without clearly defined leaders or rulers.

In this regard, the Cordillera Negra Huaylas population of the Late Intermediate Period combined a remarkable number of hydrological technologies, which included dams, reservoirs, canals, and irrigated terraces with a patent lack of internal socio-political hierarchy. Ethnohistoric evidence [53,54] for Huaylas hierarchical social organization is likely a reflection of later social changes under Inca and Spanish hegemony, rather than the state of affairs during much of the Late Intermediate Period. Instead of institutionalized elites for the Huaylas we have a much more fluid situation in which few chiefs are born into power, rather there was a favoring of transient leadership with political power vested in the existing corporate lineage groups otherwise known as the *ayllu*.

Indeed, two types of authority seem to have existed throughout the Andean highlands, and among the Huaylas, at this time; authority-in-death by which important personages in life, once dead became venerated kin known as *mallqui* [55] with influence over day-to-day actions, and that of religio-technical authorities, such as water adjudicators, otherwise known as *cilquiua* [56] or water *camayuc* or *cochacamayuc* [57]. Similar to the modern *varayoc* or *alcalde de agua* (water mayor) [58–61], these individuals or specialized groups were responsible for the equitable distribution of water, as well as maintenance of the existing hydrological infrastructure through recourse to community *faenas* or labor, known generically as the *minka*, which included terrace upkeep, the dredging of reservoirs, and canal cleaning among others [24].

For the Late Intermediate Period, it was usually the herders within these agropastoralist communities who exercised the greater political control, and potentially control over water resources [62,63]. This pre-eminence of herder groups is well documented for the area of the central Andes 47, where subservient lower *yunga* (500–2300 m) populations were inducted into repairing the high-altitude dams controlled by the agro-pastoralist *kichwa/suni* (2300–4100) and *puna* (4100–4800) dwellers. As such, these agro-pastoralists were followers or 'sons' of the Andean thunder-God, purveyor of water, fertility, and animals.

Consequently, irrigation water for use in farming by lower-lying agriculturalists was often dependent upon alliances and bargaining with the herder component of these unequal, moiety-organized agro-pastoralist societies. In the study area, crucial water storage infrastructure, including dams and anthropogenically altered wetlands, were located within the herding specialized uplands. Furthermore, extant evidence shows that within these Late Intermediate Period landscapes, herder economies where not just restricted to the *puna* ecozone but vied, depending on climate, environmental and landscape conditions, with agriculture to extend the pastoralist boundary altitudinally downwards to cover a large chunk of the agricultural *suni* (3500–4100) ecozone, this served to increase the available pasturage in certain areas by as much as 50% [64].

1.2. Survey and the Study Area

The research area lies between 89°77′ to 90°04′ North and 171°70′ to 171°81′ within the Cordillera Negra, in the Ancash Province of Peru (Figure 1). Here, the Cordillera Negra mountain-range reaches a maximum height of c. 5200 m. More specifically, the research area encompasses part of the northern section of the Cordillera Negra bounded by the Chaclancayo River in the south and the Uchpacancha River to the north, comprising the Pamparomás watershed, centered on the eponymous town.

The eastern peaks of the Cordillera Negra divide this area from the north–south running intermontane Santa Valley, forming the eastern boundary of our study region. Towards the west the limit of the area of research is set at 2300 m, the altitudinal division between the *yunga* and *kichwa* eco-zones. Downstream, both these highland rivers join the Colcap and Loco rivers to form the coastal Nepeña River down to the Pacific.

Geologically the upper Cordillera Negra is composed of Tertiary Volcanic Calipuy formations interspersed with pockets of earlier Cretacian geology, such as the Inca, Chulec, and Pariahuanca/Pariatambo formations; these are mainly solid Andesite conglomerates that are known for their hard though brittle nature (Figure 2). The predominance of Andesite in this cordillera makes for a rock-type with low permeability averaging 10–20 m [2,65], which permits the formation of substantial natural water basins or lakes; the fracturing of the brittle Andesite in turn generates a higher degree of permeability through crack porosity of the rock. Fresh Andesite has an 8% porosity, while weathered Andesite has porosity between 10–20% [66].

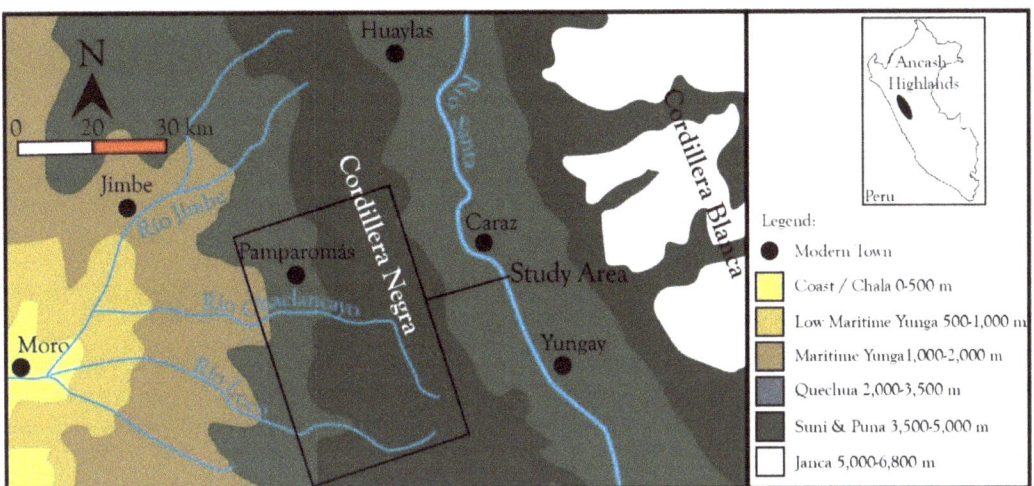

Figure 1. Map of the study area.

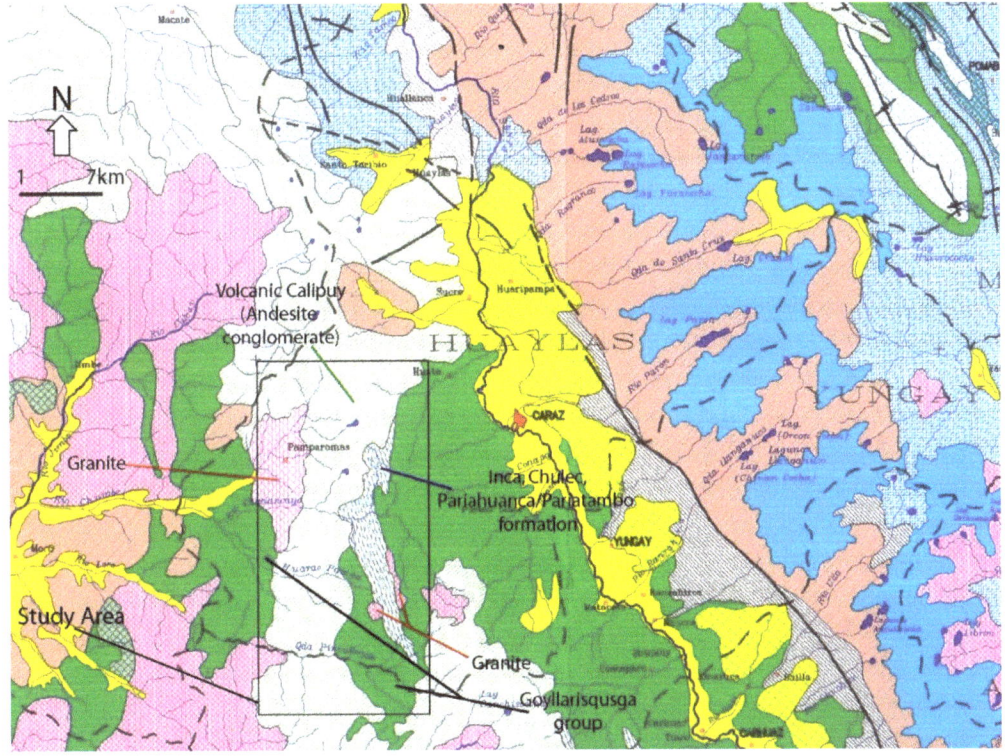

Figure 2. Geology of the study area showing the main groups, formations, and types [67].

Andesite porosity reinforces geological water storage through the replenishment of underground aquifers that then feed into *puquios* or natural springs. The presence of a large area of igneous granite around the modern town of Pamparomás, on the lower limit of the survey area along the Chaclancayo River, provides an important quarrying source for stone construction material, especially for the dams and other hydrological features found in the area. Seismically, it is an area of considerable activity as can be evidenced by the numerous fault-lines throughout these still-young fold mountains. This is a key factor in the creation of fissures and the weathering of the underlying Andesite rock of this region.

This region of the Cordillera Negra also represents the intersection between two major recognized eco-zones: the *páramo* and *puna*. The *Páramo* is a cold and humid landscape scarred by ancient glacial action; the average temperatures gravitate around 12 °C with extreme shifts between −23 °C and 23 °C. The often-water-logged soils are acidic (pH 3.7–5.5) and composed of humic black to dark brown earth. As the elevation increases, rock and sand proportions increase leading to a general decrease in water retention. Low atmospheric pressure, intense ultra-violet radiation, extremes of heat loss and accumulation lead to the endemic plant suites whose growth is slow and with low productivity. If severely disrupted, this ecosystem can take a long time to recover [68].

Similarly, the *puna*, is traditionally described as a harsh, usually dry Alpine meadow tundra located above the treeline; its soils are also highly humic, although with restricted sedimentation [69]. Our area is on the boundary of the *puna* type identified as humid [70]. The humid *puna* has daily temperature fluctuations between 25 and −20 °C; ground frost occurs some 300 days a year. The upper *puna* or *puna brava* has less sedimentation and is increasingly rugged, cold, and harsh. Traditionally thought of as a setting for a predominantly herding economy, the humid *puna* nevertheless has a long tradition of agricultural cultivation, usually of bitter potatoes *(Solanum x juzepczukii* and *Solanum x curtilobum)* [71].

Both the *páramo* and *puna* are what have been described as alpine type ecozones where it is, 'summer every day and winter every night' [72]. In the case of the Cordillera Negra, precipitation averages a yearly 500 mm, which is sufficient for rainfed agriculture and for the replenishment of existing *páramo*/*puna* wetlands. Nevertheless, the main problem in the Cordillera Negra is, and has always been, the lack of a reliable water supply throughout the year [73]. This means that, if unimpeded, the seasonal rains flow swiftly downslope to the sea. Natural lakes and springs, whilst common to the area [67,74], are not normally very large and are insufficient for local needs.

Prehispanic people appear to have tackled this longstanding problem through the construction of a series of dams and reservoirs that spanned side-valleys and ravines possibly as part of an integrated management of water resources that stretched across whole tributaries and covered the Andean highland section of drainages.

Viewed as whole-of-tributary systems, they potentially significantly increased the water-holding capacity of the affected areas and, in turn, held back a large proportion of the rich highland sediments that would otherwise have been eroded downwards to the coast. Furthermore, scholars have been at pains to highlight the importance that herding and subsequently the *puna* and *páramo* ecozones had in the past, with some stating that the herding economy was at least equal to, or greater, than that of agriculture [75].

1.3. Hydraulic Infrastructure and Watershed Management

Elsewhere, I have described how it is useful to think of water technology in the Andes in terms of 'dry' or 'wet'; in which 'dry' designates those in which water flows intermittently, for instance terraces, and 'wet' are those features in which water is almost always present, such as dams [3]. The same article provides dense descriptions of the main different types of hydrological features found in the Andes. For our purposes here, I will summarize the main features of the technologies pertinent to our study area (Table 1).

A crucial aspect of these hydrological features is that they are not present in isolation, rather they are found as packages across the Cordillera Negra, likely creating an interlaced system that links water management strategies from the herding-intense *puna* down through the mixed herding-farming *suni* and subsequently to the predominantly agricultural *kichwa*. In end-effect we appear to have a seamless system of water management linking these agro-pastoralist communities across the gradient of the Cordillera Negra. Nevertheless, this seamlessness is possibly the result of centuries of technological accumulation across watersheds.

Within our study area, I surveyed a total of over 100 archaeological sites, including settlements, tombs, and sacred standing stones. Within this total, there are 21 specific hydrological features and areas (Figure 3). These hydrological features and groups of features were the lynchpin towards harnessing the regions scarce water resources leading to greater economic productivity. A cursory appraisal of the survey data on Figure 3 suffices to show the potentially integrated nature of hydrological technology found in the Pamparomás watershed.

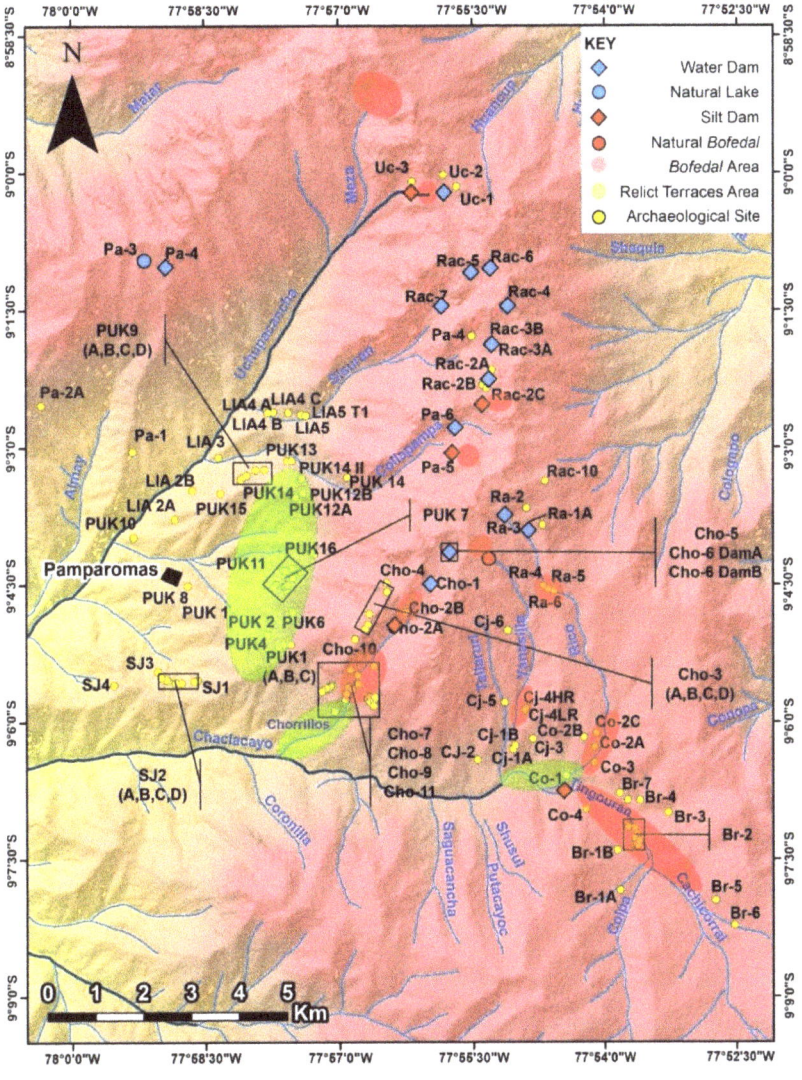

Figure 3. Map of the study area showing the location of hydrological features, including relict terrace areas and wetlands (*bofedal*).

The Pamparomás watershed is essentially fed by two large rivers the southern Chaclancayo and the northern Uchpacancha. These two water sources are mentioned in an important local historical document pertaining to a long-running court case over land and water rights between Dominican monks and the local community dating to the end of the 18th and beginning of the 19th century [76]. Here, the Uchpacancha River is known as 'Río Grande' or big river, with three other rivers, including the Chaclancayo and its three main tributaries, making up the main water sources for the Pamparomás watershed area (Figure 4 & Table 2).

Figure 4. Map of the Pamparomás area showing the main water sources, *Anexo de los Religiosos de Predicadores de Sto. Domingo con los Indios de Pamparomás* (Huaráz, setiembre 1803).

Table 1. Types and categories of hydrological features referred to here.

	Category	Type	Description
Hydrological Feature	Dry	Terraces	Built steps or benches creating cultivation platforms on sloped terrain
		Irrigation canals	Canals distributing water to fields and land features
	Wet	Water dams	Artificial lake
		Water reservoirs	Artificial pond
		Silt dam	Large artificial silt trap or check dam
		Silt reservoir	Small artificial silt terrace or small check dam

Table 2. The dam and reservoir infrastructure in the study area by the principal river.

	Site Code	Site Name	Altitude (m)	Surface Area (m^2)	Notes
Uchpacancha River	Pa 3	Tayapucro	4250	286	*Estancia*/lake
	Pa 4	Racracocha	4350	7850	Water dam
	Uc 2	Agococha/Negrahuacanan	4525	68,750	Water dam
	Uc 3	Tsaquicocha	4300	16,000	Silt dam
	Rac 2	Sacracocha	4590	35,000	Water dam
	Rac 3	Huaytacocha	4500	22,500	Water dam
Sisuran River	Rac 4	Iscaycocha	4575	13,392	Water dam
	Rac 5	Yanacocha Macho	4725	32,291	Water dam
	Rac 6	Yanacocha Hembra	4725	21,875	Water dam
	Rac 7	Alichococha	4325	17,500	Water dam
	Pa 5	Tsaquicocha	4625	1027	Silt dam
Collapampa River	Pa 6	Carhuacocha	4550	38,392	Water dam
	Rac 1	Huancacocha	4425	17,500	Silt dam
	Cho 1	Yanacocha	4550	55,468	Water dam
	Cho 2	Oleron Cocharuri	4200	53,125	Silt Dam
	Cho 6	Orconcocha	4660	35,000	Water dam
	Cho 7	Putacayoc/Kaukayoc	3900–4050	150,000	Silt reservoirs
	Cho 8	Llanapaccha	3600–3900	170,000	water reservoirs, canal to Pukio, terraces
Chaclancayo River	Cj 1	Nununga	3800	340	Water reservoirs
	Cj 4	Represa Decisión	3890	480	2x Silt Reservoirs
	Co 1	Collpacocha	3825	284,375	Silt dam
	Co 2	Intiaurán	3985	75	Silt reservoir
	Ra 1	Ricococha Baja	4485	20,625	Water dam
	Ra 2	Ricocochoa Alta	4560	18,750	Water dam

Extrapolating from the modern map of the area (Figure 3) I can identify the other two smaller rivers in the 1803 map as the modern Sisuran and Collapampa rivers. In toto, these four collect their waters from natural lakes or anthropogenic dams located in the high-altitude *puna/páramo* headwater zone directly in respect to the Uchpacancha, Sisuran, and Collapampa rivers or indirectly, as in the case of the Chanclancayo, which gathers it from three smaller tributaries—the Chorillos, Rico, and Colpa.

In fact, the hydraulic system in the study area can be divided into two large blocks, with the Sisuran and Collapampa feeding into the larger Uchpacancha drainage constituting one of these and the Chaclancayo and its tributaries the other. In this respect, the Uchpacancha includes ten water and three silt dams, while the Chaclancayo has four and two, respectively. The large number of water reservoirs positioned above the circum-Pamparomás basin would have sustained the *kichwa* and *suni* ecozone agriculture of this area. Nevertheless, the more undulating Chaclancayo drainage also includes the huge Collpacocha silt dam [Co 1], as well as large areas of silt reservoirs making it a particularly productive pastoralist landscape.

Throughout the system, other smaller streams and springs feed into all these rivers, but these constitute the hydrological mainstay of the watershed and, thereby, the areas in which past populations built their hydrological infrastructure. In some instances, these waters fed directly onto terrace systems, and three large areas of abandoned or relict terraces have been identified. It is probable that there were others, but they were most likely dismantled by sub-modern and modern cultivation.

Terrace agriculture is no longer practiced in the area; yet, in the past, these terrace systems were very well integrated into the larger water management system of the area. In this regard, water from the Chorrillos side-valley located in the Chaclancayo drainage, especially from the huge Yanacocha water dam [Cho 1] was diverted by means of an irrigation canal to the Pukio terraces located above the modern town of Pamparomás Figure 5.

Photo left: Chorrillos-Pukio Canal as it negotiates the Yurakpecho-Shunak Massif from silt reservoir [42], Cho 7E

Photo right: Detail of Chorrillos-Pukio Canal flowing west towards Pukio. Note that canal has a stone-lining

Figure 5. Stone-lined Chorrillos-Pukio canal.

The crucial point to understand about the construction of this hydraulic infrastructure is the likely interconnectedness of the different features through time. Essentially, beginning in the late Middle Horizon (AD 600–1000) and Late Intermediate Period (AD 1000–1400), the local populations invested heavily in hydraulic engineering to offset increased water insecurity.

It is probable that, through a long process of accretion, these local Huaylas generated an integrated system of water, silt and land management that stretched from the top of the cordillera down to the confluence of the Uchpacancha and Chaclancayo rivers at the site of Tincu near the village of Ullpán (c. 1400 m), thereby, combining herding and farming across this whole altitudinal range. Even so, significant local tributary differences existed. In this regard, altitude is a key determinant in how the hydraulic system of this valley were constituted such that the four main types of 'wet' hydrological features invariably occurred in a particular order:

1. water dams,
2. silt dams,
3. silt reservoirs, and
4. water reservoirs.

These structures also share a common constructive technique given that they are built of granite blocks on the outside along both faces, while the area between the wall is infilled with a compacted coarse gravel and soil/clay admixture (Figure 6). The stone used in the construction of these features are usually rough-cut, although two, Collpacocha [Co 1] and Yanacocha Hembra [Rac 6], have better cut stone in their construction, perhaps suggesting a later construction date, possibly linked to the Inca expansion into the area.

Figure 6. Profile of the Estanque Dam [Ti 1] showing the inner and outer stone walls, including the central soil and stone infill.

For instance, Collpacocha [Co 1] is located alongside a small Inca administrative site [77]. This construction technique lends the structures robustness and durability, but importantly not fixed rigidity, allowing the structures to resist seismic movement, a common occurrence in this part of the Andes. By way of comparison, modern micro-dams in the area are built using steel rods and cement creating rigid structures, which are all too susceptible to fracture during seismic events [78].

1.3.1. Water Dams

On an individual basis, all these structures are found at different altitudinal heights, with water dams located above 4150 m, within the *puna/paramo* ecozone. At this altitude, there is little sedimentation from the surrounding geology. Even today, the dammed lakes have little accumulated sediment. It is also likely that the dams were built to augment already pre-existing glacial lakes located within Andesite rock basins. Of all the lakes identified in the study region only one, Itchicocha, was natural providing an example of how these lakes looked like before human intevention. These Prehispanic structures are gravity dams, which use their weight and bulk to provide the wherewithal to hold back water, as such they are roughly triangular with a broader base and narrower crest, mimicking the triangular nature of water pressure.

Water pressure is proportionally greater at the base of a dam, whilst that at the top it is negligible. Normally, the base of modern gravity dams is equal to, or greater than, 0.7-times the height of the dam. Ancient dams typically exceeded this ratio, as seen in all the dams within the study area. Constructively, the gravity dams usually anchor or seal at least one, and where possible both, of their walls onto the natural rock outcrops that exist in this rugged landscape. This is a simple way of giving greater stability and strength to the structure.

The water dams in the study area usually have a varying number of sluice outtakes (known locally as *desfogues*), whilst a few of the dams have a single sluice located at the base of the structure allowing for a single on-off unregulated flow, likely through the use of a stone-plug. It is possible that the base-level sluice was also used to rid the basin of excess sediment, although it is likewise likely that this would not have been all that effective leading to gradual sedimentation of these structures, especially the lower-lying silt dams (see below).

Some of the more complex structures, such as Yanacocha Hembra [Rac 6], have two sluices, one at the base and one at the crest of the wall, while other examples, such as Ricococha Baja [Ra 1A], have up to six sluices (Figures 7 and 8), allowing for a greater regulation of water flow from the dam, this is similar to the Prehispanic Yanascocha dam registered in the Central Andes by Frank Salomon [17], reiterating that these type of constructions were likely ubiquitous across large stretches of the Andes, especially along the drier Pacific-facing cordilleras.

Figure 7. External face of Ricococha Baja [Ra 1] showing six sluices.

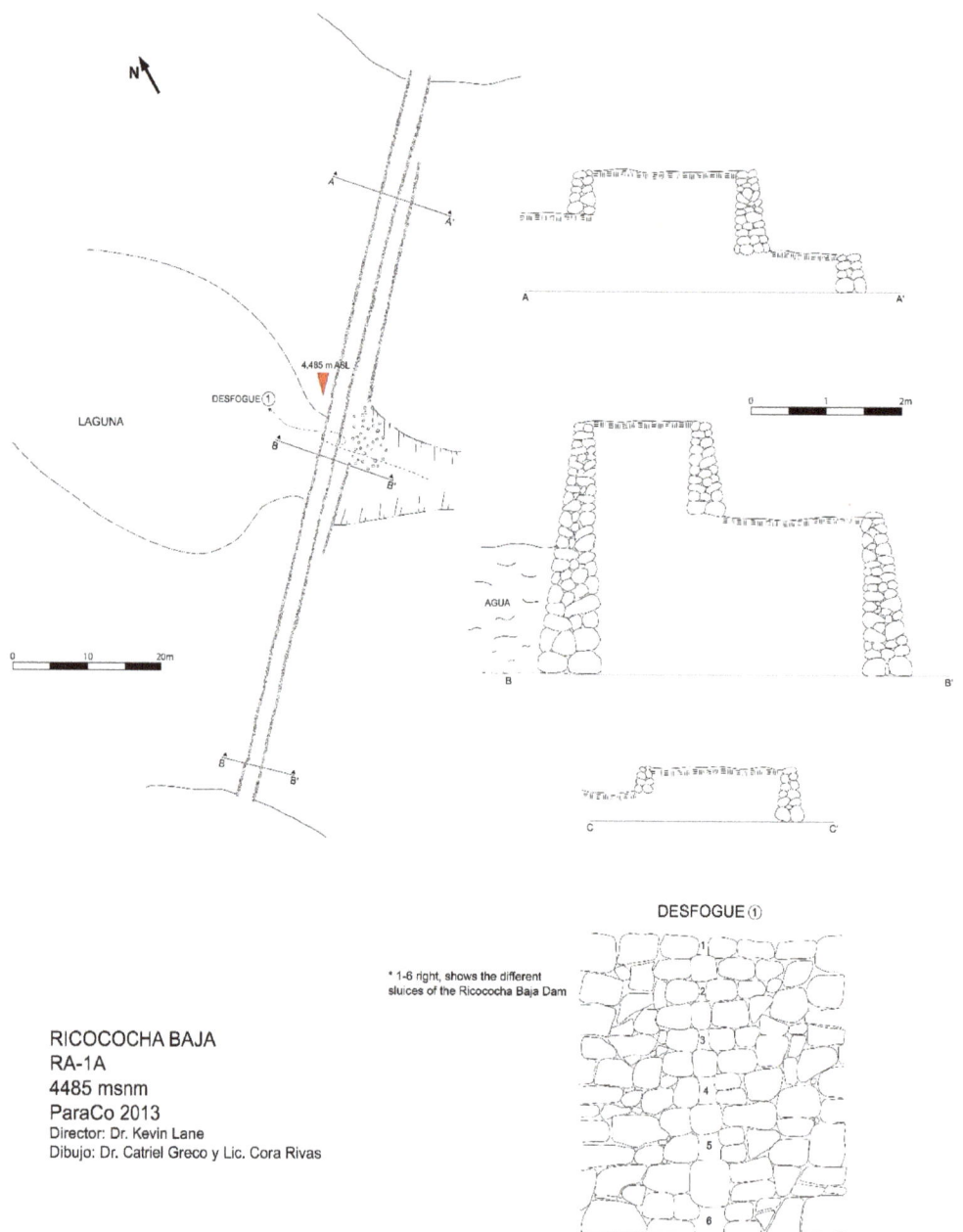

Figure 8. Plan and profile of Ricococha Baja [Ra 1].

1.3.2. Silt Dams

Ethnohistoric evidence shows that, during the Prehispanic period, there was anthropogenic wetland or *bodedal* management [56]. This has been validated both archaeologically, e.g., [14], and anthropologically, e.g., [79]. In principle, the basic concept behind this is that, through wetland management and enlargement, large areas of the *puna* can be turned into a rich plant biota for camelids, especially the more delicate and softer plant-eating alpaca (*Lama pacos*) [80].

The difference is striking, while a Ha. of normal *puna* pasture is adequate for between 1 and 2.25 animals [81], a Ha. of *bofedal* is sufficient for 3.25 animals [82]. In the study area, I have two main types of *bofedal* or wetland creation: silt dams and silt reservoirs. Both are integral elements to the Prehispanic agropastoralist economies of the area.

In this regard, silt dams bear striking similarities to normal water dams in that they are roughly the same size, constructed of double-faced walls, in-filled with compacted earth, gravel and clay, making an impermeable barrier behind which sediment and water are stored, sometimes they are also anchored or sealed onto natural rock outcrops, although this is not always the case (Figures 9 and 10). The central section of most of the dams is further reinforced by step-like walling, most likely because of the added pressure present in this segment of the structure.

The main difference with water dams is the altitude at which these structures appear, between 3800 and 4450 m. With the sole exception of the small silt dam at Tsaquicocha [Pa 5], located at 4625 m, all the other silt dams are altitudinally lower than their water cousins and in three cases, Tsaquicocha [Uc 3] on the Uchpacancha River, Oleron Cocharuri [Cho 2], and Huancacocha [Rac 1] on the Sisuran River, they are located directly below important water dams, respectively, Negrahuacanan [Uc 2], Yanacocha [Cho 1], and Sacracocha [Rac 2], as part and parcel of an integrated water and soil management system.

Figure 9. Huancacocha silt dam [Rac 1].

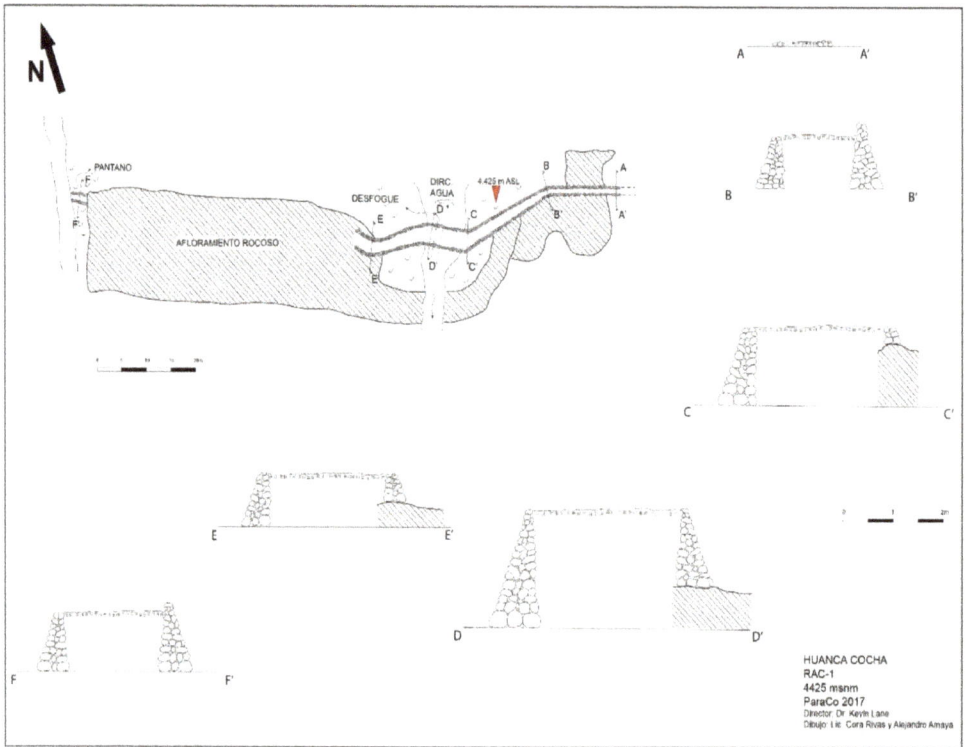

Figure 10. Plan of the Huancacocha silt dam [Rac 1].

Nevertheless, it is also possible that these structures represented an earlier construction phase of water dams at a lower altitude, and it was only after they had sedimented over, effectively becoming silt dams, that this new role was then envisaged for them. Only further research, especially the dating of the lake sediments, e.g., by isotope ratios or pollen, will resolve this issue. That said, it is likely that the construction of these silt dams, whether intentional or not would have been preceded by a period when the area behind the dam wall would have contained an ever-decreasing water basin. This is evident from the soil-sampling data from Collpacocha (Co 1, see below).

Although these features in the area have been interpreted as secondary erosion dams [83] to counteract the effects of periodic landslides known as *huaicos*, they had more than this one function. Silt dams, like the silt reservoirs that I will describe later, were more than simply features to stop erosion flooding; rather, the silt platforms that they created allowed better pasture to flourish while retaining rich deposits of mineral salts that could also animals ingest through soil-munching. This underscores one of the reasons why silt dams were located altitudinally in areas where sedimentation naturally occurred.

Other differences with water dams existed, for instance, silt dams are not necessarily anchored onto rock outcrops, and they usually have only one discernible outtake sluice located along the base of the structure, usually at its centre. The basic principle governing the silt dam is that of geologic water storage [84]; in this case, the accumulated soil basin acting as an aquifer in which water is stored, filtered and purified through the soil. Since the soil also acts as a barrier to water seepage, the sluice should be viewed rather as a 'sieve' drain that siphons excess water out of the basin, whilst the soil retains enough saturated moisture for the growth of *bofedal*-type conditions for animals.

Silt dams were the result of a slow process of soil accretion or varve-formation [85] through years of careful nurture. Therefore, it is highly probable that some of the larger silt dams retained a small spectrum of water until this was eventually silted over. This seems to have been the case with Collpacocha [Co 1] where soil auguring revealed that the area nearest the dam wall was a lake for a considerable period of time [18]. Likewise, the overflowing conditions experienced today on many of these silt dams, such as Collpacocha [Co 1] and Huancacocha [Rac 1], are not indicative of the way they would have looked during the Prehispanic period.

It is likely that periodic partial de-silting of the structures occurred, with this excess mineral-rich silt possibly reused on nearby terraces and fields for use in agriculture. De-silting has been evidenced in similar structures, such as in the Indian *gabarband* silt-traps [86], providing a useful analogy for what might have happened in the Andes. Nor is the idea of soil removal a new one in Andean studies, it had already been documented ethnohistorically in the Sixteenth Century [87] and has been suggested as a wide-ranging practice during the Prehispanic period and subsequently [88].

In total, our study area evidences five silt dams: Oleron Cocharuri [Cho 2]; Collpacocha [Co 1]; Tsaquicocha [Pa 5]; Huancacocha [Rac 1]; and Tsaquicocha [Uc 3], covering an area totalling 372,027 m^2 of anthropogenically enhanced *bofedal* sufficient to maintain a herd of over 1000 animals for a year, although we also have to consider that these silt dams and their wetlands would likely only have been used at the end of the dry season (July–September) when these might have represented some of the last extant good pasturage before the highland rains.

1.3.3. Silt Reservoirs

Altitudinally below the silt dams, there are a series of irregular, accidented open areas or *pampas*, especially on the Chorrillos and Rico side-valleys, within which, a series of small, horseshoe-shaped walls were built (Figure 11). This elevation range is significant as, these reservoirs represent a natural progression from the slightly higher-placed silt dams, while also impinging into what has traditionally been viewed as agricultural land along the upper *suni* range, but which, during the Late Intermediate Period (AD 1000–1450), seems to have been given over to herding [64]. These features trap silt and water behind the wall, thus, creating small *bofedal*-like wetlands.

I refer to these features as 'silt reservoirs', from the local Spanish term for them—*reservorios de limo*. They occur below the silt dams and above the agricultural terraces and fields, between *c*. 3900 and 4400 m, ranging between 7 and 20 m in length, with an average height that ranges between 0.6 and 2 m. These features are similar to erosion-controlling check terraces [89], but are here used to generate small compact wetlands for pasture. They tend to be thicker at the centre of the wall, varying between 0.6 and 1.2 m, although some of the larger constructions reach widths of up to 1.6 m across the center. As with the water dams that they resemble, albeit not in scale, they have a drainage sluice along the central base of the structure, from which excess water seeps out.

Analogous to the other stone features reviewed here, these silt reservoirs are also built with a double-faced stone wall, in-filled with compacted rock and clay. They are set perpendicular to water flow, impeding its uninterrupted movement; this acts as a silt trap and as an erosion barrier to the annual hill-wash episodes and high energy highland discharges that periodically occur in these largely tree-denuded landscape. In turn, the entrapment of silt permits the creation of small *bofedal* micro-environments behind the walls.

Figure 11. Two silt reservoirs from Putacayoc-Kaukayoc [Cho 7].

Like the silt and water dams, these silt reservoirs utilise, where possible, the natural rock outcrops, although, as with silt dams, this is not always the case. In total, across the study area, the silt reservoirs cover an approximate area of 300,415 m^2, representing over 300 Ha. of *bofedal*. Together with the silt dams above, this totals in the region of *c.* 700 Ha. of improved pasture reiterating yet again the importance of these types of technology for the provision of pasture at a local level.

1.3.4. Water Reservoirs

Next, there are what Andean scholars have defined as water reservoirs, likely translating from *reservorio* the Spanish for pond or tank, a small water catchment structure [15,88] similar in principle to the water dams but at a much-reduced scale. Reservoirs have been documented on the coast usually in close association with irrigation canals and the agricultural fields they feed into [90]. Reservoirs in the highlands follow a similar pattern [91], with the added take that these agricultural fields can also be terraces, such that reservoirs serve to regulate water into agricultural terrace systems. This close association between cultivated fields and reservoirs means that these features are normally located at a lower altitude, within the *kichwa* and *suni* ecozones.

In our study area, reservoirs appear between 2500 and 3800 m, hugging the limits of viable cultivation. These reservoirs come in a variety of shapes and sizes; round, ovular or roughly rectangular, contorting themselves to the available space, while varying between 10 to 15 m in relative diameter or length, with an average depth of between 1.5 and 3 m.

Construction is of a rough coursed-stone internal wall and either a similar external wall or an earthen embankment (Figure 12).

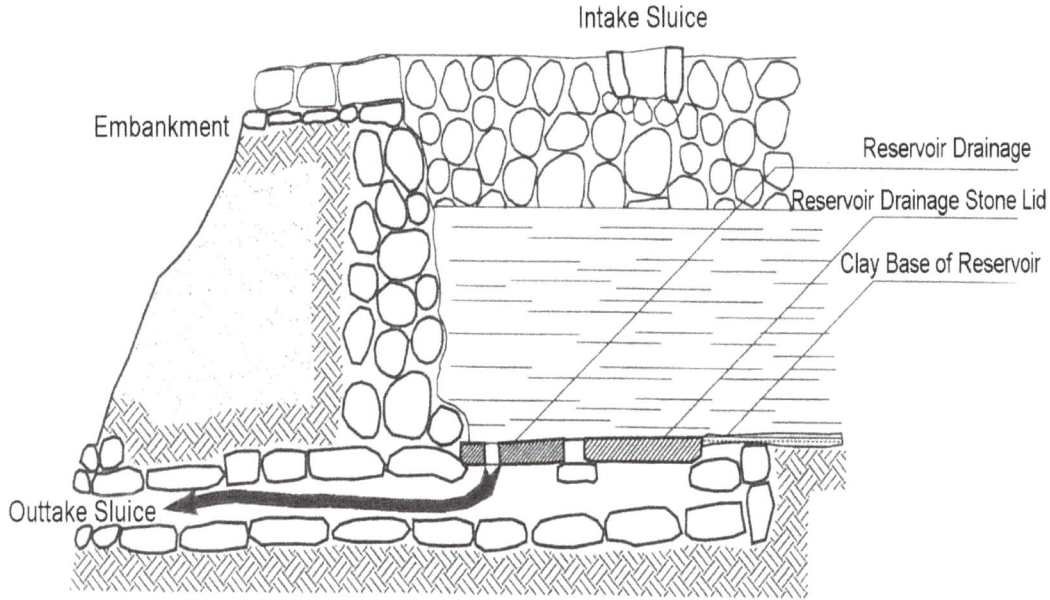

Figure 12. Sketch of the internal plan of a reservoir adapted from [91].

Reservoirs are fully enclosed structures so that walls bound the structure on all sides. In turn, a canal, either earthen or stone-walled, acts as a feeder channel, diverting water towards the reservoir. An outtake sluice is located at the bottom of the reservoir near to an outtake canal that channels the water from here to the irrigated fields.

As with all hydraulic structures, they are difficult to date, especially if they are found disassociated from datable archaeological sites. Nevertheless, the Prehispanic reservoir of Nununga [Cj 1] (Figure 13) is closely associated to mortuary structures or *chullpas* [Cj 2] with late Middle Horizon and Late Intermediate Period (AD 900–1400) material providing a potential *terminus ante quem* for the construction and use of this reservoir. Likewise, in the Chorrillos side-valley, dating of ceramic material found in the adjacent settlements and tombs gives us a similar use date (late MH-LIP). In this side-valley, the water reservoirs are directly placed alongside the agricultural terrace systems of Llanapaccha [Cho 8], thereby, providing water to them.

Figure 13. Water reservoir of Nununga [Cj 1].

1.3.5. Agricultural Terraces

Finally, below the silt terraces there are agricultural terraces. As previously mentioned, these are all now in disuse, but two major types exist in their abandoned state within the study area: sloping field terraces and bench terraces. In all cases, while overall cultivation land is lost through terrace construction, this is ameliorated by the deeper soil matrix created by their construction [15]. Aside from this greater potential productivity, terraces also control erosion and humidity, creating a more sustainable micro-climate amenable to a greater variety of crops, such as maize [75].

Sloping field terraces are the most common type of terracing found in the Andes, usually in side-valleys away from rivers or intermittent streams [15]. Sloping field terraces are built along the natural slope with soil accumulation occurring behind a retainer wall. Although the cultivation zone itself is sloped, the terrace wall acts to level the slope through soil accumulation, allowing for the control of erosion, water run-off, and moisture retention.

Sloping field terraces are usually partitioned into discrete parcels incorporating side walls; these can act as conveyors or deflectors of water flow.

These terraces are sometimes irrigated by overhead feeder canals, although the vast majority sustain themselves on the yearly rains, localized springs or slope seepage. These features are, therefore, not regimented into equidistant parallel strips but can occur haphazardly along a mountain flank. The vast majority of these terraces have been documented for areas above 3500 m. In the Central Andes, at this height, there is usually sufficient rainfall for at least one crop a year. In the study area, sloping field terraces have been found at the top of the Chaclancayo Valley around the Huaylas-Inca site of Intiaurán [Co 2] where the land opens up into a wide undulating expanse of *suni-puna* located between 3800 and 4300 m.

Nevertheless, the study area has larger expanses of abandoned bench terraces. Bench terraces convey a classic image of Andean highland agriculture in which close-fitting stone retaining walls, usually between 1 and 5 m in height, step upwards along the valley sides usually arranged in parallel vertical rows. The walls slope inward, presenting a level planting surface that is typically fed by some form of integrated canal irrigation system.

The terraces of Llanapaccha [Cho 8B], for instance, incorporated a system of main canals fed by catchment reservoirs built alongside the Chorrillos River, from where water was diverted to feed the terraces (Figure 14). Internally, water from the three main canals, offset at the top, middle, and lower parts along the terraced slope would have cascaded downwards throughout the whole system.

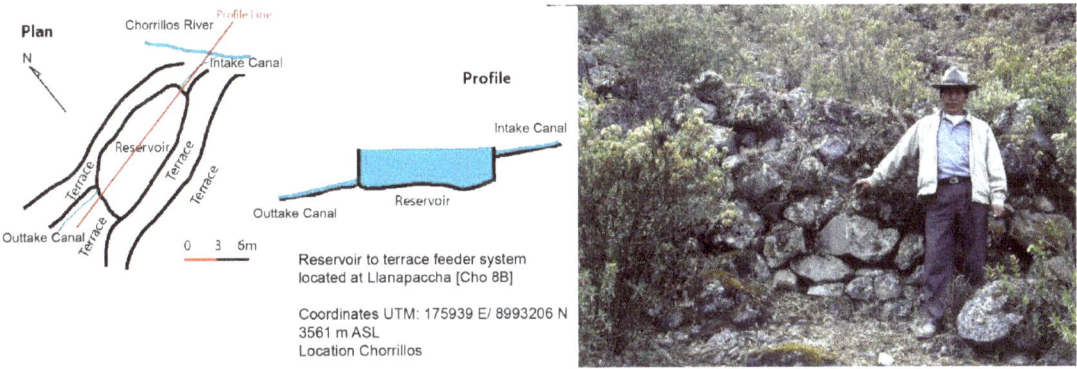

Figure 14. Right image—Bench terraces of Cho 8B Llanapaccha; Left image—Plan of reservoir and bench terraces at Cho 8B Llanapaccha.

Excavation at one of the bench terraces at Intiaurán [Co 2] revealed that the terrace had a base of small to large cobbles over which a richer humic soil level was present. The first level allowed for good drainage of the terrace, while the second provided the soil matrix necessary for successful agriculture. This is similar to what other researchers have documented across the Andes [88], e.g., [92–94]. Bench terracing is found at the lower end of the Chorrillos side-valley, where the Rico Valley joins the Chaclancayo and in a wide area above the modern town of Pamparomás, this last was fed by an important stone-lined canal that connected Chorrillos to the Pukio area (see above).

As can be seen, this area of the Cordillera Negra during, at least, the Late Intermediate Period (AD 1000–1400) and likely earlier, manifested a highly intricate management of water that straddled both herding and farming zones as part of integrated, complex agropastoralist Huaylas communities. Given that the Huaylas settled the Cordillera Negra from at least the Fortaleza River in the South to the Santa in the North, this water management system and the concomitant agropastoralist organisation unpinning it most likely extended itself across this whole mountain range.

Evidence from further north seems to bear this out [21], and, while extensive research to the south is still lacking in this respect, preliminary unpublished surveys by research colleagues seem to show the same pattern emerging in the southern Aija Valley [95].

2. Discussion

The above review details the strong technological interconnectedness within a bounded area across relatively short rivers and side-valleys highlighting the rich hydrological suite employed by the ancient Huaylas populations to counteract the persistent effects of water insecurity in the Cordillera Negra during the Late Intermediate Period and subsequently.

Furthermore, it was this insecurity that must have precipitated these technological investments in the first place. Fluctuating water availability in the face of worsening climatic and environmental conditions, possibly from the late Middle Horizon (AD 600–1000) onwards, necessitated further and sustained technological investment, and it was through these cumulative actions that resilience was generated within the ecological system providing water security to the communities in the area [96].

Resilience within ecological systems is affected by both slow-moving and fast-moving variables in the social sphere that can impact, negatively or positively, these same systems [97]. These social variables can be as important as ecological ones in the long-term stability and equilibrium (or not) of the system, with human ability to adapt to these variable conditioning how a given socio-ecological system bounces back to an appropriate state of equilibrium after disturbance [97].

In this sense, the increasing relevance of the social in ecological systems pushes us to consider the interrelatedness between resilience studies and the wider political ecology behind human decisions in regards to their environment [98–100]. In this regard, it is then possible to chart the development of hydraulic technology and the variables determining its success and failure within the context of the Cordillera Negra.

While Middle Horizon (AD 600–1000) Wari or Wari-inspired changes in food production through recourse to hydraulic technology might well have provided the impetus towards the wholescale technological transformation of the Cordillera Negra landscape at this time; it is evident that it was only under the Late Intermediate Period (AD 1000–1400) that small-scale communities move towards stabilising water security played out in its most technologically complex expression.

As previously stated, against a backdrop of increasingly dry and erratic weather patterns commencing towards AD 900, Huaylas communities in the Cordillera Negra negotiated increasing water insecurity through recourse to technology. Herein, it is possible to see a series of slow-moving social and ecological variables that underpinned human strategies in the region, these included the ecological spectrum, climate change, accompanying land degradation, and changes in the water regime, with social variables that emphasised the emergence of community-centred, specialised agropastoralism employing accretional and incremental technological transformation of the landscape to abet wetlands and provide the water security necessary for successful farming and herding production across the vertically stacked ecozones of the region.

From a political ecology standpoint, these small-scale, acephalous communities had little centralised leadership or outside interference, with water and engineering managed and controlled at the local level. Nevertheless, no system is entirely stable [101], and against the slow-moving variables described above, there would have been other fast-moving social and ecological variables that would have affected the resilience of the ecological system, such as persistent drought [2] due to fluctuating patterns of ENSO (El Niño–Southern Oscillation) [102] and SASM (South American Summer Monsoon) [41] during the early Late Intermediate Period, combined with the social variables of rising internecine warfare and raiding as the struggle for resources became more acute throughout the Late Intermediate Period.

This was especially so post-AD 1250, after which there was a general move by human groups to more-easily defended ridge and hill-top settlements [103,104]. It is likely that, this human migration to higher areas, especially in the *suni* and *puna* ecozones, further reinforced the socio-economic power of herders within these societies, given that these areas were much better suited to an intensive, specialised pastoralism than any other economic activity.

The incorporation of the Huaylas region into the Inca Empire (AD 1400–1532) bought new social variables into the ecological system, mainly regarding a more extractive political economy. In the short-term, this led to further investments in hydraulic technology in the area, especially around the important wetland of Collpacocha [Co 1] associated with the nearby Inca administrative site of Intiaurán [77] and elsewhere in the study area [105].

Nevertheless, the Inca empire likely brought with it greater political stability, largely ameliorating the effects of internecine warfare and raiding, which, in turn, would have helped guarantee water security and with it a more secure economic return from the region's ecological system. It is during this period that local population numbers probably reached their Prehispanic peak, with the study area being part of the Inca *guaranga* (nominally 1000 households) of Mato [106], later the Spanish colonial district of San Luis de Macati, modern-day Macate [25].

At this time and throughout much of the Sixteenth Century, Huaylas was still considered the richest and most populated of the north-central provinces, a reflection of the prosperity and importance that this area held throughout the Inca period and subsequent early colonial era [107]. Even so, Spanish colonisation of the Andes and the Huaylas brought with it new social variables that strongly impacted the prevailing resilience of the study area's ecological system.

The social (and biological) variables that adversely affected Huaylas resilience were principally two: disease and depopulation. Conservative estimates for the study area put population decline between AD 1534 and 1629 at 63% [18], although the final figure was likely higher given that the Andes experienced considerable population decline already prior to the arrival of the Spanish in AD 1532. European diseases predated the arrival of Francisco Pizarro by at least eight years, decimating in the region of 25–50% of the indigenous population before the Spanish even made landfall in the Central Andes [108].

Even in the Nineteenth Century, travellers to Pamparomás noted the abject poverty as well as reduced state and population of the area [109]. These two factors played havoc on the prevailing ecological system as human depopulation led to the wholesale abandonment of huge swathes of the highlands, especially the upper *suni* and *puna* areas, which were largely emptied of people and animals. A declining population required significantly less water, leading to the neglect of much of the ancient installed hydrological capacity of the region, including the water dams, anthropogenically enhanced wetlands and terrace systems.

Only since the mid-twentieth century and, in particular, since the 1970s, has the population recovered sufficiently to start re-stressing the ecological system, a system that is now ill-equipped from a technological standpoint to ensure resilience and water security. There is a danger that new pressures on the system, such as the introduction of water-intense cash crops, for instance avocado, coupled with the greater human use of water due to changing hygiene habits together with climate change will tilt the prevailing ecological system into irreversible crisis, as ecological system transformation leads to ill-adaptation and eventual collapse.

A governmental and NGO-led political ecology emphasising the construction of steel and concrete micro-dams [110] offers short term resilience and a partial solution to water security given that these modern dams, often located on top of, and destroying, their Prehispanic ancestors have a shelf life of some 100 years; considering that endemic seismic activity and a lack of maintenance in the region could reduce this to effectively between 20 and 30 years. New thinking is urgently required if we are to break this cycle of water insecurity and undue stress on the resilience of the region's ecological system.

3. Conclusions

To conclude, the sheer ubiquity of ancient water engineering systems in the Cordillera Negra is countermanded by their almost complete abandonment in the present. On the basis of a recently funded project, I aim to start reverting this condition through the rehabilitation of part of these ancient systems, while respecting modern rural economic practices that emphasis farming, rather than herding.

Using the past as our model, the underlying aim is to reinstall some of the pre-existing Prehispanic resilience into the modern water management systems of this increasingly water insecure region. Indeed, water insecurity is understood by experts as the single-most important human threat resource in the face of climate change and ever-retreating tropical glaciers [111]. In the non-glaciated Cordillera Negra, this is a particularly pernicious concern, a sentiment echoed by local communities and populations.

While both old and new dams control water and soil erosion, feed subterranean aquifers, and provide water for people's livelihoods, modern concrete micro-dams are expensive, rely on non-local expertise, and have a curtailed functional lifespan. When they break down, the funds to repair them are not readily available, leading to a significant decrease in existing water levels for the impacted communities.

By way of contrast, the ancient dams of the area were the product of century-long engineering projects that seamlessly integrated these technologies with their immediate landscape. They were based on local know-how and were easily maintained by the community. Crucially, for these cash-strapped regions, these stone and clay constructions are cheap to build and upkeep. They are also resilient, after close to 1000 years of negligible maintenance, most of them are still standing and potentially functioning.

In this regard, our rehabilitation project is rooted in a respect for community and local knowledge and allied to flexible modern engineering to shore up the effectiveness of these ancient structures. I contend that modern engineering can only provide partial solutions to increasing water insecurity in the Andes, and that the marriage between past and present knowledge and technology can deliver a better, locally informed answer to future water security and climate change in the Cordillera Negra.

Thinking ahead, ancient dams are common to the Andean highlands; therefore, success on this rehabilitation project will provide a further cheap, easily applicable, community-based solution to water scarcity across large areas of these hydrologically challenged highlands. Let us hope it does.

Funding: Funding for this research was provided by Alexander von Humboldt Foundation (2010–2011), Anthony Wilkins Fund (2001–2003), British Academy (2008–2010), Chadwick Fund (2001–2003), Crowther-Beynon Fund (2001–2003), Dorothy Garrod Fund (2001–2003), Gerda Henkel Foundation (2021–2023), Leverhulme Trust (2008–2010), Ridgeway-Venn Fund (2001–2003), and the (2001–2004).

Institutional Review Board Statement: Not applicable.

Informed Consent Statement: Not applicable.

Conflicts of Interest: The author declares no conflict of interest.

References

1. Kosok, P. *Life, Land, and Water in Ancient Peru; An Account of the Discovery, Exploration, and Mapping of Ancient Pyramids, Canals, Roads, Towns, Walls, and Fortresses of Coastal Peru with Observations of Various Aspects of Peruvian Life, Both Ancient and Modern*; Long Island University Press: New York, NY, USA, 1965.
2. Ortloff, C.R. *Water Engineering in the Ancient World: Archaeological and Climate Perspectives on Societies of Ancient South America, the Middle East, and South-East Asia*; Oxford University Press: Oxford, UK, 2010.
3. Lane, K. Water Technology in the Andes. In *Encyclopaedia of the History of Science, Technology, and Medicine in Non-Western Cultures*; Selin, H., Ed.; Springer: New York, NY, USA, 2014; pp. 1–24.
4. Mark, B.G.; Mckenzie, J.M. Tracing Increasing Tropical Andean Glacier Melt with Stable Isotopes in Water. *Environ. Sci. Technol.* **2007**, *41*, 6955–6960. [CrossRef]
5. Olarte Navarro, B. La Cuenca Del Río Chillón: Problemática y Potencial Productivo. *Ing. Ind.* **2007**, *25*, 53–68. [CrossRef]

6. Tremolada, P.; Villa, S.; Bazzarin, P.; Bizzotto, E.; Comolli, R.; Vighi, M. POPs in Mountain Soils from the Alps and Andes: Suggestions for a 'Precipitation Effect' on Altitudinal Gradients. *Water Air. Soil Pollut.* **2008**, *188*, 93–109. [CrossRef]
7. Covey, R.A. Multiregional Perspectives on the Archaeology of the Andes during the Late Intermediate Period (c. A.D. 1000–1400). *J. Archaeol. Res.* **2008**, *16*, 287–338. [CrossRef]
8. Lanning, E.P. Current Research: Highland South America. *Am. Antiq.* **1965**, *31*, 139–140.
9. Lane, K. Through the Looking Glass: Re-Assessing the Role of Agro-Pastoralism in the North-Central Andean Highlands. *World Archaeol.* **2006**, *38*, 493–510. [CrossRef]
10. Pärssinen, M. *Tawantinsuyu: El Estado Inca y Su Organización Política*; Travaux de l'Institut Francais d'Études Andines; IFEA/Fondo Editorial de la Pontificia Universidad Católica del Perú/Embajada de Finlandia: Lima, Peru, 2003.
11. Mächtle, B.; Eitel, B. Fragile Landscapes, Fragile Civilizations—How Climate Determined Societies in the Pre-Columbian South Peruvian Andes. *Catena* **2012**, *107*, 62–73. [CrossRef]
12. Morehart, C.T.; Millhauser, J.K.; Juarez, S. Archaeologies of Political Ecology: Genealogies, Problems, and Orientations. *Archaeol. Pap. Am. Anthropol. Assoc.* **2018**, *29*, 5–29. [CrossRef]
13. Robbins, P. *Political Ecology: A Critical Introduction*; Critical Introductions to Geography; Blackwell Publishing: Oxford, UK, 2004.
14. Lane, K. Water, Silt and Dams: Prehispanic Geological Storage in the Cordillera Negra, North-Central Andes, Peru. *Rev. Glaciares Ecosistemas Mont.* **2017**, *2*, 41–50.
15. Denevan, W.M. *Cultivated Landscapes of Native Amazonia and the Andes*; Oxford Geographical and Environmental Studies Series; Clark, G., Ed.; Oxford University Press: Oxford, UK, 2001.
16. Scarborough, V.L. *The Flow of Power: Ancient Water Systems and Landscapes*; School of American Research Press: Santa Fe, NM, USA, 2003.
17. Salomon, F. Collquiri's Dam: The Colonial Re-Voicing of an Appeal to the Archaic. In *Native Traditions in the Postconquest World: A Symposium at Dumbarton Oaks, 2nd through 4th October 1992*; Hill Boone, E., Cummins, T., Eds.; Dumbarton Oaks Research Library and Collection: Washington, DC, USA, 1998; pp. 265–293.
18. Lane, K. Engineering the Puna: The Hydraulics of Agro-Pastoral Communities in a North-Central Peruvian Valley. Ph.D. Thesis, University of Cambridge, Cambridge, UK, 2006.
19. Vivanco Pomacanchari, C. Obras Hidráulicas de Etapa Prehispánica En Huaccana, Chincheros—Apurímac. *Arqueol. Soc.* **2015**, *30*, 315–333. [CrossRef]
20. Palacios, J. *Agua: Ritual y Culto En Yañac (Ñaña): La Montaña Sagrada*; Universidad Peruana Unión: Lima, Peru, 2017.
21. Combey, A. *Dynamiques Spatiales et Mobilités Dans La Cordillère Noire: Permanences et Évolutions, Vallon De Capado, Ancash, Pérou Horizon Ancien (900–300 Av. J.-C.)–Horizon Récent (1460–1532 Apr. J.-C.)*; Sorbonne: Paris, France, 2018.
22. Lanzelotti, S.L. Indicadores Para El Reconocimiento de Represas Arqueológicas. *Relac. Soc. Argent. Antropol.* **2011**, *XXXVI*, 177–196.
23. Wittfogel, K.A. *Oriental Despotism: A Comparative Study of Total Power*; Yale University Press: New Haven, CT, USA, 1957.
24. Lane, K. Engineered Highlands: The Social Organisation of Water in the Ancient North-Central Andes (AD 1000–1480). *World Archaeol.* **2009**, *41*, 169–190. [CrossRef]
25. Aibar Ozejo, E. La Visita de Guaraz En 1558. *Cuad. Semin. Hist. Inst. Riva Agüero* **1968**, *9*, 5–21.
26. Brush, S.B.; Gulliet, D.W. Small-Scale Agro-Pastoral Production in the Central Andes. *Mt. Res. Dev.* **1985**, *5*, 19–30. [CrossRef]
27. Zimmerer, K.S. Wetland Production and Smallholder Persistence: Agricultural Change in a Highland Peruvian Region. *Ann. Assoc. Am. Geogr.* **1991**, *81*, 443–463. [CrossRef]
28. Mayer, E. *The Articulated Peasant: Household Economies in the Andes*; Westview Press: Boulder, CO, USA, 2002.
29. Rasmussen, M.B. *Andean Waterways: Resource Politics in Highland Peru*; University of Washington State: Seattle, WA, USA; London, UK, 2015.
30. Grant, J.L.; Lane, K. The Political Ecology of Late South American Pastoralism: An Andean Perspective A.D. 1000–1615. *J. Polit. Ecol.* **2018**, *25*, 446–469.
31. Kolata, A.L. *Tiwanaku and Its Hinterland*; Smithsonian Institution Press: Washington, DC, USA, 1996.
32. Schreiber, K.J. *Wari Imperialism in Middle Horizon Peru*; Anthropological Papers of the Museum of Anthropology; University of Michigan Press: Ann Arbor, MI, USA, 1992; Volume 28.
33. Calaway, M.J. Ice-Cores, Sediments and Civilization Collapse: A Cautionary Tale from the Lake Titicaca. *Antiquity* **2005**, *79*, 778–790. [CrossRef]
34. Arnold, T.E.; Hillman, A.L.; Abbott, M.B.; Werne, J.P.; McGrath, S.J.; Arkush, E.N. Drought and the Collapse of the Tiwanaku Civilization: New Evidence from Lake Orurillo, Peru. *Quat. Sci. Rev.* **2021**, *251*, 106693. [CrossRef]
35. Marsh, E.J.; Contreras, D.; Bruno, M.C.; Vranich, A.; Roddick, A.P. Comment on Arnold et al. "Drought and the Collapse of the Tiwanaku Civilization: New Evidence from Lake Orurillo, Peru" [Quat. Sci. Rev. 251: 106693]. *Quat. Sci. Rev.* **2021**, *251*, 107004. [CrossRef]
36. Sandweiss, D.H.; Maasch, K.A.; Landazuri, H.A.; Leclerc, E. Comment on Arnold et al. (2021): CP in the MCA? *Quat. Sci. Rev.* **2021**, *269*, 107053. [CrossRef]
37. Kellerhals, T.; Brütsch, S.; Sigl, M.; Knüsel, S.; Gäggeler, H.W.; Schwikowski, M. Ammonium Concentration in Ice Cores: A New Proxy for Regional Temperature Reconstruction? *J. Geophys. Res.* **2010**, *115*, D16123. [CrossRef]

38. Neukom, R.; Luterbacher, J.; Villalba, R.; Küttel, M.; Frank, D.; Jones, P.D.; Grosjean, M.; Wanner, H.; Aravena, J.-C.; Black, D.E.; et al. Multiproxy Summer and Winter Surface Air Temperature Field Reconstructions for Southern South America Covering the Past Centuries. *Clim. Dyn.* **2011**, *37*, 35–51. [CrossRef]
39. Thompson, L.G.; Mosley-Thompson, E.; Davis, M.E.; Zagorodnov, V.S.; Howat, I.M.; Mikhalenko, V.N.; Lin, P.N. Annually Resolved Ice Core Records of Tropical Climate Variability over the Past ~1800 Years. *Science* **2013**, *340*, 945–950. [CrossRef]
40. Baker, P.A.; Seltzer, G.O.; Fritz, S.C.; Dunbar, R.B.; Grove, M.J.; Tapia, P.M.; Cross, S.L.; Rowe, H.D.; Broda, J.P. The History of South American Tropical Precipitation for the Past 25,000 Years. *Science* **2001**, *291*, 640–643. [CrossRef]
41. Bird, B.W.; Abbott, M.; Vuille, M.B.; Rodbell, D.T.; Stansell, N.D.; Rosenmeier, M.F. A 2300-Year-Long Annually Resolved Record of the South American Summer Monsoon from the Peruvian Andes. *Proc. Natl. Acad. Sci. USA* **2011**, *108*, 8583–8588. [CrossRef]
42. Schreiber, K.J. Conquest and Consolidation: A Comparison of the Wari and Inka Occupations of a Highland Peruvian Valley. *Am. Antiq.* **1987**, *52*, 266–284. [CrossRef]
43. Vining, B.; Williams, P.R. Crossing the Western Altiplano: The Ecological Context of Tiwanaku Migrations. *J. Archaeol. Sci.* **2020**, *113*, 105046. [CrossRef]
44. Finucane, B.; Maita Agurto, P.; Isbell, W.H. Human and Animal Diet at Conchopata, Peru: Stable Isotope Evidence for Maize Agriculture and Animal Management Practices during the Middle Horizon. *J. Archaeol. Sci.* **2006**, *33*, 1766–1776. [CrossRef]
45. Browman, D.L. Origins and Development of Andean Pastoralism: An Overview of the Past 6000 Years. In *The Walking Larder: Patterns of Domestication, Pastoralism and Predation*; Clutton-Block, J., Ed.; Unwin-Hyman: London, UK, 1989; pp. 256–268.
46. Duviols, P. Huari y Llacuaz: Agricultores y Pastores: Un Dualismo Prehispánico de Oposición y Complementaridad. *Rev. Mus. Nac.* **1973**, *39*, 153–191.
47. Rostworowski, M. *Conflicts over Coca Fields in XVIth-Century Peru*; Studies in Latin American Ethnohistory and Archaeology; Marcus, J., Ed.; University of Michigan Press: Ann Arbor, MI, USA, 1988; Volume 21.
48. Orlove, B.S. Native Andean Pastoralists: Traditional Adaptations and Recent Changes. In *Contemporary Nomadic and Pastoral Peoples: Africa and Latin America*; Salzman, P.C., Ed.; College of William and Mary: Williamsburg, VI, USA, 1982; Volume 17, pp. 95–136.
49. Salzman, P.C. *Pastoralists: Equality, Hierarchy, and the State*; Westview Press: Oxford, UK, 2004.
50. Lane, K. The State They Were in: Community, Continuity and Change in the North-Central Andes, 1000AD–1608AD. In *Socialising Complexity: Structure, Integration and Power*; Kohring, S., Wynne-Jones, S., Eds.; Oxbow: Oxford, UK, 2007; pp. 76–99.
51. Fleming, D. Can We Ever Understand the Inca Empire? In *28th Northeast Conference on Andean Archaeology and Ethnohistory*; State University of New York: New York, NY, USA, 2016.
52. Evans-Pritchard, E.E. *The Nuer: A Description of the Modes of Livelihood and Political Institutions of a Nilotic People*; Oxford University Press: Oxford, UK, 1940.
53. Zuloaga Rada, M. *La Conquista Negociada: Guarangas, Autoridades Locales e Imperio en Huaylas Perú*; IEP: Lima, Peru, 2012.
54. Espinoza Soriano, W. Etnia Guaylla (Ahora Huaylas). *Investig. Soc.* **2013**, *17*, 179–190. [CrossRef]
55. Isbell, W.H. *Mummies and Mortuary Monuments: A Postprocessual Prehistory of Central Andean Social Organization*; University of Texas Press: Austin, TX, USA, 1997.
56. Guaman Poma de Ayala, F. *Nueva Crónica y Buen Gobierno*; Fondo de Cultura Economica: Lima, Peru, 1993; Volume 1–3.
57. de Avila, F. *Ritos y Tradiciones de Huarochirí*, 2nd ed.; Travaux de l'Institut Francais d'Études Andines; IFEA, Banco Central de Reserva Del Peru, Universidad Particular Ricardo Palma: Lima, Peru, 1999; Volume 116.
58. Paerregaard, K. Complementarity and Duality: Oppositions between Agriculturists and Herders in an Andean Village. *Ethnology* **1992**, *31*, 15–26. [CrossRef]
59. Mitchell, W.P.; Guillet, D. *Irrigation at High Altitudes: The Social Organization of Water Control Systems in the Andes*; Ehrenreich, J.D., Ed.; Society for Latin American Anthropology Publication Series; American Anthropological Association: Washington, DC, USA, 1994; Volume 12.
60. Gelles, P.H. *Water and Power in Highland Peru: The Cultural Politics of Irrigation and Development*; Rutgers University Press: New Brunswick, NB, USA, 2000.
61. Trawick, P.B. *The Struggle for Water in Peru: Comedy and Tragedy in the Andean Commons*; Stanford University Press: Stanford, UK, 2003.
62. Zuidema, R.T. Kinship and Ancestor Cult in Three Peruvian Communities. Hernández Príncipe's Account of 1622. *Bull. Inst. Fr. DÉtudes Andin.* **1973**, *2*, 16–33.
63. Gose, P. Segmentary State Formation and the Ritual Control of Water under the Incas. *Comp. Stud. Soc. Hist.* **1993**, *35*, 480–514. [CrossRef]
64. Lane, K.; Grant, J.L. A Question of Altitude: Exploring the Limits of Highland Pastoralism in the Prehispanic Andes. In *The Archaeology of Andean Pastoralism*; Capriles, J.M., Tripcevich, N., Eds.; University of New Mexico Press: Albuquerque, Mexico, 2016; pp. 139–157.
65. Pérez-Flores, P.; Wang, G.; Mitchell, T.M.; Meredith, P.G.; Nara, Y.; Sarkar, V.; Cembrano, J. The Effect of Offset on Fracture Permeability of Rocks from the Southern Andes Volcanic Zone, Chile. *J. Struct. Geol.* **2017**, *104*, 142–158. [CrossRef]
66. Jamtveit, B.; Kobchenko, M.; Austrheim, H.; Malthe-Sørenssen, A.; Røyne, A.; Svensen, H. Porosity Evolution and Crystallization-Driven Fragmentation during Weathering of Andesite. *J. Geophys. Res. Solid Earth* **2011**, *116*, B12204. [CrossRef]

67. INRENA. *Base de Datos de Recursos Naturales e Infrastructura: Departamento de Ancash-Primera Aproximacion*; INRENA/Ministerio de Agricultura: Lima, Peru, 2000; p. 166.
68. Luteyns, J.L. Páramos: Why Study Them? In *Páramo: An Andean Ecosystem under Human Influence*; Balslev, H., Luteyns, J.L., Eds.; Academic Press: London, UK; New York, NY, USA, 1992; pp. 1–14.
69. Luteyns, J.L.; Churchill, S.P. Vegetation of the Tropical Andes: An Overview. In *Imperfect Balance: Landscape Transformations in the Precolumbian Americas*; Lentz, D.L., Ed.; Columbia University Press: New York, NY, USA, 2000; pp. 281–310.
70. Custred, G. Las Punas de Los Andes Centrales. In *Pastores de Puna: Uywamichiq Punarunakuna*; Flores Ochoa, J.A., Ed.; Instituto de Estudios Peruanos (IEP): Lima, Peru, 1977; pp. 55–85.
71. Young, K.R.; León, B.; Herrera-MacBryde, O. Peruvian Puna. In *Centres of Plant Diversity: A Guide and Strategy for Their Conservation*; Davis, S.D., Heywood, V.H., Herrera-MacBryde, O., Villa-Lobos, J., Hamilton, A.C., Eds.; WWF/IUCN: Cambridge, UK, 1997; Volume 3, pp. 470–476.
72. Hedberg, O. *Features of Afroalpine Plant Ecology*; Acta Phytogeographica Suecica; Almqvist & Wiksells Boktryckeri AB: Uppsala, Sweden, 1964.
73. Tosi, J.A. *Zonas de Vida Natural En El Peru*; Instituto Interamericano de Ciencias Agricolas de la OEA, Zona Andina: Lima, Peru, 1960.
74. ONERN. *Inventario Nacional de Lagunas y Represamientos*; ONERN: Lima, Peru, 1984.
75. Treacy, J.M. *Las Chacras de Coporaque*; Estudios de la Sociedad Rural; Instituto de Estudios Peruanos: Lima, Peru, 1994.
76. Anonymous. Anexo de Los Religiosos de Predicadores de Sto. *Domingo con los Indios de Pamparomás, Archivo Comunitario de Pamparomás:* Pamparomas, 1774.
77. Lane, K.; Contreras Ampuero, G. An Inka Administrative Site in the Ancash Highlands, North-Central Andes. *Past Newsl. Prehist. Soc.* **2007**, *56*, 13–15.
78. Llosa Larrabure, J. *Elaboración e Implementación de Un Programa Nacional de Adaptación al Cambio Climático, Con Énfasis En Zonas Seleccionadas de La Sierra Centro y Sur Del País*; CONCYTEC/OAJ—Contrato de Subvención No064: Lima, Peru, 2008.
79. Palacios Rios, F. Pastizales de Regadío Para Alpacas En La Puna Alta (El Ejemplo de Chichillapi). In *Comprender la agricultura campesina en los Andes Centrales: Perú-Bolivia*; Morlon, P., Ed.; IFEA/CBC: Lima, Peru, 1996; pp. 207–212.
80. Verzijl, A.; Guerrero Quispe, S. The System Nobody Sees: Irrigated Wetland Management and Alpaca Herding in the Peruvian Andes. *Mt. Res. Dev.* **2013**, *33*, 280–293. [CrossRef]
81. CEDEP. *Experiencia de Repoblamiento de Alpacas en La Sierra de Ancash*; CEDEP, Lima as Internal Memo, Centro de Estudios para el Desarrollo y la Participación: Lima, Peru, 1997.
82. Browman, D.L. High Altitude Camelid Pastoralism of the Andes. In *The World of Pastoralism: Herding Systems in Comparative Perspective*; Galaty, J.G., Johnson, D.L., Eds.; Guilford Press: New York, NY, USA, 1990; pp. 323–352.
83. Freisem, C. Vorspanische Speicherbecken in Den Anden: Eine Komponente Der Bewirtschaftung von Einzugsgebieten Das Beispiel Nepeñatal—Peru. Diploma Thesis, Technische Universität Berlin, Berlin, Germany, 1998.
84. Fairley, J.P. Geologic Water Storage in Precolumbian Peru. *Lat. Am. Antiq.* **2003**, *14*, 193–206. [CrossRef]
85. Leet, L.D. *Physical Geology*; Prentice Hall: Engelwood Cliffs, NJ, USA, 1982.
86. Possehl, G.L. The Chronology of Gabarbands and Palas in Western South Asia. *Expedition* **1975**, *17*, 33–37.
87. Cieza de León, P. *Crónica Del Peru: Segunda Parte*, 3rd ed.; Colección Clásicos Peruanos; Franklin Pease, G.Y., Ed.; Pontificia Universidad Católica del Peru/Academia Nacional de la Historia: Lima, Peru, 1996.
88. Donkin, R.A. *Agricultural Terracing in the Aboriginal New World*; Viking Fund Publications in Anthropology 56; University of Arizona Press: Tucson, AZ, USA, 1979.
89. Zuccarelli Freire, V.; Meléndez, A.S.; Rodriguez Oviedo, M.; Quesada, M.N. Erosion Control in Prehispanic Agrarian Landscapes from Northwestern Argentina: El Alto-Ancasti Highlands Case Study (Catamarca, Argentina). *Geoarchaeology* **2021**, 1–21. [CrossRef]
90. Netherly, P.J. The Management of Late Andean Irrigation Systems on the North Coast of Peru. *Amercian Antiq.* **1984**, *49*, 227–254. [CrossRef]
91. Venturi, F.; Villanueva, S. *El Manejo Del Agua en la Cordillera Negra: Estudio Elaborado Por Encargo Del Programa Cordillera Negra en Los Distritos de Huaylas y Huata*; Programa Cordillera Negra, Programa de Lucha Contra la Pobreza en Zonas Rurales de la Region Chavin: Huaraz, Peru, 2002.
92. Bonavia, D. Investigaciones Arqueológicas En El Mantaro Medio. *Rev. Mus. Nac.* **1967**, *35*, 211–294.
93. Schjellerup, I. Andenes y Camellones En La Región de Chachapoyas. In *Andenes y Camellones en el Peru Andino: Historia, Presente y Futuro*; de la Torre, C., Burga, M., Eds.; CONCYTEC: Lima, Peru, 1986; pp. 133–150.
94. Keeley, H.C.M. Soils of Prehistoric Terrace Systems in the Cusichaca Valley, Peru. *Br. Archaeol. Rep. Int. Ser.* **1985**, *232*, 547–568.
95. Herrera, A. *CROPP Report on Archaeological Field Reconnaissance in the Pescado River Valley, La Merced District, Aija Province, Cordillera Negra, Conducted January 2nd to January 6th 2021*; CROPP: Bogota, Colombia, 2021; pp. 1–18.
96. Holling, C.S. Engineering Resilience vs. Ecological Resilience. In *Engineering within Ecological Constraint*; Schulze, P.C., Ed.; National Academy Press: Washington, DC, USA, 1996; pp. 31–43.
97. Beymer-Farris, B.A.; Bassett, T.J.; Bryceson, I. Promises and Pitfalls of Adaptive Management in Resilience Thinking: The Lens of Political Ecology. In *Resilience and the Cultural Landscape: Understanding and Managing Change in Human-Shaped Environments*; Plieninger, T., Bieling, C., Eds.; Cambridge University Press: Cambridge, UK, 2012; pp. 283–299.

98. Peterson, G. Political Ecology and Ecological Resilience: An Integration of Human and Ecological Dynamics. *Ecol. Econ.* **2000**, *35*, 323–336. [CrossRef]
99. Turner, M.D. Political Ecology I: An Alliance with Resilience. *Prog. Hum. Geogr.* **2013**, *38*, 616–623. [CrossRef]
100. Kull, C.A.; Haripriya, R. Political Ecology and Resilience: Competing Interdisciplinarities? In *Interdisciplinarités entre Natures et Sociétés*; Hubert, B., Mathieu, N., Eds.; PIE Peter Lang: Brussels, Belgium, 2016; pp. 71–87.
101. Lentz, D.L. Introductions: Definitions and Conceptual Underpinnings. In *Imperfect Balance: Landscape Transformations in the Precolumbian Americas*; Lentz, D.L., Ed.; Columbia University Press: New York, NY, USA, 2000; pp. 1–13.
102. Sandweiss, D.H.; Andrus, C.F.T.; Kelley, A.R.; Maasch, K.A.; Reitz, E.J.; Roscoe, P.B. Archaeological Climate Proxies and the Complexities of Reconstructing Holocene El Niño in Coastal Peru. *Proc. Natl. Acad. Sci. USA* **2020**, *117*, 8271. [CrossRef]
103. Langlie, B.S.; Arkush, E.N. Managing Mayhem: Conflict, Environment, and Subsistence in the Andean Late Intermediate Period, Puno, Peru. In *The Archaeology of Food and Warfare*; VanDerwarker, A.M., Wilson, G.D., Eds.; Springer: Basel, Switzerland, 2016; pp. 259–289.
104. Arkush, E. *Hillforts of the Ancient Andes: Colla Warfare, Society, and Landscape*; University Press of Florida: Gainesville, FL, USA, 2011.
105. Lane, K. Hincapié En Los Andes Nor-Centrales: La Presencia Inca En La Cordillera Negra, Sierra de Ancash. In *Arquitectura Prehispánica Tardía: Construcción y Poder en Los Andes Centrales*; Lane, K., Luján Dávila, M., Eds.; UCSS/CEPAC: Lima, Peru, 2011; pp. 123–170.
106. Ampuero, F.; Yupanqui, I. Información Hecha Por Francisco de Ampuero y Doña Ines Yupanqui, Su Mujer, Vecinos de La Cuidad de Los Reyes, Sobre La Recompensa Que Piden Se Les Haga Del Repartimiento de Guíalas, Cuyas Guarangas Disfruto Contarguacho. *Rev. Mus. Nac.* **1976**, *XLII*, 272–288.
107. Levillier, D.R. *Gobernantes de Perú: Cartas y Papeles Siglo XVI: El Virrey Garcia de Hurtado de Mendoza, Marques de Cañete*; Documentos del Archivo de Indias; Colección de Publicaciones Históricas de la Biblioteca de Congreso Argentino: Madrid, Spain, 1926; Volume XIII.
108. Cook, N.D. *Born to Die: Disease and New World Conquest, 1492–1650*; University of Cambridge Press: Cambridge, UK, 1998.
109. Raimondi, A. *El Peru, Tomo I, Parte Preliminar*; Imprenta del Estado: Lima, Preu, 1874.
110. Junta de Desarrollo Distrital de Pamparomás. *Proyecto: Uso Productivo Del Agua y Desarrollo Agroecológico de La Microcuenca Chaclancayo*; Junta de Desarrollo Distrital de Pamparomas: Pamparomas, Peru, 2000.
111. Gleick, P.H.; Ajami, N.; Juliet, C.-S.; Cooley, H.; Donneley, K.; Fulton, J.; Ha, M.-L.; Herberger, M.; Moore, E.; Morrison, J.; et al. *The World's Water Volume 8: The Biennial Report on Freshwater Resources Paperback*; Island Press: Washington, DC, USA, 2014.

Case Report

Hydraulic Engineering at 100 BC-AD 300 Nabataean Petra (Jordan)

Charles R. Ortloff [1,2]

[1] CFD Consultants International, 18310 Southview Avenue, Los Gatos, CA 95033, USA; ortloff5@aol.com
[2] Research Associate in Anthropology, University of Chicago, Chicago, IL 60637, USA

Received: 4 November 2020; Accepted: 8 December 2020; Published: 12 December 2020

Abstract: The principal water supply and distribution systems of the World Heritage site of Petra in Jordan were analyzed to bring forward water engineering details not previously known in the archaeological literature. The three main water supply pipeline systems sourced by springs and reservoirs (the Siq, Ain Braq, and Wadi Mataha pipeline systems) were analyzed for their different pipeline design philosophies that reflect different geophysical landscape challenges to provide water supplies to different parts of urban Petra. The Siq pipeline system's unique technical design reflects use of partial flow in consecutives sections of the main pipeline to support partial critical flow in each section that reduce pipeline leakage and produce the maximum flow rate the Siq pipeline can transport. An Ain Braq pipeline branch demonstrated a new hydraulic engineering discovery not previously reported in the literature in the form of an offshoot pipeline segment leading to a water collection basin adjacent to and connected to the main water supply line. This design eliminates upstream water surges arising from downstream flow instabilities in the two steep pipelines leading to a residential sector of Petra. The Wadi Mataha pipeline system is constructed at the critical angle to support the maximum flow rate from a reservoir. The analyses presented for these water supply and distribution systems brought forward aspects of the Petra urban water supply system not previously known, revising our understanding of Nabataean water engineers' engineering knowledge.

Keywords: Petra; Nabataean; water systems; hydraulic analysis; CFD; canals; reservoirs; pipelines; flow stability

1. Introduction

The history of Petra's monumental architecture and historical development has been described by many authors [1–6]. Some scholars have concentrated on technical and location aspects of water supply and distribution systems within Petra [7–21], while other surveys [16,22–27] have concentrated on the water control and distribution technology available to Nabataean water engineers from Roman and other eastern and western civilizations through trade and information transfer contacts during Petra's expansion period (100 BC–AD 300) period. This paper was designed to add further depth to the hydraulic engineering technology used in the design and operation of Petra's three major pipeline water supply systems serving the urban center of Petra: The Siq system sourced from the Ain Mousa spring, the Ain Braq system, and the Wadi Mousa system.

Petra's openness to foreign influence is demonstrated in the city's monumental architecture that reflect elements of Greek, Persian, Roman, and Egyptian architectural styles integrated into Nabataean monuments [1–6,28,29]. Later Roman occupation of Petra past 106 AD exhibits Roman pipeline technologies employed to expand the marketplace, the Paradeisos Pool Complex [8,9], and city precincts responding to increased water demands for an expanding population as the city's status advanced as a key trade and emporium center. Petra's ability to manage scarce water resources to provide constant potable water supply for its permanent population of ~20,000 citizens with reserves for large caravan arrivals and drought periods was vital to its centuries of prominence as the nexus of a trade network between African, Asian, and European cities for luxury goods. While knowledge of hydraulic technologies from foreign sources was available, the rugged mountainous terrain, distant spring water sources, and brief rainy periods posed unique water supply challenges that required technical innovations consistent with the site's ecological and geophysical constraints. Effective utilization of scarce and seasonally intermittent water supplies required technical expertise to ensure optimum system functionality from the use of distant interconnected spring-supplied pipeline transport and distribution systems (Figure 1) that guided water to the city's densely populated urban core and surrounding agricultural districts.

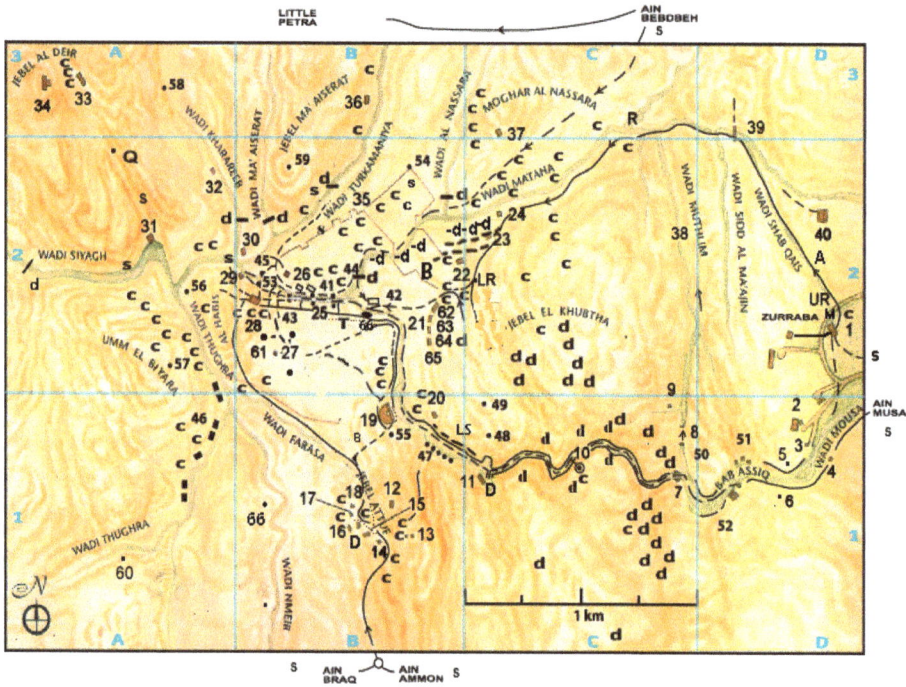

Figure 1. Site map of Petra: Site names corresponding to numbers given in the Appendix A.

Examination of three of Petra's major water conveyance pipeline systems using computational fluid dynamics (CFD) analysis permitted discovery of the rationale behind design selections utilized by Nabataean hydraulic engineers. Solution of fluid dynamics equations using CFD finite-difference methodology together with models of existing pipeline structures [30] within Petra's urban center graphically demonstrate internal water flow patterns within pipelines that revealed damaging hydraulic phenomena that would be observationally familiar to Nabataean engineers from past trial-and-error experience. What the CFD solutions revealed were the problems and their solutions used to guide their pipeline design/selection progress to avoid system failures. CFD results demonstrating flow patterns within pipelines showed what Nabataean hydraulic engineers intended (or avoided) in their hydraulic designs and give insight into their civil engineering knowledge base. Given the low survival rate of documents and descriptions of hydraulic phenomena from ancient authors and from the descriptions of water engineering phenomena from the few surviving ancient documents, the descriptions are given in prescientific terms with little correlation to modern terminology to understand the ancient author's meanings of hydraulic phenomena. Given this lack of descriptive material to interpret the water engineering base underlying their water system designs and functions, the analysis of internal flow patterns within pipelines using CFD methodology presents a viable recourse to calculate, visualize, and analyze hydraulic engineering problems Nabataean hydraulic engineers encountered and solved in their piping network designs. Within the archaeological remains of Petra's pipeline systems lies insight into Nabataean technical processes and the knowledge base available to their water engineers that served as the foundation for their engineering decision making. Use of CFD analysis provided graphic displays of problems encountered and solutions developed by Nabataean water engineers and helped to understand what lay behind field observations of archaeological remains relevant to Petra's water supply and distribution systems.

Additionally, descriptive terms from ancient texts describing hydraulic phenomena given in prescientific terms unfamiliar to western notations can now be associated with actual phenomena through CFD modeling of the flow patterns through water conveyance structures.

2. The Water Infrastructure of Petra

To begin the discussion of Petra's water system development and progress toward utilizing all possible water resources to meet population requirements and the greatly increased water demands from arriving caravans (which according to the ancient author Strabo, consisted of many hundreds of camels and hundreds of accompanying support personnel), the solid and dashed lines of Figure 1 detail the known water supply and distribution pipeline systems leading water into the city's urban core. Numbered locations denote major site features listed in the Appendix A. Shown are major and minor catchment dams and multilevel stepped dams (-d, d, D), cisterns (c), water distribution tanks (T), and springs (s). The (-d) dams located across streambeds are stone barrier structures built to hold large quantities of rainfall runoff water for redistribution to urban structures and local agricultural areas. Dams denoted (d) are minor catchment structures that stored water in mountainous areas to limit descent and flooding of Petra's low altitude urban center and have channels leading water to cisterns. Solid lines represent original pipeline paths originating from distant springs which in many cases now show only barren channels that once contained interlocked terracotta piping elements. The superimposed grid system (A, B, C; 1, 2, 3) defines 1.0-km^2 grid boxes to enable the location of numbered Appendix A features. Figure 1 represents the present state of knowledge of the location of pipeline structures from personal field exploration and survey results from previous site explorers [7–16,18,21]. Due to erosion and soil deposition landscape degradations over ~2000 years, as well as reuse and pillaging of piping elements by later inhabitants of the area, many terracotta pipeline sections are now missing from their original holding channels, obscured by subterranean placement, or are yet to be discovered. Extrapolation of pipeline continuance between available visible pipeline segments then represents the pipeline paths shown in Figure 1 and is, at best, a first approximation to the entirety of the total pipeline system as much of the site remains unexcavated and subject to landscape change from millennia of soil erosion and deposition events caused by flood episodes, landslide events, and frequent earthquakes. Further hidden subterranean pipelines buried for defensive security purposes may exist to supplement the known pipelines shown in Figure 1. Recent on-site discoveries appear to add new information in this regard: For example, Site 66 in Figure 1 represents a large newly discovered (in 2016) platform with a centrally located temple structure yet to be excavated to determine its role in Petra's administrative and ceremonial life [28]. The water supply system serving this newly discovered complex remains to be discovered but surely exists given that all ceremonial structures presently known are accompanied by elaborate water systems for both aesthetic display and supply of potable water.

Additionally, as elements of major pipelines were intentionally built subterranean and as yet undiscovered due to limited excavations performed in the urban center and distant reaches of Petra, Figure 1 can be considered a first approximation of the total piping network. Some major monuments located at high altitude exclude the possibility of canal water supplies from springs located at lower levels; in this case, channels directing rainfall capture to cisterns located nearby provided the water supply to high-level sites.

Petra's urban core lies in a valley surrounded by rugged mountainous terrain. Figure 2 demonstrates a section of the mountain terrain behind the ~10 m-high Q'asar al Bint temple (Feature 29, Appendix A) located at the base of the Jebel al Deir mountain and gives a perspective on the rugged mountainous terrain encompassing the urban core of Petra. Seasonal rainfall runoff passes into the valley through many streambeds and was drained out primarily through the Wadi Siyagh streambed in antiquity (Figure 1, A-2). Although present measurement of the Wadi Siyagh streambed slopes now precludes drainage resulting from water-transferred sediment deposits over millennia, this path was the main drainage channel during the city's existence. Water flow into the city [10] originated from captured rainfall runoff into cisterns from Wadi al Hay to the north, Wadi al Hudayb to the south, and from the watershed area supplying runoff into the Wadi Mousa streambed. Surface flows from these sources passed through the Siq (Figure 1, D-1 to C-1; Appendix A Feature 10) which is the ~2 km long narrow entry passageway into the urban center of Petra (Figures 1 and 2A). The Siq is a natural, narrow passageway through the Jebel el Khubtha mountain range, varying from 5 to 10 m in width; the north and south walls of the Siq are near vertical with maximum heights up to 80 m.

By construction of a diversion dam and tunnel shown in Figure 3 (Figure 1; Feature 8 in Appendix A) in the first century AD, flood water from the Wadi Mousa River was deflected from the Siq entrance and led to Wadi Mataha (Figure 1, C-2, 3) that linked up with the section of the Wadi Mousa streambed within the city center; water then drained into Wadi Siyagh (Figure 1, A-2) through a channel passing through the city center. This dam and tunnel construction served to divert the silt-laden floodwater away from entry into the Siq that would compromise the structural integrity of many of the monuments within Petra's urban core region. At present, the Wadi Siyagh drainage path is partially blocked from silt deposits, causing flooding of Petra's urban center during the rainy season. As the chemical composition of Petra's stone monuments contain different salt varieties [31], exposure to flood water causes destructive flaking from monument walls; this effect is apparent from observation of many of Petra's urban sector monuments. Runoff water from the southern sector Wadi Thughra, Nmeir, Farasa, and northern sector Wadi Kharareeb, Ma'aiserat, and Turkamanya gullies (Figure 1) additionally drained rainfall runoff into the Wadi Siyagh away from the city center [10]. In sector B-2 of Figure 1, a local depressed low area lies between the Jebel el Khubtha Mountain and sites to the west. Currently, a modern bridge crosses over the southern narrow end of this area. If the southern narrow drainage end of this basin was dammed in antiquity, rain water or a Wadi Mataha diversion channel could produce an internal shallow lake within city confines that would serve local gardening, nearby ceramic and metal working workshop areas as well as city aesthetic beautification purposes consistent with Nabataean plans to utilize all available water before disposal into the Wadi Siyagh drainage. As mentioned in Strabo ([32]), " ... the inside parts (of Petra) having springs both for domestic purposes and for watering gardens ... ", which would be consistent with an internal shallow lake within city precincts whose borders and groundwater could serve as gardening areas and for specialty crops. At this stage of research, this feature awaits verification for its existence in ancient times.

Figure 2. (**A**) The Q'asar al Bint Temple at the base of Jebel al Deir Mountain. (**B**) Narrow part of the Siq passageway; note the figure located on the passageway for scale of the height of Siq near vertical walls.

Figure 3. Nabataean drainage tunnel leading water away from the Siq entry to the Wadi Sayagh through channels extending from and parallel to the Wadi Mataha streambed (Figure 1, solid line path through 2-D, 3-C, 2-C sectors).

Spring-fed water supply to the city before the first century BC was through a slab-covered, open channel dug into the floor of the Siq, conducting water from the distant Ain Mousa spring into the urban center of the city. A trace of this system in the vicinity of the south Nymphaeum along Colonnade Street may exist from GPS data [33]. This channel perhaps extended as far as the Temenos Gate (Figure 1, B-2; Feature 43 in Appendix A) later built in early Roman occupation times past AD 106 to honor the visit of Hadrian in 131 AD. First century BC and first century AD Nabataean builders replaced this early Siq channel water supply system with a pipeline system along the north wall of the Siq supplied by a ~14-km pipeline carrying water from the Ain Mousa spring. Details of most of this pipeline system from the Ain Mousa spring are largely unknown due to later urban construction overlay except for a pipeline segment before the Siq entrance. First century BC and early century AD construction revealed a more integrated approach to site water system management as demonstrated by new and novel features. The major Siq northern wall channel supporting a pipeline (Figure 4) was an additional feature eliminating earlier use of the Siq floor channel (Figure 1; B-C-D-1 sectors). New features included surface cisterns to capture rainfall runoff, deep underground cisterns (one located in the eastern part of the site, Appendix A Feature 28, Figure 1), multiple pipeline systems sourced by different springs and storage reservoirs, floodwater control through diversion dams and tunnels, and pipeline supply system redundancy to ensure water delivery from multiple spring and high-level reservoirs—notably the massive Zurraba reservoir that supplied lower level smaller distribution and water storage reservoirs such as the M reservoir (Figure 1, D-2 and Figure 5B. The Figure 5B M reservoir no longer exists due to construction expansion of the nearby town and may have been one of several other distribution reservoirs sourced by networks of canals from the Wadi Mousa spring.

Figure 5A, Appendix A Feature 1, shows the high-level Zurraba reservoir supplied by a branch channel from the Ain Mousa spring that supplied water to high population concentration parts of the city. This reservoir also served as a drought remediation measure and served to provide large water supply storage and delivery for large caravan arrivals into the city. The lower level M reservoir (Figure 5B, Figure 1, 2-D) was connected to the upper level Zurraba reservoir through a pipeline; this lower reservoir provided water for occupation zones close to the Siq entrance. Due to modern road building and urban construction obliterating many of the ancient pipeline connections between reservoirs, the totality of the reservoir network pipeline system sourced by the Wadi Mousa spring remains to be determined.

Major pipelines, primarily on the north wall of the Siq, provided sediment particle filtration and removal basins (Figure 6) and served a sophisticated hydraulic function particular to partial pipeline flow as described in detail in a later section. The Siq south-side water open channel, perhaps sourced by a south-side spring, likely served as the water supply for camels and horses accompanying personnel passing through the Siq. The north-side Siq pipeline had four open basins supplying potable water for personal use (typical of Figure 6); this feature was an integral part of the hydraulic design of the pipeline as discussed in detail in a later section.

Figure 4. Channel cut to support pipeline water supply to the city center along the north wall of the Siq passageway. Pipeline continuation extends to the Outer Siq and further on to the city center.

(**A**)

Figure 5. *Cont.*

Figure 5. (**A**) Upper level Zurraba reservoir (Figure 1, 2D), (**B**) Lower-level M reservoir.

Figure 6. Typical open basin between segments of the Siq pipeline.

The new hydraulic features developed at later times reflected the need to bring high-quality potable water into the city center and serve city hillside occupation zones above the valley floor. Late first century BC and early first century AD developments demonstrated continual evolution of the Petra water system and reflected application of acquired technologies from exterior sources integrated with indigenous hydraulic engineering innovations to respond to the water needs of the city as a sustainable population center. Contingency water needs resulting from caravan arrival and water restocking anticipating transit to distant cities in North Africa and Mediterranean cities were vital considerations to maintain Petra's importance as a vital trade emporium. Beyond serving as a transit center for foreign goods to distant areas through known trade routes, the Nabataeans had a monopoly on incense trade originating from Southern Arabia—a vital and profitable product that the ancient world valued for ceremonial functions.

In summary, the means to capture and store a fraction of rainfall runoff through dams and cisterns, to build flood control systems, and build pipelines and channels to deliver water from distant springs to provide a constant water supply to the city was vital to understand Nabataean contributions to hydraulic engineering and water management practice. While water storage was vital to the city's survival, springs internal and external to the city (Ain Mousa, Ain Umm Sar'ab, Ain Braq, Ain Dibdiba, Ain Ammon, Ain Beidha, and Ain Dibidbeh) supplied water channeled and/or piped into the city to provide the main water supply (Figure 1). Several spring-supplied reservoirs exterior to the city (M–Zurraba, Figure 5A,B) provided additional water storage reserves. No accounts of other lower level, near city reservoirs exist at present due to modern urban construction; thus the complete reservoir water storage system remains unknown. Additional on-demand use of reservoir water for industrial ceramic and metal working workshop areas located at the base of the Jebel el Khubtha Mountain and for ceremonial use to supply triclinium rituals for tomb celebratory functions are indicated in [4,10,12]. Of the many springs used in antiquity, the Wadi Siyagh spring near the quarry (Figure 1, Appendix A Feature 31) remains functional for local inhabitants living in remote sections of the ancient city together with the Ain Mousa spring which now provides water for the modern town of Wadi Mousa and the water display structure in the town center. Water taken from the open channel within this structure has ritual use as evidenced by members of holy sects collecting water in containers for ceremonial functions.

3. Details of the Siq Pipeline System

The main city water supply of ancient Petra originated from the Ain Mousa spring about 7 km east of the town of Wadi Mousa (Figure 1, D-1), combined with the waters from the minor Ain Umm Sar'ab spring. The flood bypass tunnel (Figure 3, Appendix A Feature 8) at the Siq entrance, together with an entry dam and elevated paving of the Siq floor, reduced flooding into the urban center from rainy season runoff into the Wadi Mousa River. Part of the early slab-covered channel water supply to the urban core of Petra now lies under hexagon slab pavement in front of the Treasury construction attributed to Aretas IV (9 BC–AD 40). Recent excavations of this area revealed an empty tomb structure below the front part of the Treasury (Figure 7), although the intended use remains conjectural.

Later construction of the early Nabataean Siq water channel was replaced by the Siq pipeline water supply system supplied from the Ain Mousa spring source. The four open surface water basins between Siq piping segments (Bellwald, principal Siq excavator, personal communication) typical of those shown in Figure 6, give indication as to the hydraulic technology available to Nabataean water engineers as further discussion indicates. It is of note that in several locations along the ~2-km length of the Siq pipeline, top portions of the pipeline have been opened up, revealing heavy sinter deposits on pipeline bottoms that taper off about halfway up the pipeline side walls. This observation would indicate that a partial flow existed in at least some portions of the pipeline by observation of the sinter deposit patterns. If a partial flow in Siq pipeline sections was the original design intent of Nabataean water engineers, then opening up the top part of the pipelines made no difference in the pipeline flow rate and may have had the additional benefit of flow observance and access to cleaning sinter from pipeline inner surfaces. Additionally, if partial flow in the Siq pipeline was the design intent, then the Siq pipeline system was not full-flow pressurized as an air space existed over the partial flow, thus reducing the possibility of pressurized water leakage at pipeline joints. Details of the design intent of the Siq pipeline based upon these preliminary observations are next considered to proceed toward hydraulic analysis of the Siq pipeline and its Outer Siq continuation to the city center.

Figure 7. The Treasury, located at the Siq exit.

With increasing water needs for increasing population in early centuries AD, an elevated ~1-km long, north-side pipeline extension (Figure 8; LS, Outer Siq, Figure 1 1-B, C) extended the utility of the Siq pipeline system [14–16,25,27] to supply potable water to further reaches of Petra's urban core. This pipeline extension provided water supply to structures located in the Figure 1 B-2 district, as well as providing pipeline water supply to the Nymphaeum (Appendix A Feature 42, Figure 1). Further pipelines emanating from the Nymphaeum to the Temple of the Winged Lions (Figure 9), major tombs fronting the Wadi Mataha housing structures, and elite palace structures in early stages of excavation [18,19] existed, although details await further excavation. Continuance of water flow through a pipeline to a bridge across the Wadi Mataha led water to the south side of the site as further discussion details.

Figure 8. Elevated Outer Siq pipelines located ~6 m above the current site surface. These pipelines lead to an extension of the Siq pipeline to the city center. Several meters below the current site surface is the original paved floor of the area in front of the Treasury.

Figure 9. Temple of the Winged Lions [4].

To fully analyze the design intent of the Siq pipeline, note first that within pipelines, destructive hydraulic instabilities may exist [34]. These include transient pressure waves, flow intermittency, internal pipeline hydraulic jumps, transient turbulent drag amplification zones, partial vacuum regions, and subcritical to supercritical (and vice versa) flow transitions—these instabilities largely depend upon pipeline slope and internal pipeline wall roughness—all of which affect flow stability and pipeline flow rate. Such hydraulic problems affecting the design of pipeline systems required development of advanced technologies to produce stable pipeline flows that matched a sizeable fraction of the Ain Mousa spring source input. The total pipeline design from the Ain Mouse spring source to the Siq and through the Outer Siq needed to be designed to produce the maximum flow rate the total pipeline system could sustain. Further design considerations involved limiting full flow conditions within pipelines (pipeline cross-section fully occupied by water) to limit hydrostatic pressure within a pipeline. This design consideration limited leakage at the thousands of terracotta pipe-joint-connections along long lengths of pipelines. For a typical pipe element length of ~0.35 m and total pipeline length over ~14 km from the Ain Mousa spring origin location to the end of the Siq and Outer Siq pipeline with a total height drop well over ~50 m (data derived from the contour *Map of Petra*), high hydrostatic pressure in a full flow pipeline would increase the likelihood of pipeline element connection joint leakage at~42,000 pipeline joint connections. Solutions to mitigate many of the design and operational problems provide insight into Nabataean water management expertise.

CFD calculation [30] of full-flow volumetric flow rates in 14 cm diameter(D) piping were made for internal wall roughness $\varepsilon/D > 0.01$ (where ε is the mean-square internal pipeline roughness height); this ε/D value holds for roughness typical of terracotta internal pipeline walls, demonstrating casting corrugations and erosion usage pitting [35] for typical Reynolds number Re ~10^5 values. For pipelines at 2-, 4-, and 6-degree declination slopes for 1.0 and 3.0 m supply head, Figure 10 reveals that past ~400 m pipeline length, an increase of input head does not substantially increase the flow rate [17]. Past this length range, even a 3× head increase does not substantially increase flow rate. The implication for the ~14-km long Ain Mousa spring-supplied pipeline to the Siq and Outer Siq extension is that sophisticated hydraulic engineering had to be in place to match the estimated [12,13] maximum Ain Mousa spring output of~1000–4000 m^3/day; (~0.01–0.04 m^3/s) through the Siq pipeline without invoking any of the flow instability problems previously alluded to. Flow transmitted through the Siq pipeline at the maximum Ain Mousa output flow rate clearly required knowledge of flow conditions in the pipeline before the Siq entrance (which is likely full flow due to internal pipeline wall roughness flow resistance considerations and low pipeline slope) coupled with knowledge of Siq and Outer Siq pipeline designs that mitigate flow resistance and flow instabilities that would propagate upstream into the subcritical supply flow pipeline.

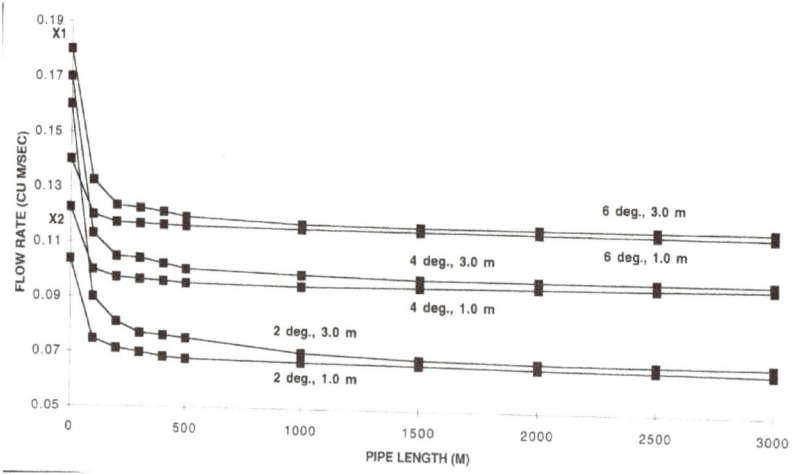

Figure 10. Flow rate in full flow pipelines as a function of supply head (m), pipeline slope (deg.), and pipeline length (M) for a given ε/D value.

The internal pipeline flow rates shown in Figure 10 are based upon a low pipeline wall roughness of ε/D ~0.01; higher ε/D ~0.07 values that account for pipe element connection socket roughness reduce flow rates substantially, leading to Figure 10 curves being shifted downward by a factor of ~3 for typical Re ~10^5 Reynolds number values. On this basis, the observed Siq rough internal wall pipeline at a 2-degree slope permitted a 0.023 m^3/s flow rate and, at a one-degree pipeline slope, permitted a 0.017 m^3/s flow rate corresponding to ~1.5 and ~1.1 m/s mean velocity in the 14-cm internal diameter pipeline with corresponding flow rates of ~80 (0.02 m^3/s) and ~60 m^3/h (~0.02 m^3/s) compared to the Ain Mousa flow rate of 0.01 to 0.04 m^3/s. Thus, even high head values from the Ain Mousa spring to the Siq entrance pipeline had little effect of increasing the flow rate given the long 14-km supply pipeline as Figure 10 implies. The additional ~2-km Outer Siq pipeline lengths to the 14-km Ain Mousa spring to Siq exit length added additional flow resistance due to additional pipeline internal wall roughness resistance; all these elements had to be considered for a substantial fraction of the Ain Mousa spring flow rate to enter the total pipeline length under maximum flow rate conditions without instabilities and leakage problems. For lower flow rate conditions due to Ain Mousa spring seasonal rainfall absorbance conditions, only a trickle flow would go through the Siq pipeline, emphasizing the need for available storage reservoir water supplements to maintain potable water supply for the city's ~20,000 population and caravan trade. Any excess flow beyond the Siq extraction flow rate from the Ain Mousa spring was likely used to recharge upper and lower level reservoir systems for on-demand water usage. Here, normal flow to the Siq may have been temporarily stopped to serve lower reservoir filling—this being done by pipeline blockage—and for upper reservoir filling, pipelines from the Ain Musa spring likely served this purpose. Given the ~10,000 m^3 capacity of the main Zurraba reservoir (Figure 5A) and that of several additional minor reservoirs, the additional Ain Mousa flow rate capacity not used for the total Siq pipeline system served to provide additional reservoir water storage for on-demand and drought remediation use as well as for pipeline maintenance shut-offs where alternate pipeline water supplies were available from reservoirs. In summary, to this point it appears that a sizeable fraction of the Ain Musa spring flow rate could be transmitted through the Siq pipeline with provision for system shut-off to refill lower reservoirs for drought condition water availability, caravan water supply, and pipeline maintenance; however, there is more to this story to demonstrate the engineering that this water system contains.

Along the Siq north-side pipeline, several typical open basin features (Figure 6) are found at intervals along the pipeline. Each basin permits an accumulation of water serving as a head tank for its downstream piping length segment. Figure 10 reveals that hydrostatic head substantially influences flow rate for pipeline lengths less than ~500-m, and more so for shorter pipeline lengths. The head is reset at the origin of each of the piping lengths between basins so that flow through each piping segment is only sensitive to its local head tank head value. This modification led to partial flow (pipeline cross-section partially occupied by water) in Siq pipeline segments reducing overall flow resistance. As pipeline flow extending ~14 km from the Ain Mousa spring to the Siq entry point was mostly subcritical full-flow due to the high pipeline internal wall roughness flow resistance and low slope, the lowered resistance of Siq pipeline segments at ~3-degree hydraulically steep slopes between basins induced partial flow in distal portions of the ~14-km Ain Mousa spring supply pipeline to the Siq pipeline. The slightly lowered flow resistance could help to raise the overall system flow rate due to lowered back pressure and partial flow frictional wall resistance decrease. In simpler terms, lowering the frictional flow resistance in the end regions of a pipeline permitted a higher flow rate to be achieved through the entire pipeline system.

The four Siq north-side open basins (Bellwald, Siq principal excavator, personal communication) additionally served to provide drinking water along the Siq narrow passageway (Figure 2A) and access for cleaning of silt particles entrained in the flow to enhance potable water quality. Atmospheric pressure in the air space above the partial flow water surface in the Siq pipeline segments reduced pipeline joint leakage as would occur under a pressurized, full-flow condition design. To achieve a piping design based on a 14-cm pipe inner diameter at the declination angle of the Siq, ideally partial, critical flow at unit Froude number (Fr = 1) at critical depth (y_c/D~0.8–0.9) would provide the maximum flow rate the piping could convey. As open basins were sequentially lower than their predecessor upstream basins due to the declination slope of the Siq pipeline, each basin water height supplied head for its adjoining downstream piping section. Thus, partial flow existed throughout most of the Siq piping sections where flow entering a ~3-degree slope pipe section from a basin transitioned to a partial-flow, critical depth. This result indicates that the Froude number for the Siq pipeline flow is on the order of unity at close to the permissible ~96 m^3/hr flow rate. As the main purpose of the Siq water supply system was for the maximum flow rate to the urban center, any flow resistance changes that would increase flow rate were part of Nabataean water engineers' design intent. This was apparently achieved to gain a larger portion of the Ain Mousa flow rate to be used to supply the urban center's water needs. The thick sinter build-up in the (now visible) open bottom-most part of Siq piping supports the deposition pattern resulting from partial flow conditions existing in the Siq pipeline segments.

In the area immediately west of the Treasury, the Outer Siq pipeline (Figure 1, LS, B, C, and Figure 8) continued the Siq pipeline and was set at a slightly lower angle than the supply Siq pipeline. Given the Siq pipeline flow rate, a further consideration arises as to the stability of the flow in piping sections within the Siq as well as for water arriving to the Nymphaeum (Figure 1, Appendix A Feature 42) through the further Outer Siq pipeline extension subject to the input Siq flow rate. Any flow resistance from the Outer Siq pipeline extension would have a small effect on the upstream flow rate given the near-critical flow rate sections in the Siq pipeline that limit upstream influence. For water delivery from the Outer Siq portion of the total pipeline, if water delivery was pulsating into open basins within the Nymphaeum, then spillage and sloshing occurs amplifying flow instabilities that would compromise the aesthetic display of fountains and open pools in the Nymphaeum and induce forces within adjoining distribution pipelines promoting joint separation and leakage. For an examination of flow stability within the Outer Siq pipeline extension, CFD models were made for a ~1220-m section of the piping extension with a higher declination slope than that in the Siq pipeline; this length represents the approximate distance from the exit of the Siq pipeline to the Nymphaeum (Feature 42, Figure 1). Previous research [10] has indicated a further south-side Nymphaeum structure across the Wadi Mousa streambed serving the marketplace area with an extension of the south arm of the Siq pipeline; traces of elevated reservoirs providing pressurized water to the south Nymphaeum are described in [25].

FLOW-3D computer solutions (Figure 11) for 0.305(A), 0.610(B), 1.54(C), and 3.05(E) m/s water velocities were made with observed ε/D values typical of the interior surface of piping elements at the higher declination slope of the Outer Siq segment; here water velocity is expected to be higher than in the supply Siq pipeline. Figure 11(D) results are for a smooth interior pipe element wall at 1.54 m/s velocity to demonstrate the effect of wall roughness on flow patterns at the same water velocity as 1.54 m/s (C). For an illustrative example, water at ~96 m^3/hr was accepted into the Outer Siq piping—average pipeline water velocity was then ~1.73 m/s. Given the flow was slightly supercritical at this velocity, the flow approximated critical depth for Cases (B), (C), and (D). For flows close to critical, the normal depth was close to the critical depth [35]. As partial flow existed in the extension section (similar to that in the Siq piping) and had approximately the ~96 m^3/h flow rate at about the steeper declination angle as the Siq piping, near critical flow conditions existed throughout the Outer Siq extension pipeline. Figure 11 shows the transition to lower water depths from the entry boundary condition for all supercritical (B), (C), (E) and (D) cases, primarily due to the steeper declination angle. For the subcritical Case (A), water was at a low level due to low flow velocity. Figure 11, Case (B) at ~1.2 m/s approximated the flow effect at a mean velocity of ~1.75 m/s and indicated that only minor, random height excursions occurred in water flowing within the Outer Siq pipeline—this indicated that the water supply system through the Outer Siq extension to the Nymphaeum was stable, which was advantageous for proper operation of the Nymphaeum water display system. At twice this velocity, Figure 11(C) indicates large height fluctuations in the water surface, indicating unstable flow delivery occurring in the pipeline system. This indicates that the flow rate selected from the Ain Mousa spring by Nabataean engineers was a deliberate design choice to achieve stable flow in the total system. For Case (E), the high flow rate case shows susceptibility to a hydraulic jump creation due to the amplified supercritical flow interaction with pipeline internal roughness. This case serves as limit on the declination angle that could be implemented by Nabataean engineers. Here Cases (A) and (B) likely serve to give the declination angle of the Outer Siq extension pipeline due to the small flow disturbances indicated; Cases (C) to (E) would require a mandatory settling basin design to ensure smooth flow on to city center locations. Such an intermediate settling basin was observed in practice most likely to smooth the passage pipeline flow on to the city center and Nympheaum indicating a safety measure to ensure smooth flow for any of the Cases shown for any of the different declination angles shown that supported higher flow rates.

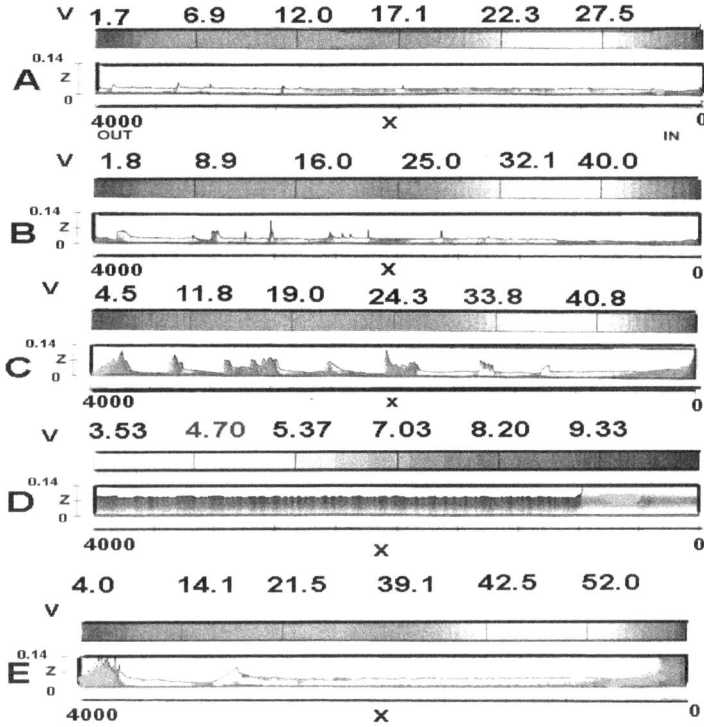

Figure 11. Internal pipeline flow stability results relevant to the Outer Siq pipeline for Cases **A** to **E**; right (IN) to left (OUT) flow direction for all cases.

Comparison of Figure 11, Cases (C) and (D) indicates that internal pipeline wall roughness amplified large-scale turbulent eddies that contributed to flow unsteadiness; this effect was included in the Case (B) result. Tracking the distal end of the Outer Siq extension pipeline revealed it was intercepted by a head reset open basin, indicating a further feature to ensure flow stability for flow continuing onto further pipeline branches and the Nymphaeum.

In summary, the total Siq system was designed to optimally transfer close to the Ain Mousa spring output flow rate of ~96 m^3/hr with partial flow conditions in several pipeline segments. This design largely eliminated large hydrostatic pressure conditions associated with full flow designs that would cause pipeline joint leakage at high hydrostatic pressures. Additionally, open basins along the Siq pipeline permitted removal of sediment debris that would ultimately clog the Siq system and destroy its functioning. In total, the various hydraulic flow regimes designed into the Siq pipeline (Fr < 1 subcritical, Fr > 1 supercritical, and Fr = 1 critical) and its Outer Siq extension necessary to capture a large fraction of the Ain Mousa spring output flow rate indicates an advanced knowledge of hydraulic science.

4. The Wadi Mataha Pipeline System

The water supply systems of Petra employed springs (s, Figure 1) both within (Ain Siyagh) and outside city limits (Ain Dibidbeh, Ain Mousa, Ain Braq, and Ain Ammon) to bring water to city monuments through pipelines. Reservoirs fed by the Ain Mousa (M–Zurraba, Figure 1) as well as other springs and reservoirs (no longer in existence due to modern urban expansion) provided supplemental water on-demand for arriving caravans or other utilitarian purposes. Pipelines composed of short interlocking terracotta pipeline elements, with sealed at joint connections by hydraulic cement, were typical of pipeline construction. Nabataean pipe elements were typically ~0.35-m long with an inner diameter range of ~14 to 23 cm, as observed from sample pipeline sections on display at the Petra on-site museum. As technologies developed over centuries, many different pipeline types developed with improved features to reduce internal pipeline wall frictional resistance. One special design of Nabataean tailored internal wall roughness minimized drag resistance ([25], pp. 268–271) according to modern hydraulic engineering theory and laboratory tests—if intentional from early observations and tests, this hydraulic science development occurred some ~2000 years before its "official" discovery in western hydraulic science.

Later Roman modifications to existing Nabataean water systems in the Great Temple area [20,21] (Figure 12) include small diameter lead piping and Roman standardized pipe diameter designs [22] for Roman extensions of the marketplace and Cardo Street areas, as well as for the Paradeisos Pool areas. The Roman standards for what a city under its control should have (fountains, baths and other water display structures) were an inherent part of Roman reconstructions present in post-106 AD Petra together with standardized piping size elements and lead piping. Excavations in the Cardo area show standardized Roman pipelines grafted onto existing Nabataean pipelines originating from Siq pipeline extensions to this area that expanded the marketplace area with fountains. For the Wadi Mataha pipeline, construction details of the channel on the Jebel al Khubtha mountain face (C-2, Figures 1 and 13, A–C) carry partial pipeline remains permitting analysis of its design and function.

Figure 12. The Great Temple.

The presence of pipeline segments, as opposed to an open channel conveying water (in the Wadi Mataha pipeline for example, Figure 13, A–C), is evident by noting that flow accelerating on a hydraulically steep open channel encountering bends would form surface height excursions and hydraulic jumps, causing channel overflow. For an open channel design, only a trickle flow from the upper reservoir could be contained in the channel without spillage. As water conservation was a prime concern, open channel systems subject to evaporation and spillage were not considered viable compared to pipeline systems. Some outer slab-covered sections of the Wadi Mataha channel are evident from and typical of Nabataean methods to provide a defense against water heating from direct sunlight exposure. Additionally, by surrounding embedded pipelines in channels with stabilizing soil to limit pipeline motion caused by water frictional forces and flow instabilities, leakage could be reduced at pipeline connection joints. Figure 10 governs admissible flow rates for full flow conditions such as would apply to pipeline flows originating from the M–Zurraba reservoir; for partial flow conditions as indicated in Figure 11, theoretical means are available for flow rate estimates [35,36]. Note also that open channels carrying water from distant springs were not a design option due to evaporation loss of scarce water supplies.

The pipeline along the eastern face of Jebel al Khubtha originated from a pipeline branch originating from the M–Zurraba reservoir complex. Typical full flow maximum flow rate estimates from Figure 10 are on the order of ~0.05 m^3/s, which would satisfy on-demand water supply requests; lesser flow rates would be possible by using a flow control valve originating from the reservoir. Piping from the high-level reservoir source proceeds to a lower reservoir elevated above the Sextus Florentinus Tomb area(Appendix A Feature 22, Figure 14), then proceeds southward toward the royal compound area [19] (Figure 15, Figure 1, 2-C), then on to monumental royal tomb architecture (Figure 15) near the Palace Tomb and a nearby fountain [10], then further on to the Nymphaeum for further pipeline water distribution to other parts of the city together with multiple basin/reservoir areas created by multiple dams along the Wadi Mataha streambed (-d, Figure 1).

(**A**)

Figure 13. *Cont.*

(B)

(C)

Figure 13. (**A**) The Wadi Mataha pipeline channel on the Jebel al Khubtha mountainside. (**B**) Continuation of the Wadi Mataha pipeline above the Sextus Florentinus Tomb (Figure 14). (**C**) Continuation of the Wadi Mataha channel is above workshop areas.

Figure 14. The Sextus Florentinus Tomb below the Wadi Mataha pipeline channel.

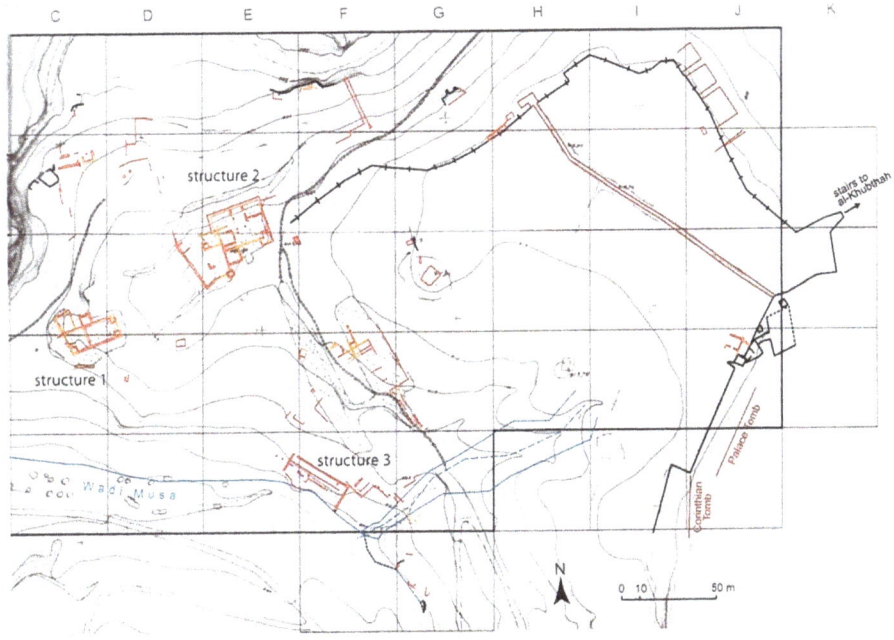

Figure 15. Royal compound structures 1, 2, and 3 in the Wadi Mataha area [18].

This flow network provided water to cisterns associated with ceramic and metal workshop areas and for celebratory triclinium functions associated with monumental royal tomb structures (Figure 1, Appendix A 62, 63, 64, and 65; **B** in region 2-B) and additionally made available upper reservoir water as a drought remediation measure. Recent research [18,19] indicates that several royal compounds exist (Figure 15) in the terminal pipeline area supplied by the Wadi Mataha pipeline (Figure 1, **B**, region 2-C). In conjunction with water from the Outer Siq pipeline extension directed into this area, ample water supply to royal compounds, even in drought periods, was assured as royal compound water needs for both consumption and aesthetic water displays dictated. Pipeline connections to the multiple royal compounds await excavation, but given nearby multiple water sources from the Wadi Mataha and Ain Mousa springs through the Outer Siq pipeline extension, as well as the upper reservoir water supply, abundant water supplies from multiple sources would be available to serve multiple palace structures. Water continued by pipeline past the Nymphaeum southward to city center areas and ultimately was transferred by a pipeline across a bridge to the commercial marketplace and fountain area district to supplement Siq pipeline water to the marketplace area. Final water discharge from all pipeline sources was ultimately led to the Wadi Siyagh natural drainage path. The present-day bridge in the Q'asar al Bint area contains visible ancient pipeline elements remaining after modern modifications to the earlier Nabataean bridge; this pipeline permitted water to cross over the bridge to sites on the south-side marketplace, Great Temple (Figure 12) and al Bint (Figure 2A) areas. Multiple pipeline water supplies to the same site areas would permit pipeline cleaning outages given redundant water supplies from other parts of the city, as well as a pedestrian transit path across the Wadi Mataha streambed.

Given the slope (~2.8-degrees) and construction detail of the entire Wadi Mataha pipeline, the question arises as to the engineering behind the selection of a pipeline slope that would yield the maximum flow rate given a variety of possible design options. Given the necessity to construct an elevated basin carrying the pipeline on the near vertical face of the Jebel al Khubtha mountainside (Figure 13, A–C) and the surveying required for the pipeline slope over a considerable length, a preliminary engineering analysis was necessary preceding construction. Analysis of alternate pipeline slope choices that have the potential to maximize the flow rate, while minimizing leakage from the many connecting joints between piping elements, was vital to a successful design. Details of flow patterns within the pipeline connecting the upper to the lower reservoir were analyzed by computer simulation to provide insight into the decision-making rationale underlying the Nabataean water engineers' choice of the most efficient pipeline slope.

Starting from an elevation of ~910 masl (meters above sea level) near the Mughur al Mataha and Jebel al Mudhlim areas (Figure 1, point **R**), the pipeline on the face of Jebel al Khubtha extends ~1.5 km to the lower reservoir near, but offset and above, the Sextus Florentinus Tomb (Figure 14) area located at ~770 masl. The upper water source is from the M–Zurraba reservoir connection supplied by a Wadi Mousa spring branch possibly augmented by a local spring. A ~15 m lower reservoir elevation above the tomb base at 770 masl yields an observed pipeline declination slope of ~2.8 degrees measured from the horizontal. While this slope was the ultimate choice of Nabataean hydraulic engineers, an alternate shallower slope choice of one-degree declination would require a yet higher elevation reservoir, requiring a high support base built onto the near-vertical Jebel el Khubtha mountainside. While this construction had the benefit of allowing the pipeline to extend further to the south to provide water to urban habitation areas, this design option was redundant given water supplies through the Siq and Outer Siq providing water to the same area of the city. For a one-degree declination slope option (slope measured from the horizontal origin point **R** of the pipeline as it switches to the western face of the mountain as shown in Figure 1), the elevation of the lower reservoir would be much higher above the ground surface, requiring an extensive base construction support. Beyond structural difficulties involved in this construction, computer modeling of water flow patterns in a long pipeline at one-degree declination slope revealed a problem. To expose this problem, a FLOW-3D CFD model (Figure 16, Case A) consisted of a left-most upper reservoir connected to a lower right-most reservoir by a pipeline at a one-degree declination slope; CFD calculations qualitatively duplicated hydraulic phenomena present in the pipeline flow from the upper reservoir **UR** to the lower reservoir **LR**. Figure 16, Case A results demonstrated the pipeline cross-section was fully occupied by water (full flow) given the submerged pipeline exit condition into the lower reservoir. A design with the pipeline exit above the lower reservoir water level would still support mostly full-flow conditions over most of its length, but be subject to ingested air slugs moving counter to the flow direction ingested at the pipeline exit, causing unstable flow conditions [34]. A volumetric flow rate of ~0.05 m^3/s entered the pipeline from the upper reservoir; this was the permissible full-flow rate for ε/D ~0.1 pipeline internal roughness over long pipeline distances (Figure 10) at a one-degree declination slope. Increasing the upper reservoir head did not result in a proportionate increase in volumetric flow rate for long pipelines over 400–500 m length as indicated by Figure 10. CFD results for Figure 16, Case A indicated full-flow for a submerged outlet pipeline under high hydrostatic pressure near the end reaches of the pipeline—this increased the likelihood of pipeline joint leakage due to full flow conditions under high hydrostatic pressure. Based on the combination of leakage problems and the difficulty of building a high-elevation reservoir with a steep connecting pipeline to the Nymphaeum, hydraulic engineers rejected the one-degree declination slope pipeline design as impractical.

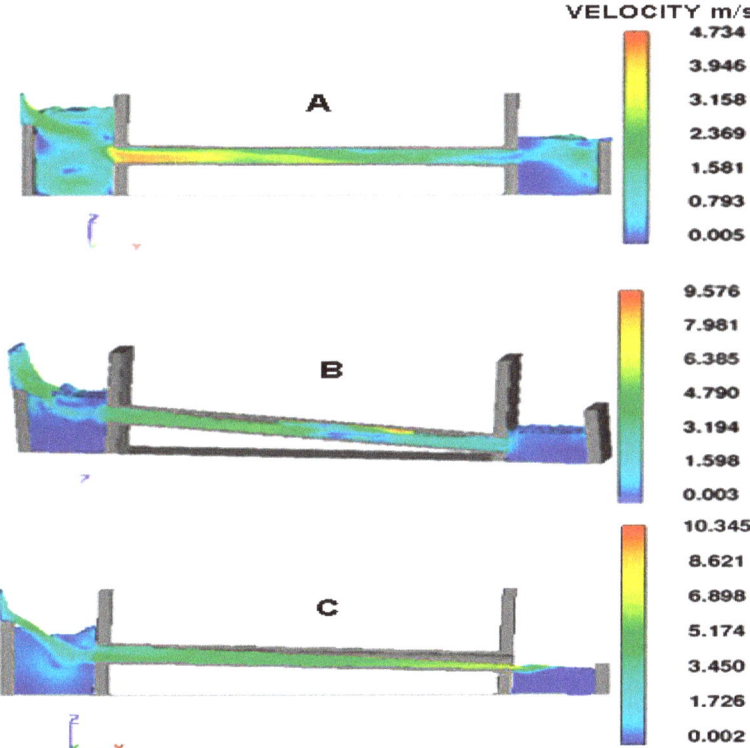

Figure 16. FLOW-3D computational fluid dynamics (CFD) results for Cases **A–C**. The hydraulic jump shown(directly under **B**) in Case **B** causes pressure oscillations, inducing separation forces at pipeline connection joints amplifying the potential for joint leakage and does not support the highest flow rate possible despite its steeper slope.

A further design option involved a steeper pipeline slope choice of four degrees with a volumetric flow rate adjusted to 0.08 m^3/s. The intuitive (but incorrect) observation that the steeper the pipeline slope, the faster the water velocity will be serving the on-demand rapid water delivery maximum flow rate requirement best, is an observation that may have guided the steeper slope choice. The steeper slope design option would result in a lower reservoir height from ground level but would require an additional length of piping to carry water on to the tomb structures (Figure 1, Appendix A site numbers 62, 63, 64, and 65) as water would be delivered to ground level cisterns distant from these structures. Further piping to the (North) Nymphaeum and the nearby sites including the Temple of the Winged Lions (Figure 9) required pipeline construction which, at low slope, may have further impeded the flow on to further water distribution centers across the lower bridge to south-side sites. Based upon the easier design option construction of a low-level reservoir on the el Khubtha mountainside, this option appears to have had positive benefits for workshop areas north of the monumental tomb structures; however, hydraulic analysis of this design option reveals a problem. For a steep pipeline slope transitioning from a mild slope, the high velocity supercritical internal flow developed a hydraulic jump with an upstream partial vacuum above the partial flow water surface (Figure 16, Case B). The flow height approached normal depth [35] for supercritical flow in a hydraulically steep pipeline. This result, approximated as flow resistance, was amplified by the high partial flow velocity that converted flow from supercritical (Fr > 1) to subcritical (Fr < 1) through a hydraulic jump within the pipeline as Figure 16, Case B calculation shows. This effect was largely caused by pipeline wall roughness and the high-water velocity used for Case B. The hydraulic jump is indicated below **B** in Case B. Froude number (Fr) was defined as Fr = $V/(g\ D)^{1/2}$, where V is the average flow velocity, g the gravitational constant (9.82 m/s^2), and D is the hydraulic depth [36]. While partial-flow in the upper supercritical portion of the pipeline helped eliminate pressurized full flow conditions that promoted leakage, the full-flow, post-hydraulic jump region within the downstream reaches of the pipeline promoted air entry into the pipeline exit to counter the partial vacuum region that promoted unstable flow pressure variations that caused pipeline joint leakage. Since water in the lower reservoir may have undergone sloshing and transient height changes during this process, the hydraulic jump position could be oscillatory due to sporadic air ingestion slugs countering the partial vacuum region from the pipeline exit downstream of the hydraulic jump. This effect contributed to sloshing in the lower reservoir and pipeline forces that compromised connection pipeline joint stability. While a design option existed to have the pipeline outlet above the lower reservoir water level under free fall conditions, this only amplified air ingestion to counter the partial vacuum region and flow instability. The computer solution shown in Figure 16, Case B, verified the formation of a hydraulic jump showing the interaction between the incoming supercritical partial flow delivered by the submerged pipeline to the lower reservoir. The hydraulic jump established a pressurized, full-flow water region extending downstream from its location in the pipeline to the reservoir. Provided the lower reservoir maintained constant height by limiting flow into piping directed toward the Nymphaeum and basin regions, the hydraulic jump position could be stabilized; nevertheless, the full-flow, pressurized water region constituted a potential pipeline leakage and flow instability problem. Given problems associated with the steep-piping slope choice (regardless of the pipeline exit submerged or with a free-fall delivery condition), flow instabilities arose from an oscillatory hydraulic jump position. On this basis, this design option was not implemented as an oscillatory hydraulic jump would cause pressure fluctuations inducing pipeline joint leakage.

Given the ~2.8-degree pipeline slope declination design option selected by Nabataean engineers and given an allowable input flow from the upper reservoir of 0.062 m^3/s for full-flow conditions into a long stretch of the 3.5 km pipeline upstream of point **R** in Figure 1, three criteria had to be met for acceptance of this design option: (1) An easy-to-build and maintain low reservoir that would provide observation of pipeline exit flows to check for flow problems; (2) a stable, high volume flow rate with an atmospheric pressure airspace above the (partial flow) water surface to eliminate hydrostatically pressurized connection joint leakage; and (3) easy access to the tomb structures for ritual events. Regarding these conditions, Figure 16, Case C provides insight into design criteria that satisfies conditions (1) to (3).

Theoretically, a flow approximating critical flow at Fr ≈ 1 would yield the highest flow rate possible with flow occupying a fraction of the cross-section of the pipe with an air space above the water surface. Calculations for a 0.062 m^3/s full-flow volumetric flow rate based on a ~2.8-degree slope and upper reservoir hydrostatic head values from one to three meters indicated that average velocity in the pipeline was on the order of V ~1.5 m/s with (g D)$^{1/2}$ ≈ 1.5 m/s, leading to Fr ≈ 1.0, indicating that a near critical flow condition was approximated at the ~2.8-degree declination slope. Given the near-critical flow observed at a ~2.8-degree declination slope (corresponding to the observed slope) and observing computer results (Figure 16, Case C) that indicate that this condition would produce an air space over the water free surface approximating y_c/D ≈ 0.8, an approximate critical flow condition existed at the ~2.8-degree declination slope. The significance of this is that the 0.062 m^3/s volumetric flow rate was the maximum value that this long pipeline design could sustain due to pipeline internal wall frictional effects and that the critical flow water height in the pipeline, although close to full-flow conditions, left an atmospheric pressure airspace over the water surface over a long length of the pipeline to limit pressurized leakage. Additionally, the presence of a critical flow eliminated the upstream influence from downstream disturbances thus promoting a stable flow within the critically sloped pipeline. Therefore, among the choices of the pipeline slope available to Nabataean engineers, the ~2.8-degree declination slope was selected. This slope option produced the maximum transport volumetric flow rate possible, which is what an on-demand water system is set to accomplish.

5. The Ain Braq and Theater Area Pipeline Continuance System

The upper reaches of Ain Braq water system (Figure 1) consisted of a channel cut into bedrock and surface soils that earlier supported a pipeline; its continuance to the plateau north of Wadi Farasa (Figure 1) by a supply line continues to the B-3 area shown in Figure 1 and supplied water to sites 12 to 18 (Appendix). A branch of this system may have extended to the eastern reaches of the housing district south of the Great Temple, as further excavations may indicate. A pipeline originating from the theater frontage area (Figure 17), given the land contours shown in ([37], p.136), would support a canal path to this area and sites along its path. The plateau shown in the Weis reference (Abb.5, p.136) [37] indicates many water structures consistent with this water supply system.

Figure 17. Theater with a seating capacity of 5000 to 6000 persons. Frontal pipelines extend to elevated basins above the south bank marketplace area to support fountains with (a probable) pipeline extension to areas above the Great Temple.

A survey of the area from the Ain Braq, Ain Ammon springs (Figure 1) revealed channels cut into mountain sides that supported a pipeline water supply from the two main springs to sites shown in Figure 1. Some traces of an attempt to bridge a large valley by a siphon are indicated in Figure 18. This design choice originated from a spring originating above the Wadi Mousa town that was likely abandoned early due to the necessity of constructing a large siphon across a deep valley to reach city precincts. The main Ain Braq and Ain Ammon springs supported a pipeline that sent water to the Lion Fountain (Figure 19, Appendix A Feature 14) and sites 12 to 18.

Figure 18. Remains of an abandoned early pipeline channel that would employ a large siphon to cross a valley originating from the spring source located north of the modern town.

Figure 19. The Lion Fountain.

Figures 1 and 20 show surface water distribution features in the Wadi Farasa area (B-1, Figure 1), together with the subterranean pipeline basin pit **A** connecting to the distribution junction **E**. Further exploration of possible contour pipeline paths threaded through mountainous areas to lower sites is necessary to establish the water supply source to **D**, the Lion Fountain (Figure 19), the Soldier's Tomb area, and associated water basin structures sources by a branch of the Ain Braq pipeline. The contour map of the hill area south of the Great Temple (Figure 12) area indicated the possibility of a water supply from a Southern Siq channel passing by the theater frontage (Figure 16), but erosion changes in the landscape over millennia and multiple earthquake landscape disturbances have made the water supply connection to **D** difficult to establish.

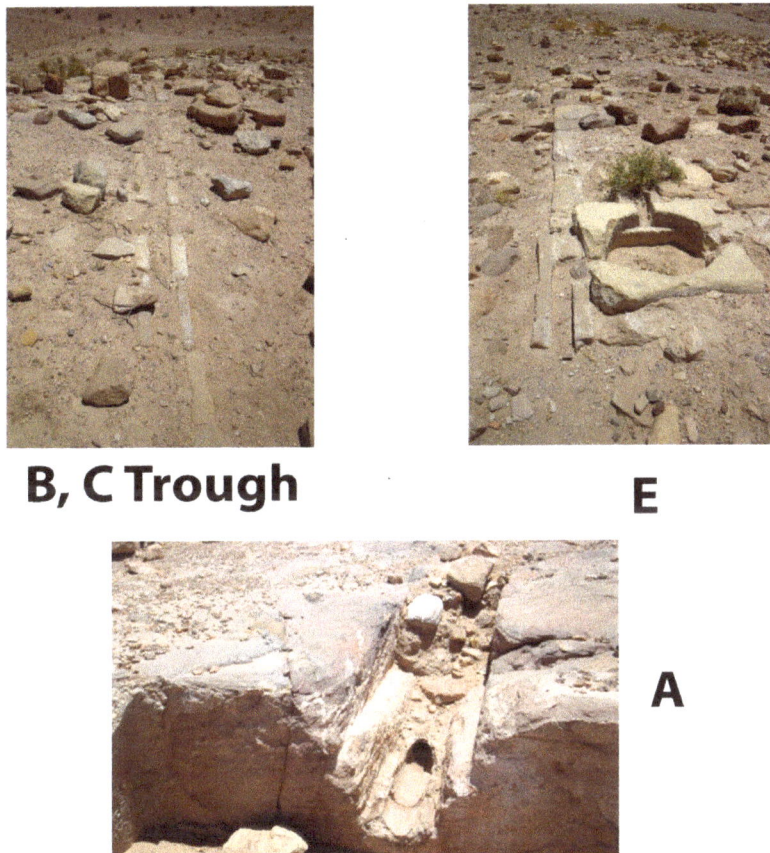

Figure 20. Details of pipeline components located above the Great Temple area. Pipeline (**A**) is immediately adjacent and connected to (**E**) and empties into a pit basin.

Of interest is the water supply to the Great Temple area. Starting from the hilltop junction (Figure 1) above the Great Temple, a pipeline component assemblage exists (Figure 20). While likely sourced by a Siq extension pipeline from the theater frontage area, there may have been a branch pipeline originating from the main Ain Braq, Ain Ammon pipeline as indicated in the B-3 area of Figure 1. Independent of confirmation of this water source to the hill area above the Great Temple, the assemblage of water transfer components (Figure 20) indicated water transfer to downhill housing areas with a probable pipeline water transfer to the Great Temple area. Combining excavated details of the Figure 20 components, a FLOW-3D computer model is surmised in Figure 21. One of the pipelines **B** from **E** led water to the lower Zhanthur housing district ([37], p. 136), while another pipeline branch

C also served the same purpose; a possible branch of the **C** pipeline may have led to the upper reaches of the Great Temple and lower Paradeisos area [8,20,21] as further excavations may verify. On site conversations with Brown University personnel during excavation of the Great Temple area indicated that a pipeline element was found heading downhill to the Great Temple, but was now covered over from excavation soil from the Great Temple renovation activity. Independent of verification of this pipeline water supply to the Great Temple's upper levels, much can be learned from existing Figures 20 and 21 components that give insight into the hydraulic engineering knowledge base of Nabataean water engineers.

The FLOW-3D computer model (Figure 21) composed of Figure 20 components utilized an input water velocity consistent with Figure 10 for full flow conditions into pressurized pipeline **D**. The water supply input into **D** was from the theater district (or from an Ain Braq branch pipeline) with a level portion of the pipeline existing before pipeline branches **E–C** and **E–B** continued water flow on to steeper terrain to supply downhill housing destinations and possibly the Great Temple upper level areas. A basin pit **A** shows a connection pipeline adjacent to and connected into the **E** pipeline.

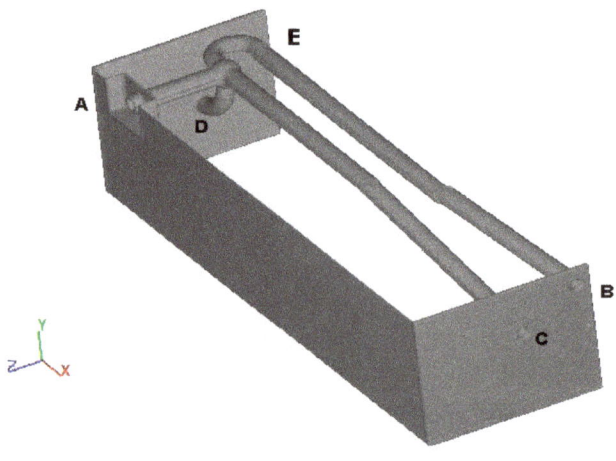

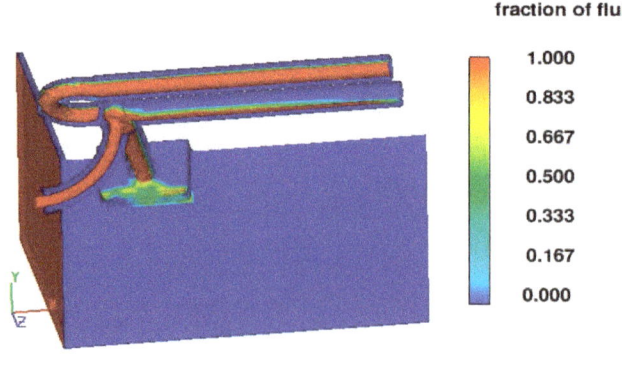

(A)

Figure 21. *Cont.*

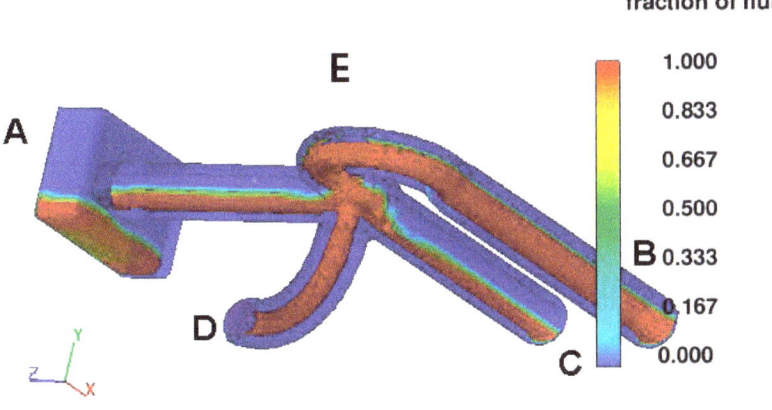

(B)

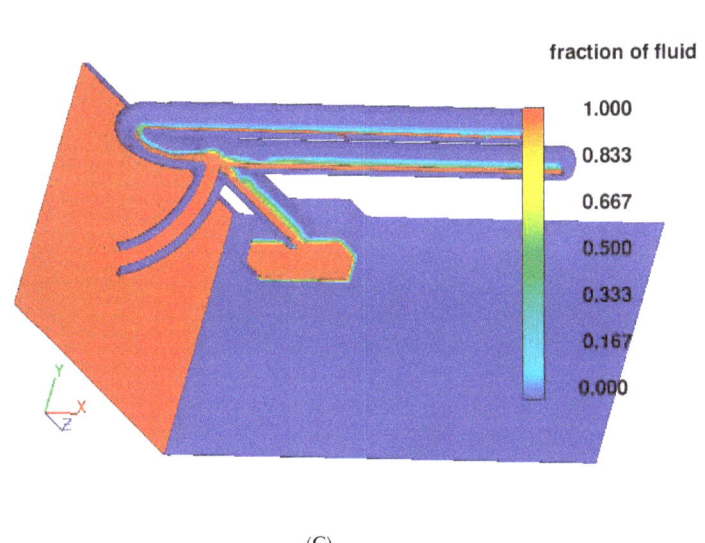

(C)

Figure 21. FLOW-3D CFD model of the pipeline connections constructed from Figure 20 components and downhill exposed pipeline elements. (**A**) Incipient filling of pit basin **A** due to hydraulic jump instability (or an **E–B**, **E—C** blockage) causing transient back flow conditions that are damped by water flow into the **A** pit basin. (**B**) Partial filling of pit basin **A** after pipeline instabilities are damped; note partial flow in high slope pipelines **E–B** and **E–C**. (**C**) Alternate view of Figure 21B conditions.

For one hydraulic function, the basin pit pipeline **A** (Figure 21) served to divert flow away from the dual downhill pipeline **E–B** and **E–C** branches if these were blocked during repair or if the input flow at **D** had flow surges past what the remaining downstream pipelines could accommodate. The blockage then diverted water to basin pit **A** in Figure 21. The flow into the basin pit continued until a pressure balance of water accumulated in basin pit **A** balances with the pressure in water supply pipeline **D** promoting an automatic system shut-off if the lower exit ports **B** and **C** (Figure 21) were blocked. The more sophisticated hydraulic engineering use of the Figure 21 configuration originated from the observation that subcritical full flow conditions existed in the lengthy, low slope supply pipeline **D**. Due to the hydraulically steep declination angle in dual pipelines **E–B** and **E–C**, input full flow past the junction area was converted to supercritical partial flow in both steep pipeline branches. This flow transition induced a local partial vacuum region above the partial flow water surface. As the branch pipeline exits were submerged into housing area reservoirs at their destination locations, then air was drawn into pipeline exit openings toward the upstream partial vacuum regions by flow patterns described in [34]. The intermittent air bubble stream proceeded upstream into the dual steep declination pipelines' branches to counter the partial vacuum region leading to flow intermittency and oscillatory hydraulic jumps in both the dual pipeline branches. This hydraulic phenomenon affected the flow stability of the entire system, leading to pulsating discharges into the downstream reservoirs as well as affecting the stability of input flow into **D**. The pulsating flows induced internal pressure changes that could affect pipeline joint connections causing leakage. If pipeline exit flow into housing area reservoirs was free-fall, then the flow instability condition was even more severe as there was easier access of air going upstream to partial vacuum regions. A further cause for pipeline flow instability was the creation of a hydraulic jump within the pipelines due to internal pipeline wall roughness when subject to upstream supercritical flow. Many situations arose then that caused flow instabilities, leading to pulsating flow delivery tohousing area reservoirs and pressure variations that induced pipeline joint leakage.

A cure to unstable flow conditions was anticipated by Nabataean water engineers when air was supplied through basin pit pipeline **A** as shown in Figure 21. Air led into the partial vacuum region where it occurred canceled the partial vacuum region and air bubble transfer from pipeline exits and subsequent flow instabilities in the dual subterranean pipeline branches. The hydraulic jumps still occurred in **E–B** and **E–C** pipeline branches, but with the partial vacuum regions eliminated, the hydraulic jumps and flow within pipelines remained stable. In Figure 21 **A–C**, the fluid fraction ff = 1 denotes water, ff = 0 denotes air, and intermediate ff values denote evaporative moisture levels. These figures indicate the transfer of air into the upstream reaches of **E** pipelines through the basin pit pipeline **A**. In summary, the basin pit pipeline connection served to minimize transient water surges originating from air pulses entering partial vacuum regions by providing an air passageway to **E** and elimination of transient water instabilities in the downstream high declination slope dual pipelines. This was done by transferring backup water caused by transient hydraulic jump motion into the pit basin, thus preventing its effect on flow input stability. Figure 21**A** shows incipient water transfer into the pit basin **A** due to pipeline blockage. Figure 21**C** shows continued filling of the pit basin until a pressure balance was achieved, shutting down further input water flow from **D**. Figure 21**B** shows that with air input from the connecting pipeline in **A** to the main **E–B** and **E–C** pipelines, the partial vacuum regions were eliminated, producing stable partial flow in these pipeline branches.

Excavation of the Great Temple area over many years by the Brown University team under the direction of Martha Joukowski [20,21] revealed that rainfall seepage from a hillside wall south of the temple was collected into a drainage channel [25] that led water to an underground reservoir on the east part of the site. A further subterranean channel starting from the surface entrance to the reservoir led water to the front platform part of the Great Temple site and likely served as an overflow drain channel activated when the reservoir was full. These additional water transfer features indicate that water supply to the Great Temple (and Paradeisos area) may have had an additional pipeline source from another branch of the Figure 21 assemblage, which is yet to be discovered. Given Great Temple water supplies from the extension of the theater district piping at lower levels, possible higher level water supplies originated from the collection trench above the Great Temple north wall ([25], pp. 265–267), as well as a possible connection from the Figure 21**C** channel as previously noted. Previously, this trench was assigned a drainage collection channel role from rainfall seepage to protect the foundation stability of the Great Temple, but other uses were possibilities. Given the elaborate upper level community assemblage structure that likely represented [20,21] a site leadership meeting place, access to water at higher Great Temple levels appeared a necessity. Given that discussions with Brown University personnel involved in the excavation indicated that a pipeline existed coming down from the hill above the temple, now covered with excavation soil, gives credence to the water supply from a branch canal from pipeline C shown in Figure 21 or an extension pipeline from the theater system (Figure 1). Future excavations are necessary to establish these connections to show the source of enigmatic structures that may have association with water transfer at upper levels of the Great Temple.

The Monastery Temple (Figure 22) at high elevation above the urban city center had no apparent pipeline water supply system due to the absence of high-level springs, but rather a series of nearby water collection basins that collected rainfall runoff.

Figure 22. The Monastery Temple.

6. Conclusions

Three pipeline water supply systems supplying urban Petra required different hydraulic engineering approaches to overcome problems to optimize the water supply to Petra's city center.

In each case, problems were overcome by engineering designs requiring empirical knowledge of fluid mechanics principles known in a prescientific format predating western science's "official" discovery of similar principles some ~2000 years later. Given the constraints from limited excavation data and given that the Nabataean hydraulics knowledge base is not known from surviving literature, its breadth may be gauged by use of CFD recreations of solutions that inform of thought processes and technical means available to Nabataean hydraulic engineers. As many of their solutions exhibit modern approaches, one can now attribute a role in the history of hydraulic science to Nabataean engineers.

Funding: This research received no external funding. Initial site exploration was conducted as part of Phillip Hammond's Temple of the Winged Lions NSF funded excavations; further site trips involved work with Talal Akasheh, Head of the Queen Rania Heritage Foundation, followed by individual trips to Petra over a period of five years.

Conflicts of Interest: There is no conflict of interest

Appendix A (Figure 1 Site Captions)

1	Zurraba–M reservoirs
2	Petra rest house
3	Park entrance
4	Early B-C-D Siq Covered Floor Channel
5	Dijn monument
6	Obelisk tomb and Triclinium
7	Siq entrance elevated arch remnants
8	Flood bypass tunnel and dam
9	Eagle Monument
10	Siq passageway
11	Treasury (El Kazneh)
12	High Place sacrifice center
13	Dual obelisks
14	Lion fountain monument
15	Garden Tomb
16	Roman Soldier Tomb
17	Renaissance Tomb
18	Broken pediment Tomb
19	Roman theater
20	Uneishu Tomb
21	Royal Tombs (62, 63, 64,65)
22	Sextus Florentinus Tomb
23	Carmine façade
24	House of Dorotheus
25	Colonnade Street
26	Winged Lions Temple
27	Pharaoh's column
28	Great Temple
29	Q'asar al Bint
30	Museum
31	Quarry
32	Lion Triclinium
33	El Dier (Monastery)
34	468 Monument
35	North city wall
36	Turkamaniya Tomb
37	Armor Tomb
38	Wadi drainage
39	Aqueduct

40	Al Wu'aira crusader castle
41	Byzantine tower
42	North Nymphaeum
43	Paradeisos, Temenos Gate area
44	Wadi Mataha major dam
45	Bridge abutment
46	Wadi Thughra tombs
47	Royal tombs
48	Jebel el Khubtha high place
49	El Hubtar necropolis
50	Block Tombs
51	Royal Tombs
52	Obelisk Tomb
53	Columbarium
54	Conway Tower
55	Tomb complex
56	Convent tomb
57	Additional Tomb complex
58	Pilgrim's spring
59	Jebel Ma'aiserat high place
60	Snake monument
61	Zhanthur mansion
62	Palace Tomb
63	Corinthian Tomb
64	Silk Tomb
65	Urn Tomb
66	Rectangular platform/temple
67	South Nymphaeum
—	Pipeline
- - -	Buried channel
B	Dual pipeline supplied basin area
d	Catchment dam
c	Cistern
s	Spring
D	Multi-level dam
A	Kubtha aqueduct
T	Water distribution tank
-d	Major dam

References

1. Browning, I. *Petra*; Chatto and Windus Publishers, Ltd.: London, UK, 1982.
2. Glueck, N. *Deities and Dolphins*; Strauss and Cutahy Publishers: New York, NY, USA, 1965.
3. Guzzo, M.G.; Schneider, E. *Petra*; University of Chicago Press: Chicago, IL, USA, 2002.
4. Hammond, P. *The Nabataeans: Their History, Culture and Archaeology*; Studies in Mediterranean Archaeology 37; Astrom Publishers: Gothenburg, Sweden, 1973.
5. McKenzie, J. *The Archaeology of Petra*; Oxford Press: London, UK, 1990.
6. Ossorio, F.A.; Porter, B.A. *Petra-Splendors of the Nabataean Civilization: Highlights*; White Star Publishers: Vercelli, Italy, 2009.
7. Akasheh, T.S. Ancient and Modern Water Management at Petra. *Near East. Archaeol.* **2002**, *65*, 220–224. [CrossRef]
8. Bedal, L.-A. *The Petra Pool Complex: A Hellenistic Paradeisos in the Nabataean Capital*; Gorgias Dissertations, Near Eastern Studies 4; Gorgias Press: Piscataway, NJ, USA, 2002.
9. Bedal, L.-A.; Gleason, K.L.; Schryver, J.G. The Petra Garden and Pool Complex, 2003–2005. *Annu. Dep. Antiq. Jordan* **2007**, *51*, 151–176.

10. Bellwald, U. The Hydraulic Infrastructure of Petra-A Model for Water Strategies in Arid Lands. *Historia* **2007**, *19*, 94–100.
11. Bellwald, U.; Al-Huneidi, M.; Salihi, A.; Naser, R. *The Petra Siq: Nabataean Hydrology Uncovered*; Petra National Trust Publication: Amman, Jordan, 2002.
12. Oleson, J.P. The Origins and Design of Nabataean Water-Supply Systems. In *Studies in the History and Archaeology of Jordan*; Department of Antiquities: Amman, Jordan, 1995; Volume 5.
13. Oleson, J.P. Nabataean Water Supply: Irrigation and Agriculture. In *Proceedings of the International Conference on the World of Herod and the Nabataeans*; Politis, K., Ed.; Franz Steiner Verlag: Stuttgart, Germany, 2002; Volume 2.
14. Ortloff, C.R. The Water Supply and Distribution System of the Nabataean City of Petra (Jordan), 300 BC–AD 300. *Camb. Archaeol. J.* **2005**, *15*, 93–109. [CrossRef]
15. Ortloff, C.R. Water Engineering at Petra (Jordan): Recreating the Decision Process underlying Hydraulic Engineering at the Wadi Mataha Pipeline. *J. Archaeol. Sci.* **2014**, *44*, 91–97. [CrossRef]
16. Ortloff, C.R. Three Hydraulic Engineering Masterpieces at 100 BC–AD 300 Nabataean Petra. In *Proceedings of the De Aquaductuatque Aqua Urbium Lyciae Phamphyliae Pisidiae: The Legacy of Julius SextusFrontinus*; Wiplinger, G., Ed.; Peeters Publishing: Leuven, Belgium, 2014; ISBN 978-90-429-3361-3.
17. Ortloff, C.R.; Crouch, D. The Urban Water Supply and Distribution System of the Ionian City of Ephesus in the Roman Imperial Period. *J.Archaeol. Sci.* **2002**, *28*, 843–860. [CrossRef]
18. Schmid, S.G. Die Wasserversorgung des Wadi Farasa Ost in Petra. In *Cura Aquarum in Jordanien, Proceedings of the 13th International Conference on the History of Water Management and Hydraulic Engineering in the Mediterranean Region*; DWhG Publications: Berlin, Germany, 2007; pp. 95–117.
19. Schmid, S.G.; Bienkowski, Z.P.; Ziema, T.; Kolb, B. The Palaces of the Nabataean Kings at Petra. In *The Nabataeans in Focus: Current Archaeological Research at Petra*; Nehme, L., Wadeson, L., Eds.; Supplement to Proceedings of the Seminar for Arabian Studies 42; Archaeopress: Oxford, UK, 2012; pp. 73–98.
20. Joukowsky, M. *Petra Great Temple, 1993–1997*; Brown University Press: Providence, RI, USA, 1998; Volume 1, pp. 265–270.
21. Joukowsky, M. The Petra Great Temple Water Supply. In *Proceedings of the 9th International Conference on the History and Archaeology of Jordan- Cultural Interaction throughout the Ages; Abstracts 25*; Department of Antiquities, Al-Hussein Bin Talal University Publication: Str Ma'an, Jordan, 2004.
22. Hodge, A.T. *Roman Aqueducts & Water Supply*; Bristol Classical Press: London, UK, 1992.
23. Laureano, P. *The Water Atlas, Traditional Knowledge to Combat Desertification*; Laia Libros: Barcelona, Spain, 2002.
24. Mays, L. Ancient Water Technologies. Springer Publishers: Dordrecht, The Netherlands, 2007; pp. 21–22.
25. Ortloff, C.R. *Water Engineering in the Ancient World: Archaeological and Climate Perspectives on Ancient Societies of South America, the Middle East and South East Asia*; Oxford University Press: Oxford, UK, 2010.
26. Ortloff, C.R. *The Hydraulic State: Science and Society in the Ancient World*; Routledge Press: London, UK, 2020.
27. Parr, P. La Date du Barrage du Sîq á Petra. *Rev. Biblique* **1967**, *74*, 45–49.
28. Parcak, S.; Tuttle, C.A. Hiding in Plain Sight: The Discovery of a New Monumental Structure at Petra, Jordan, Usung WorldView-1 Satellite Imagery. *Bull. Am. Sch. Orient. Res.* **2016**, *375*, 35–51. [CrossRef]
29. Markoe, G. *Petra Rediscovered: Lost City of the Natataeans*; H. Abrams Publishers, Inc.: New York, NY, USA, 2003.
30. Flow Science. *FLOW-3D/V.10.1 Users' Manual*; Flow Science, Inc.: Santa Fe, NM, USA, 2019.
31. Heinrichs, K.; Azzam, R. Investigation of Salt Weathering on Stone Monuments by use of a Modern Wire Sensor Network exemplified for Rock-Cut Monuments at Petra. *Int. J. Herit. Digit. Era* **2012**, *1*, 191–216. [CrossRef]
32. Strabo. *Geography, Books 15–16*; Harvard University Press: Cambridge, MA, USA, 2000; Volume 164, pp. 21–22.
33. Urban, T.M.; Alcock, S.E.; Tuttle, C.A. Virtual Discoveries at a Wonder of the World: Geophysical Investigations and Ancient Plumbing at Petra, Jordan. *Antiquity* **2012**, *86*, 331.
34. Lugt, J. *Vortex Flow in Nature and Technology*; John Wiley & Sons Publishers: Hoboken, NJ, USA, 1983.
35. Morris, H.; Wiggert, M. *Applied Hydraulics in Engineering*; The Ronald Press: New York, NY, USA, 1972.

36. Chow, V.T. *Open-Channel Hydraulics*; McGraw-Hill Book Company: New York, NY, USA, 1959.
37. Weis, L. Das Wasser der Nabataer: Zwishen Lebensotwendigkeit und Lexus: The Northwestern Project. In *Proceedings of the De Aquaductuatque Aqua Urbium Lyciae Phamphyliae Pisidiae: The Legacy of Julius Sextus Frontinus*; Wiplinger, G., Ed.; Peeters Publishing: Leuven, Belgium, 2014.

Publisher's Note: MDPI stays neutral with regard to jurisdictional claims in published maps and institutional affiliations.

 © 2020 by the author. Licensee MDPI, Basel, Switzerland. This article is an open access article distributed under the terms and conditions of the Creative Commons Attribution (CC BY) license (http://creativecommons.org/licenses/by/4.0/).

Case Report

Roman Hydraulic Engineering: The Pont du Gard Aqueduct and Nemausus (Nîmes) Castellum

Charles R. Ortloff [1,2]

1 CFD Consultants International, 18310 Southview Avenue, Los Gatos, CA 95033, USA; ortloff5@aol.com
2 Research Associate in Anthropology, University of Chicago, Chicago, IL 60637, USA

Abstract: The water distribution *castellum* at the terminal end of the Pont du Gard aqueduct serving the Roman city of Nemausus in southern France is analyzed for its water engineering design and operation. By the use of modern hydraulic engineering analysis methods applied to analyze the *castellum*, new aspects of Roman water engineering technology are discovered not previously reported in the archaeological literature. Analysis of the *castellum*'s 10 basin wall flow distribution pipelines reveals that when a Roman version of modern critical flow theory is utilized in their design, the 10 pipelines optimally transfer water to city precincts at the maximum flow rate possible with a total flow rate closely approximating the input flow rate from the aqueduct. The *castellum*'s three drainage floor ports serve as additional fine-tuning to precisely match the input aqueduct flow rate to the optimized 10 pipeline output flow rate. The *castellum*'s many hydraulic engineering features provide a combination of advanced water engineering technology to optimize the performance of the water distribution system while at the same time enhancing the *castellum*'s aesthetic water display features typical of Roman values. While extensive descriptive archaeological literature exists on Roman achievements related to their water systems both in Rome and its provinces, what is missing is the preliminary engineering knowledge base that underlies many of their water system's designs. The present paper is designed to provide this missing link by utilizing modern hydraulic engineering methodologies to uncover the basis of Roman civil engineering practice—albeit in Roman formats yet to be discovered.

Keywords: Roman; Pont du Gard; water engineering; castellum; aqueduct; CFD analysis; hydraulic design; critical flow

Citation: Ortloff, C.R. Roman Hydraulic Engineering: The Pont du Gard Aqueduct and Nemausus (Nîmes) Castellum. *Water* 2021, 13, 54. https://doi.org/10.3390/w13010054

Received: 18 November 2020
Accepted: 21 December 2020
Published: 30 December 2020

Publisher's Note: MDPI stays neutral with regard to jurisdictional claims in published maps and institutional affiliations.

Copyright: © 2020 by the author. Licensee MDPI, Basel, Switzerland. This article is an open access article distributed under the terms and conditions of the Creative Commons Attribution (CC BY) license (https://creativecommons.org/licenses/by/4.0/).

1. Introduction

The Pont du Gard aqueduct, built during the reign of Claudius (40–60 AD), involves many unique hydraulic engineering components and strategies [1–4] (pp. 181–188 [1]) that collectively worked to deliver water to the Roman city of Nemausus [2,3,5]—now the city of Nîmes in southern France. Water from the Fontaine d'Eure spring at Uzès was conducted to a regulation basin at Lafoux with an overcapacity diversion channel to the Alzon River (Figure 1); the aqueduct was designed to deliver 40,000 m^3/day through the ~50 km long aqueduct channel [2]. The Pont du Gard aqueduct/bridge spanning the Gardon River (Figure 2A–C) is located ~25 km from the spring source; a further 25 km extension of the aqueduct channel constructed partway through a tunnel delivered water to the basin distribution center (*castellum*) located ~17 m above the city of Nemausus. Changes to the original Roman aqueduct were made over the centuries: Figure 2B shows a 1743–1747 pedestrian walkway addition to the original Roman structure devised by French engineer Henri Pitot; further restoration was done in 1850–1855 by Napoleon III to reinforce the original structure. Figure 3 shows structural details of the original aqueduct base; the triangular base structure incorporates Roman knowledge to reduce the flowing water pressure on the foundation base. As the included angle of the aqueduct base exceeds the separation angle, turbulent, large-scale water rotational vortices keep sediment particles in suspension during passage under the bridge thus preventing sediment deposits under the bridge. Water delivered through the *castellum* basin tunnel opening (Figures 3–5) was further conducted through multiple pipelines to city reservoirs and site locations. The ~5.5 m diameter, ~1.0 m high *castellum* basin wall supported 10 *centenum-vicenum* ~30 cm terracotta inner diameter pipelines (Figure 6) with three additional pipelines of similar diameter (Figure 6) originating from the basin floor to complete the 13 pipeline distribution system. A sluice plate located at the tunnel opening to the basin regulated basin water height—this feature would prove vital to the design and function of the *castellum* as later discussion reveals.

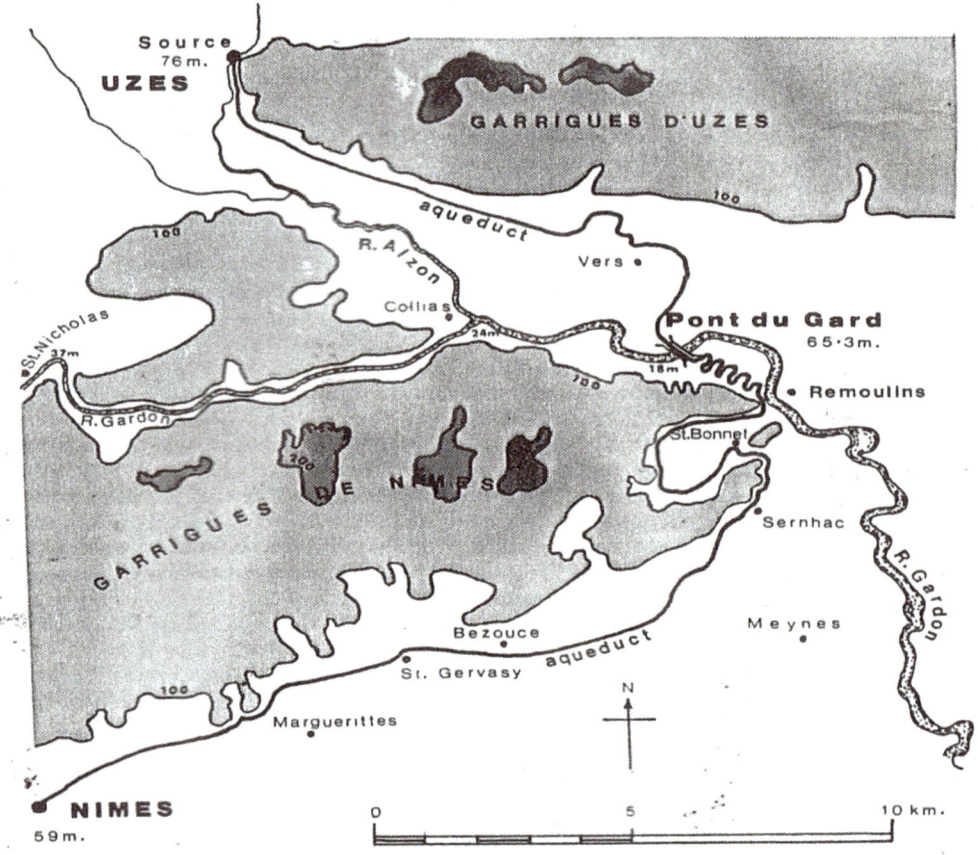

Figure 1. Map of the Pont du Gard aqueduct water system path from the Fontaine d'Eure source spring at Uzes to the destination *castellum*.

(A)

Figure 2. *Cont.*

Figure 2. (**A**) The Pont du Gard aqueduct/bridge. (**B**) Later (1743–1747) roadway passage addition to the original Roman Aqueduct. (**C**) Base structure of the Aqueduct. Note foundation triangular structures to reinforce the aqueduct base and lower river water hydrodynamic pressure forces on the base structure.

Figure 3. Front view of the Nîmes *castellum*.

Figure 4. View of the front of the Nîmes *castellum*.

Figure 5. View of structures in front of the Nîmes *castellum*.

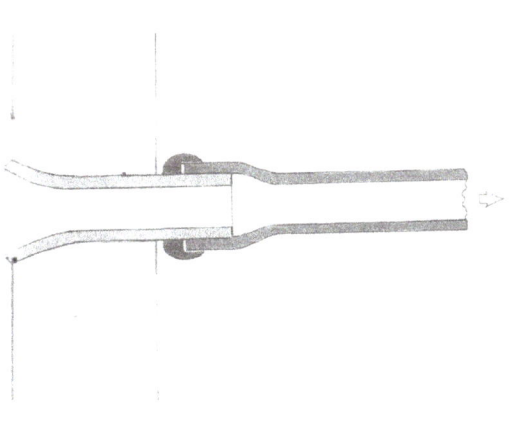

Figure 6. (**a**) Interior view of the Nîmes *castellum* showing three floor drainage ports. Port covers can be moved to adjust flow drainage flow rate output to match aqueduct input flow rate. (**b**) Typical *calyx* insert adjoining a pipeline marked with a flow rate value (in *quinaria*) for attachment to a pipeline to limit its flow rate to a prescribed value.

The Pont du Gard aqueduct/bridge crossed the Gardon River near the town of Vers-Pont du Gard and was a vital element of the ~50 km long aqueduct providing water to Nemausus. The straight-line distance between the Fontaine d'Eure spring source and the terminal distribution *castellum* was ~25 km; the final channel path selected by Roman engineers was a winding route measuring ~50 km because of construction difficulties associated with the mountainous Garrigues de Nîmes direct route. Roman surveyors selected the longer channel path to avoid difficulties associated with building numerous tunnels and bridges through mountainous terrain that would accompany the shorter length path that led directly from the spring source to the *castellum*. In addition to construction difficulties associated with the mountainous and deep gorge terrain in the northern section of the proposed aqueduct channel path, further routing changes were necessary to circumvent the southernmost foothills of the Massif Central known as the Garrigues de Nîmes. These foothills, covered in dense vegetation and indented by deep valleys, would prove difficult to cross with the shortest length water channel as they required construction of many small bridges and tunnels through a long section of hills and ravines that required a tunnel between 8 and 10 km long depending on the starting point. A diversion course around the eastern end of the Garrigues de Nîmes mountain range (Figure 1) proved to be the only practical way of transporting water from the origin spring to the city to reduce construction time and minimize labor costs. Ahead of the aqueduct/bridge, a covered continuation channel and terminal tunnel led water to the *castellum* basin through the sluice gate port shown in Figures 3–6. The aqueduct was designed and built to carry a given maximum flow rate of 40,000 m^3/day—the challenge to Roman engineers was to efficiently design the *castellum* to transport the input flow rate through a minimum number of pipelines to city destinations in the most hydraulically efficient manner. The innovative *castellum* design devised by Roman engineers to accomplish this end is described in sections to follow and gives a penetrating look into Roman hydraulic engineering practice.

In the first century AD, Nemausus was a prosperous Roman colony whose resource base consisted of Rhone Valley agricultural fields and vineyards to support trade and export to central Rome. The colony's prosperity reflected population growth from 20,000 to 40,000 over a short time span leading to designation of official city status by the central Roman administration. The original Nemausus fountain the base of Mount Cavalier did not provide the expanded city population city with its daily need of potable drinking water nor provide additional water for the baths, fountains, temples, theaters, government and commercial sector buildings and garden areas that Roman cities incorporated as standard city design practice. Based upon the need for increased water supply for the expanding population of the city, planning of an advanced design aqueduct from the Eure Uzés spring source to Nemausus anticipated the future water needs of the city. It is estimated that approximately 70% of the aqueduct channel pathway consisted of excavated stone-lined trenches with slab or arched roof covering with the remainder channel pathway in the form of short length tunnels and small bridges. As a major construction challenge, the Gardon River valley crossing required engineering design and construction innovations on a scale exceeding previous aqueduct designs to transport water across the extensive width of the river gorge area.

The planning and construction of the aqueduct during the ~19 BC time period is credited to Augustus' son-in-law and aide, Marcus Vipsanius Agrippa then serving as the senior magistrate *aedile* responsible for managing the water supply of Rome and its colonies. Espérandieu [3] writing in 1926 linked the construction of the aqueduct with Agrippa's visit to Narbonensis; later excavations [2,5,6] suggest the construction may have taken place between 40 and 60 AD. Earlier built tunnels bringing water from several local springs to the city were not considered in modification plans originated by the builders of the new aqueduct due to the greatly increased water needs required by the growing city population. Coins discovered in Nemausus' outflow pipeline catchments are no older than the reign of the emperor Claudius (41–54 AD) to more securely date construction time. On this basis, a team led by Guilhem Fabre [3] argued that the aqueduct must have been completed in the middle of the first century AD and have taken approximately fifteen years to build, employing between 800 and 1000 workers.

2. The Pont du Gard Aqueduct/Bridge Design

The Pont du Gard aqueduct/bridge has three tiers of arches, stands 48.8 m high and descends only 2.5 cm over its length of 274 m (a gradient of 1 in 18,241). The water channel from the Fontaine d'Eure spring source to the *castellum* descends by only 17 m height over its entire ~50 km length—this is indicative of the challenge in surveying precision that Roman water engineers utilized in the aqueduct design. Average slopes in meters per kilometer (and degrees per kilometer) over long sections of the channel from the Fontaine d'Eure spring at Uzès (Figure 7) to the *castellum* indicate mean constant slopes over long stretches of the aqueduct channel; local slope variations along channel path lengths were necessary due to local landscape surface irregularities and accuracy limits of surveyor's instruments [1]. The water supply system may have been in use as late as the sixth century, with some parts used for significantly longer times, but lack of maintenance after the fourth century led to accumulation of channel sinter deposits that eventually limited the water flow rate [2]. Construction details (Figures 3–5) of the *castellum* indicate 10 *castellum* basin wall ports from which 10 terracotta pipelines directed water to individual fountains, *nymphea*, baths, temples, gardens, theaters, commercial sector and administrative buildings, private homes and intermediate reservoirs around the city. Three pipeline ports (Figure 6) on the floor of the *castellum* basin served for adjustable basin drainage, flow diversion during repairs and cleaning of the *castellum* and, when partially opened by movable cover plates, served as an adjustment mechanism to guarantee that the ~40,000 m^3/day aqueduct delivery rate closely matched the design output flow rate from the 10 basin wall pipelines.

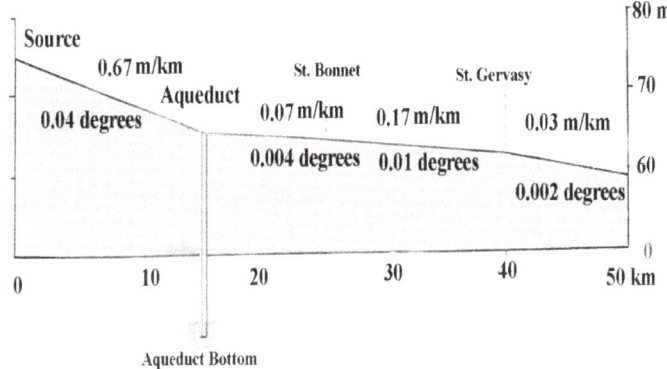

Figure 7. Average slopes of sections of the Pont du Gard aqueduct in m/km and equivalently in degrees.

The Fontaine d'Eure spring, at 76 m above sea level, is 17 m higher than the *castellum* above the city of Nemausus; this provided sufficient gradient to sustain gravity flow of aqueduct water to Neamausus. The aqueduct's average gradient is ~1/3000 but the local channel slope varies widely along its course [2,3] being as small as ~1/20,000 in some sections (Figure 7). The average gradient between the start and end of the aqueduct (0.34 m/km) is far shallower than usual for Roman aqueducts being only approximately one-tenth of the average gradient of some of the eleven aqueducts supplying Rome. The reason for the change in gradients along the water system's route is that a uniform gradient would have meant that the Pont du Gard aqueduct/bridge would have an extreme height and thus present a formidable construction challenge given the limitations of Roman construction technology. By maintaining a steeper gradient along the channel path ahead of the aqueduct/bridge (Figure 7), Roman engineers were able to lower the height of the aqueduct/bridge by 6 m to a total height of 48.7 m above the Gardon River bed. This height limit governed both the up- and downstream channel gradients of the aqueduct as well as limiting the weight of the aqueduct. Despite the weight saving design, the substantial as-constructed weight of the multi-tiered stonework created a slight depression in the middle of the aqueduct/bridge that created a slightly increased water depth in the center of the aqueduct channel. The initial gradient profile originating from the spring source before the aqueduct/bridge is relatively steep descending 0.67 m/km but descends by only 6 m over the remaining channel length to the *castellum* (Figure 7). In one channel section, the winding route between the Pont du Gard aqueduct/bridge and St. Bonnet (Figure 1) required an extraordinary degree of accuracy from Roman engineers who had to survey for a channel decline of only 7 mm per 100 m of the aqueduct.

It is estimated that the aqueduct supplied the city with a minimum of 20,000 m^3/24 h to a maximum of ~40,000 m^3/24 h and that water took 28 to 32 h to flow from the Uzés spring source to the city [2,5]. The different limits in aqueduct flow rate reflect the seasonally variable spring discharge rate dependence upon groundwater recharge from infiltrated rainwater and received groundwater from surrounding infiltration areas. The *castellum* was designed for use at the maximum 40,000 m^3/day flow rate that the Uzés spring source could supply on a steady basis—this being necessary to provide sufficient water for the large city population. Average water velocity was on the order of ~0.5 m/s according to one estimate [2] corresponding to the total maximum flow rate of ~40,000 m^3/day. Aqueduct water arrived to the *castellum*—an open, shallow, circular basin ~5.5 m in diameter by ~1.0 m deep (Figures 3–6)—and was surrounded by a (now lost) balustrade within an enclosure under a small, but elaborate, pavilion. When the *castellum* was first excavated, traces of a tiled roof, Corinthian columns and a fresco decorated with fish and dolphins were discovered in fragmentary condition. As to details of the construction of the *castellum*, adjoining curved stone slabs (Figures 3–6) lined the receiving basin inner rear wall and, for the 10 basin wall pipeline ports, large arced adjoining stone blocks were pierced by 10 circular openings to accommodate pipeline insertion with leakage prevention cementing. The three floor ports were carved through the stone floor with right angle turn passageways below the floor permitting horizontal pipelines to emerge from outside the basin (Figures 5 and 6). The basin floor has an outer rim upon which the arched basin wall blocks were placed. Although not apparent from the current state of the *castellum* remains, likely a thin layer of bitumen (or cement) between basin blocks provided leakage protection. The entry slope of the supply tunnel (0.002°, Figure 7) was vital to lower aqueduct water velocity and raise its height prior to basin entry; further flow area expansion from the narrow tunnel entrance opening into the wide basin further lowered the water entry velocity to the multiple basin wall pipelines. A (now lost) movable sluice gate (Figures 3–5) regulated the basin water height into the 10 basin wall pipelines and played a vital design role in the operation of the *castellum*. Conjecture as to the design and function of the entrance sluice gate structure [2,3] prevails with no current resolution as to the design intent of its function to regulate *castellum* water height and velocity—the resolution of this issue is addressed in subsequent sections. A series of holes penetrating the top plate of the entrance structure exist [2,3] but as to the sluice plate lifting mechanism controlling the water entrance opening height, no current information exists.

The precise control and regulation of the maximum design flow rate from the aqueduct to the *castellum* was an important Roman design consideration to avoid basin spillage from the low wall height of the *castellum* basin and overflows from the aqueduct and supply channels. As the cross section of all channels was rectangular with a constant base width, local variations in flow rate from intercepted local rain storm runoff that would cause water height changes in the low slope channels were anticipated by Roman engineers by local height increases in the channel walls where spillage was likely to occur. Although water containment in channels during rainfall events was anticipated by the channel wall height design, the use of the three *castellum* basin floor ports to dispense excess water arriving to the *castellum* over the design flow rate kept the basin from overflowing thus maintaining its aesthetics even during rainfall events.

3. The Castellum Floor Ports: Hydraulic Design Considerations

Aqueduct water entered the *castellum* through a rectangular tunnel opening, 1.2 m wide by 1.1 m high (Figures 3–6), and circular holes in the basin wall, each ~40 cm in diameter, give indication of the pipeline dimensions that directed water into the 10 basin wall pipelines. The three floor pipeline ports (Figure 6) if fully open, presuming a continuous aqueduct water supply, would induce a vortex over each floor entrance port inducing rotation of water in the basin [7]. This effect would alter the equal distribution of the water flow into the 10 wall ports. As this effect would influence the 10 pipeline design, the three-floor port opening cover plates would have been used sparingly for input flow rate regulation to produce near equalization of flow into the 10 separate pipelines. This requirement then mandated a *castellum* design that would closely regulate the aqueduct input flow rate to closely match the sum of output flow rates from the 10 basin wall pipelines with only minor flow rate corrections provided by the three-floor port opening areas. The rotating flow in each floor pipeline, given its passage through elbows below the basin floor (as Figure 5 indicates) would transition pipeline entry full flow to partial flow within the sloped pipeline extensions due to gravitational acceleration increasing pipeline flow velocity and lowered pipeline water height. Here a pipeline cross-sectional area fully occupied with water is denoted as full flow; a pipeline cross-sectional area partially occupied with water is denoted as partial flow. For high speed flow in highly sloped pipelines emanating from the *castellum* with significant internal pipeline wall roughness, a hydraulic jump may occur downstream in the pipeline. Between entry full-flow and downstream post-hydraulic jump full-flow regions, a partial vacuum region exists in the intermediate partial flow region. Unless these regions within pipelines are relieved by upper pipeline holes to admit air at atmospheric pressure, flow delivery instabilities arise as air enters the floor pipeline inlet by means of an air-entraining vortex extending from the water basin surface [7] together with air entering from the pipeline exit (for either submerged or free overfall conditions) to relieve the partial vacuum region. A further source of flow instability arises from large internal pipeline roughness slowing flow velocity and raising its height to transition partial to full flow. As upstream water length buildup occurs ahead of the full-flow hydraulic jump water region, accumulated water mass weight suddenly overcomes frictional resistance and causes the elongated water mass to rapidly transfer out of the pipeline to freefall into a reservoir. This clearing effect then restarts the creation of another full-flow region leading to periodic flow delivery to a reservoir. These effects for both high and low speed entry flow to wall pipelines cause transient, oscillatory water motion in pipelines resulting in *castellum* basin water level oscillations and unstable flow entering into pipelines. These effects are largely governed by pipeline water entry velocity, pipeline slope and diameter, internal pipeline wall roughness, pipeline segment connection joint roughness and hint of the complexity that Roman engineers contended with to produce a *castellum* design providing stable flow to city destinations. As Roman engineers had concerns about flow instabilities that induced pipeline vibrations [8] that loosened connection joints between pipeline elements to cause leakage, flow stability concerns related to *castellum* design were a major problem to be addressed and eliminated by an advanced *castellum* design. As the elimination of transient flow instability effects was an important consideration in *castellum* design, Roman engineers required design considerations to eliminate pipeline flow instabilities that could propagate upstream to the *castellum* basin and disturb the aesthetics of the basin water surface. The design considerations would therefore involve in some way the use of the controllable three-floor port opening size selection together with solutions associated with pulsating flow delivery and uneven flow rate delivery to the 10 basin wall pipelines. Further design considerations involve design of specific pipeline slopes selected by Roman engineers to limit unsteady hydraulic jump formation and transient pressure pulses to induce steady flow in pipelines.

How a design solution was accomplished by Roman water engineers to deal with and eliminate these complex interacting hydraulic effects by innovative *castellum* and pipeline designs is described in following sections. Although it has been previously suggested that the three floor ports were mainly used for flooding the amphitheater for mock naval battles (*naumachia*) [3,6], the allowable flow rate through three bottom ports alone is far lower than the input flow rate from the aqueduct as derived in a later section. The three floor ports may additionally have served the purpose of continuous flows to important sites but are inadequate, by themselves, to carry the 40,000 m^3/day aqueduct flow rate without several (or all) of the 10 basin wall pipelines simultaneously in use as a later section details. If aqueduct flow is diverted by blockage and flow diversion to the Alzon River for aqueduct cleaning and repair functions, then the bottom ports would well serve to completely drain the *castellum* basin after the water level falls below the 10 basin ports pipeline openings. Among the reasons for the use of the floor ports, the main function of the floor ports, used with partially open covers (Figure 6) during normal daily operation of the aqueduct, was to drain away excess aqueduct supply water to exactly maintain the design 40,000 m^3/day input aqueduct flow rate as apparently this rate was critical to the hydraulic design of the pipeline system of the *castellum*.

4. The 10 Castellum Wall Ports: Hydraulic Engineering Design Options

Candidates for the pipeline types emanating from the basin wall ports are the (120A) *centenum-vicenum* with a diameter of 22.83 cm and inner cross-sectional area of 409.4 cm^2 and the larger (120B) *centenum-vicenum* with an inner diameter of 29.5 cm and a cross-sectional area [3,4,9] of 686.6 cm^2. To include known Roman pipelines allowing for pipeline wall thickness of at least ~2.54 cm, then several standard Roman pipeline sizes [3] are basin entrance flow candidates. In the discussion to evaluate their merits of different flow rate measurement devices available to Roman water engineers, *calices* mounted in a horizontal pipeline section are considered as they are typical of Roman practice for flow rate measurement and can be used to regulate and/or limit flow rates when used as chokes (Figure 6b). In this figure, a typical *calyx* placed on the left is adjoined to a pipeline shown to the right. Flow direction is from left to right.

A table of *calyx* sizes and the use of a *calyx*–pipeline connection [3,10] in the 50, 80 and 100 digit sizes with diameters of 27.8, 45.5 and 57.4 cm, respectively, may have been considered by Roman water engineers to regulate the amount of water flowing in different pipelines to different destinations with different prescribed water needs. While Roman pipeline types have known standards [3,4,9], large bronze *calices* placed directly into the 10 basin wall entrance holes were a design option to match the sum of pipeline flow rates to the aqueduct flow rate input. Given that *calices* work only under full-flow conditions, their use by Roman engineers in the *castellum* pipeline entry ports may have appeared useful for precise flow rate delivery to destination sites given that full flow at basin wall pipeline entrances could be maintained by means of a horizontal pipeline elements before pipeline declination slope continuance converted full to partial flow in pipeline extensions to the lower city. This design option would require that the sluice gate was fully open and that basin wall height exceeded the observed wall height shown (Figures 3–6) to ensure full-flow entry into basin wall pipelines. The higher basin wall height design would require a higher elevation of the tunnel supply line and an even lower aqueduct slope leading to the tunnel. Sustaining entry full flow into basin wall pipelines would rely on constant water height in the elevated wall height basin just to support full-flow *calyx* usage. The advantages of full entry flow incorporating *calices* placed at the start of horizontal piping branches with markings to indicate that the output flow rate (in *quineria*) had yet a further disadvantage beyond height reconfiguration of the inlet tunnel. *Calices* used at the entrances of all 10 basin wall pipelines theoretically provided the sum of their flow rates and therefore were vital to match the aqueduct design flow rate. As *calyx* sizes appropriate for the ~30 cm inner diameter pipelines give erroneous flow rates based on the nozzle diameter rather than the square of the diameter appropriate to the cross-sectional area of the nozzle, the correct flow rate prediction would ultimately be a problem if installed due to the inaccuracy of *calyx* flow rate measurement. An analysis of *calyx* use [10] describes inaccuracies associated with large flow rate measurements. Since accurate flow rate measurements were necessary in a system design to balance aqueduct input exactly to 10 pipeline water transfer output, *calyx* flow rate inaccuracies would compromise the *castellum* design that requires precise measurements of flow rates particularly if the use of only *calices* for flow rate determination precluded the use of the three floor ports as a flow rate adjustment. If the *calyx* design option were pursued then major redesign and reconstruction of the *castellum* would follow upon flow rate proof tests using the full aqueduct design flow rate. As the existing *castellum* design is vastly different from that utilizing this design option, this indicates that Roman engineers were aware of the precision difficulty of *calyx* water measurement devices and the near impossible reduction in the lead-in aqueduct slope and tunnel elevation underlying *calyx* usage. As the existing tunnel entry slope is already small (0.002°, Figure 7), an even lower slope design would severely challenge surveying accuracy measurement capability as well as requiring significant modification of the aqueduct slopes upstream of the tunnel.

Apparently the concept of water velocity and its measurement were recognized as important to determine flow rates but such considerations were not readily available to Roman engineers due to lack of precise time measurement devices [1,3]. From these considerations, it was apparent that a *calyx*-based design option was not practical and Roman engineers would need to choose a more refined design option that eliminated problems associated with the *calyx* design option as well as flow instability problems. What then was the final Roman *castellum* design that solved all the problems mentioned to match aqueduct input and pipeline output flow rates exactly?

As no traces of actual pipelines or *calices* exist at the present *castellum* site, pipeline connection details, as well as the *exact* pipeline diameter used, remain conjectural. Nevertheless a reasonable estimate of pipeline diameter may be made for flow rate estimation based upon the *castellum* retaining wall diameters shown (Figures 3–5) and the three-floor port geometry (Figure 6). Based upon the above discussion, for the three *castellum* floor ports used to rapidly flood the nearby amphitheater for *naumachia*, this function would require the addition of several (or all) of the 10 wall outlet ports to work in conjunction with the floor ports to accommodate the 40,000 m^3/day aqueduct flow rate. The *naumachia* function would necessitate that valves were available in basin wall pipelines to redirect additional flow to the amphitheater. That such large valves were in the Roman engineer's purview has been demonstrated [9]. For the present analysis, however, it is assumed that all 10 side wall ports were in continuous use but not the three floor ports (except for fine adjustment of the input 40,000 m^3/day flow rate to wall pipelines)—this conclusion underlies pipeline flow rate results to follow.

Since the *castellum* was elevated well above the Roman city and few traces of the multiple water destinations and connection pipelines now exist, it may be assumed that pipeline lengths were on the order of a fraction (or more) of a kilometer from the *castellum* to different city destinations. A typical Roman arrangement of pipelines from the *castellum* to a lower reservoir (Figure 8) would be designed to regulate flow to destination sites. Each city destination may have had time variable flow rate requirements (particularly baths and private houses) from cisterns and stilling basins so that overflow cisterns were necessary to captures excess water flows and direct water to collection basins serving gardens, pools and storage basins that did not require steady water input. Similar designs to those shown (Figure 8) were in use at the Roman site of Pompeii [4,11]. Based on this design complexity for a city water distribution system, emphasis on stable pipeline flow delivery would be an important consideration that minimized maintenance and the use of supplemental downstream settling basins and thus was a prime consideration inherent to the *castellum* design.

The early writings of Vitruvius and Frontinus on Roman hydraulic engineering practice [12] are replete with pre-scientific notations of hydraulic phenomena related to flow velocity, flow rates, time and hydrostatic pressure that were used for water flow rate measurement. On this basis, there is much to recommend Roman hydraulic engineering practice as largely based on an observational recording basis based on pipeline slope effects on flow delivery as opposed to results derived from theoretical calculations. The basic problem in determining flow rate was the accurate measurement of time and water velocity which eluded precise Roman definitions as indicated by Roman water administrator's book descriptions. In this regard, Greek hydraulic engineers demonstrated progress in measuring average water velocity and flow rates appropriate to fountain and water outlet designs at Priene [13]. In this case, a large basin of known volume (V) was filled by water flow from a level pipeline; a time measuring device (water clock, candle burn time, as examples) provided an estimate of the time (T) to fill the basin. The flow rate (Q) for the pipeline is then Q = V/T. This same methodology could be used for different *calices* mounted on a pipeline to determine an estimate (or correction) of their *quineria* flow rate marking if indeed this was part of Roman technology yet to be elucidated.

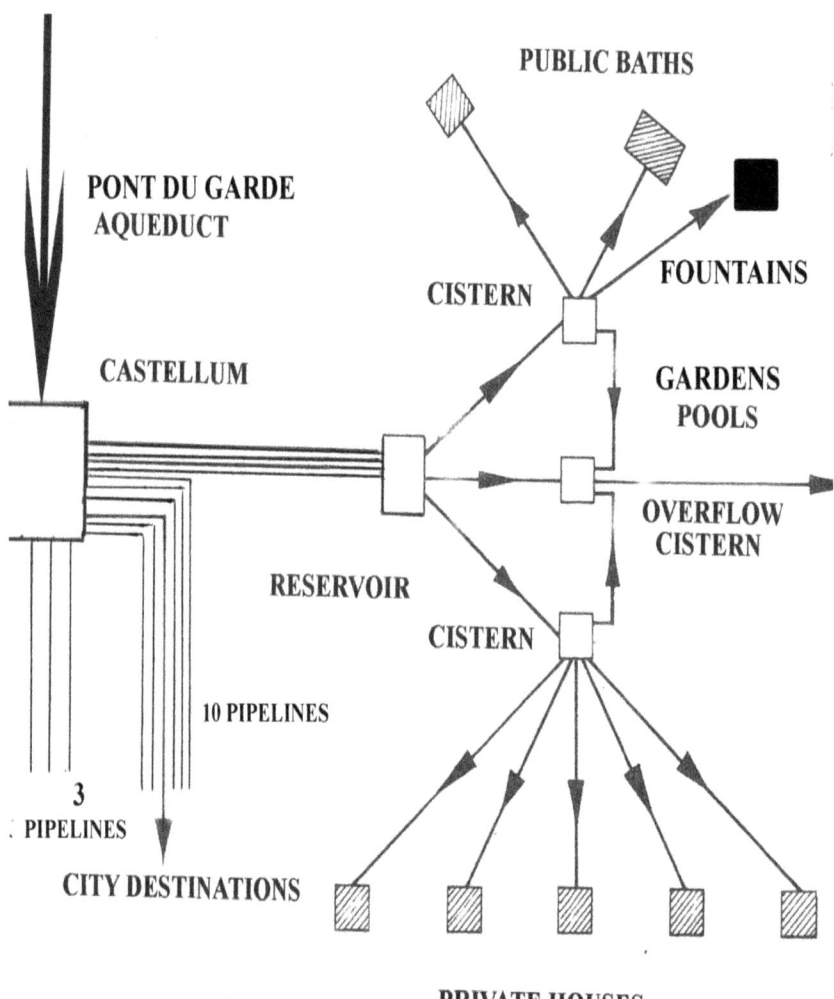

Figure 8. Aqueduct supplied water to the *castellum* showing the redistribution network structure typical of Roman cities.

To understand pipeline flow phenomena dependent upon pipeline slope and pipeline-basin attachment choices, discussion is next focused on the hydraulic positives of an alternate design choice for flow rate measurement. Here the *castellum* design contains elements of advanced Roman hydraulic technology thoughtful of the effects of pipeline slope choices and effects derived from knowing how to produce stable flow rates. From the analysis to follow, design elements noted in the actual *castellum* construction are revealed that demonstrate the design of an optimum water delivery system that effectively matched aqueduct input flow rate to the *castellum* output flow rate, produced maximum flow rates with pipelines to limit the number of pipelines used and eliminated flow instability problems.

5. Toward the Optimum Castellum Design

Many Roman pipeline designs transferring partial flow had top hole openings to eliminate partial vacuum regions [14,15] (pp. 314–320 [15]) but use of this feature for the present case is not known due to absence of pipeline remains. In the absence of pipeline top openings, transient air ingestion at the pipeline exit port, or from the basin water surface, occurs to counter a partial vacuum region that induces pulsating forces acting on pipeline joints that promote leakage. Such partial vacuum regions occur when subcritical entry full flow transitions to a partial critical (or supercritical) flow in the declination sloped pipeline continuance from an initial horizontal pipeline section; when a downstream hydraulic jump occurs due to inner pipeline roughness, then the partial vacuum region is trapped between these two separating flow conditions. This condition exists for either free or submerged pipeline exit conditions into a reservoir as for either case, atmospheric air enters the entrance and exit regions of the pipeline to counter the partial vacuum region inducing transient flow instabilities as the hydraulic jump location is unstable due to air ingestion rate differences and the transient, variable pressure and size of the partial vacuum region. A further contribution to flow instability occurs when the upstream size of the subcritical hydraulic jump region increases to a point where the weight of this region causes a flushing of the region out of the pipeline exit; when this flushing is done, the previous flow conditions restart once again inducing flow instabilities. When post-hydraulic jump subcritical pipeline flow delivery to a reservoir (Figure 8) is erratic, induced flow oscillations propagating upstream in a flow pipeline cause erratic motion of water flow to adjoining distribution basins as well to *castellum* basin water thus destabilizing smooth flow delivery and cancelling water basin aesthetics by surface wave occurrence. As flow stability considerations were known to Roman engineers, their *castellum* and pipeline designs must reflect a design that would eliminate erratic flow oscillations in pipelines and in the *castellum* basin and provide resolution of all flow instability problems through knowledge of preferential pipeline slopes in some manner. A further consideration known to Roman water engineers was that maximum pipeline flow rates are associated with critical partial flow—not full flow—conditions and that this is related in some way to the declination slope of a pipeline. Noting that the pipeline entrances are very close to the top rim of the *castellum* basin (Figure 6), this feature indicates that the *castellum* design reflects knowledge of inducing partial critical flow into pipeline entrances and this is related to achieving total maximum flow rate through the 10 pipelines equal the input aqueduct flow rate in an optimum manner. If pipelines could be designed to transmit the maximum flow rate possible, then this would reduce the need for additional pipelines emanating from the *castellum*.

As a first consideration of the observed *castellum* design, the water height entering pipelines is controlled by lowering the sluice gate to precisely control the basin entry water height to the pipelines (Figure 9). Provided this water height from the basin bottom comes up to half the pipeline diameter, then from the aqueduct input flow rate of ~40,000 m^3/day to the *castellum*, the average flow rate for a single pipeline is 1.63 ft^3/s; for 13 ports open, the average single pipeline flow rate is 1.26 ft^3/s. From hydraulic engineering theory [16], for D the pipeline diameter (~30 cm) and g the gravitational constant (9.82 m/s^2), $Q/D^2 (g D)^{1/2}$ = 0.29 for 10 open pipelines and 0.22 for 13 open pipelines. From [11], Henderson's Figures 2–12, y_c/D = 0.5. This means that the y_c critical depth entering pipelines, regulated by the height position of the sluice gate, is equal to half of the pipe diameter (Figure 12). Thus water enters the 10 wall pipeline entry ports at half the pipeline diameter height at critical (Fr = 1) Froude number [11,16–19]. The Froude number is defined as Fr = $V/(g D_m)^{1/2}$ where V is water velocity, g the gravitational constant and D_m the hydraulic depth [11–13,19]. From critical entry flow, the continuance of partial critical flow in pipelines is determined by pipelines set at the critical slope range (θ_c, Figure 12).

The physical significance of establishing critical flow conditions in pipelines [11–13,16,19] emanating from the *castellum* lies in the fact that when pipelines are set at a critical slope, this minimizes the energy expenditure to transport the flow at the highest flow rate. The lower the energy expenditure to transport pipeline flow, the less reliance on supplemental ways to increase flow rate such as an elevated *castellum* basin wall height and increased water height to provide additional hydrostatic pressure to increase flow rate. As partial critical flow can be maintained (or somewhat extended) over the entire pipeline length given smooth interior pipeline walls, atmospheric pressure exists over the partial flow from air entry into the open pipeline exit. Pipelines set at the critical slope are therefore free of hydrostatic pressure that would induce leakage under full-flow conditions. Most importantly, critical flow (Fr = 1) conditions throughout entire pipeline lengths produce the highest pipeline flow rate [10,15–19]. Pipeline designs that sustain flows equal to or close to partial critical flow (Fr = 1) over long distances would largely eliminate flow instability concerns from hydraulic jump creation as only a small water contact area with the interior pipeline wall roughness exists under partial flow conditions thus lowering flow resistance effects that would lead to internal hydraulic jump formation. As critical flow is maintained in critically sloped pipelines, no upstream influence from downstream resistance elements (pipe bends, contractions, bifurcations, hydraulic jumps) can propagate upstream to disturb basin water height and stability conditions [10,16–19]. Such considerations related to the effects of pipeline slope to produce stable flows must have been known to Roman engineers from observation of the many hydraulic engineering projects they implemented; in this regard, several elements of the pipeline system at Ephesus [14] constitute a prime example. With the advantage of flow stability derived from a critical flow design, the use of downstream accumulators, water towers, settling basins and open basin reservoirs and settling tanks (Figure 8) used to stabilize flow conditions between pipeline exit flows to different destinations with specific flow rate needs can be minimized. In the discussion to follow, the English unit system is used as this underlies many of the empirical hydraulic relations used in the analysis.

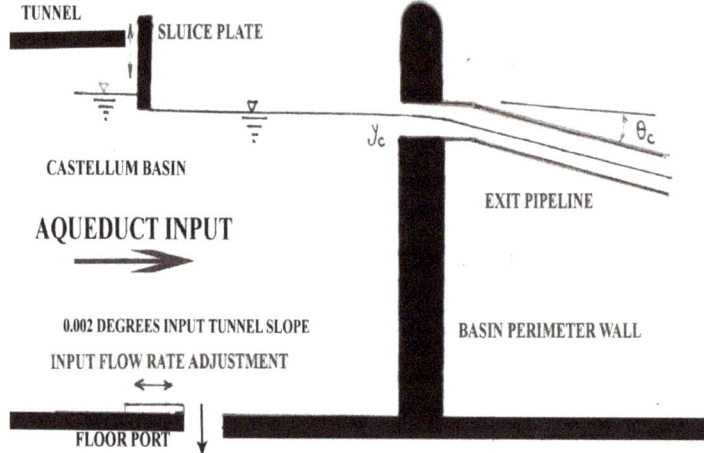

Figure 9. Basin exit wall flow dynamics (not to scale).

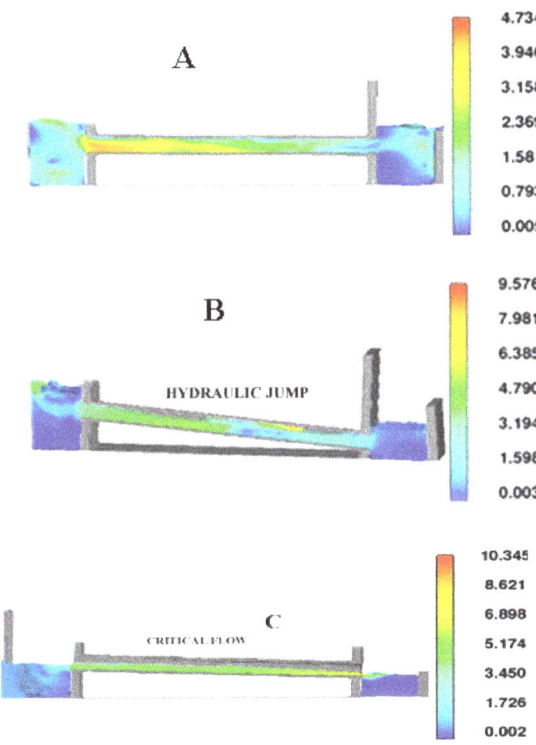

Figure 10. FLOW-3D CFD calculation results for pipeline Cases (**A–C**). Velocity in m/s.

Figure 11. The Maison Carré Roman Temple in central Nîmes.

Figure 12. The Roman amphitheater located in central Nîmes.

The critical, partial flow entry velocity for 10 pipes is $V_c = (g\, y_c)^{1/2} = 4.01$ ft/s and 5.8 ft/s for the 13 ports open case. The Froude number is $Fr_{10} \approx 1$ for the 10 port open case and $Fr_{13} \approx 1.3$ for the 13 port open case. The important conclusion is that wall port entry Froude numbers are either near critical ($Fr \approx 1$) or slightly supercritical ($Fr > 1$). Again, the importance of this design feature, as regulated by the sluice gate height position (Figure 12), is that with critical ($Fr = 1$) and near supercritical ($Fr \approx 1$) entry port flows that continue critical (or near critical) in pipelines, this eliminates downstream resistance influence that may induce upstream flow instabilities and destabilize the input flow rates to the pipeline ports. This positive effect is induced with smooth interior wall pipelines with little if any connection joint roughness—an option likely available for optimum water transfer flow conditions to the city. This is a most important design feature of the actual *castellum* and its sluice gate regulation mechanism. In modern hydraulic terminology, upstream influence derived from downstream resistance obstacles does not occur for $Fr = 1$ critical or $Fr > 1$ supercritical flows [10,15–19]. The θ_c angle (Figure 10) is derived from the Manning equation [19] where n is an empirical resistance constant indicative of a likely worst case internal pipeline wall roughness (given here as $n = 0.034$) and connection joint roughness accumulated from the thousands of piping connection sections of ~0.5 m length that comprise long pipelines from the *castellum* to destination sites. Here R_h is the hydraulic radius given by the cross-sectional area of the critical flow (A_c) divided by its wetted perimeter. The critical pipeline angle θ_c is given by:

$$\theta_c = \tan^{-1}(n\, V_c/1.49\, R_h^{2/3})^2 = \tan^{-1}(n\, Q/1.49\, A_c R_h^{2/3})^2 \qquad (1)$$

Substituting, for the 10 basin wall ports open case, $\theta_c \approx 4.6°$ and for the 13 port open case, $\theta_c \approx 7.1°$ with an average value of ~ 5.8°. For the 3 basin floor pipelines closed, pipeline slopes emanating from the *castellum* should have declination slopes in the range of ~4.6° (or somewhat higher). For all 13 ports open, the pipeline declination slopes should be a bit higher at ~7.1° to maintain partial critical flow. As a later section demonstrates, the 10 basin wall pipelines set at θ_c are sufficient to transfer the 40,000 m³/day (1.41×10^6 ft³/day) aqueduct flow rate indicating the minimal use of basin floor pipelines for fine-tuning of the input flow rate to 40,000 m³/day. Thus the three basin floor ports are considered closed (or partially open) to fine tune the y_c water height and design critical flow rate at pipeline entrances. When final destination site locations are distant from the *castellum*, pipelines emanating from the basin should have slopes in the ~4.6° < θ_c < ~7.1° range. The higher slope value is for high internal pipeline roughness over longer pipeline lengths that likely induce a downstream hydraulic jump (Case B, Figure 9) unless the use of smooth pipeline interior walls for longer pipelines is in place to maintain critical or near-critical flow advantages. Based on the surveying accuracies obtainable by Roman engineers [1,3,4,9] such slope accuracies are easily within their surveying capabilities.

6. Pipeline Destination Types Served by the Castellum

As critical pipeline slope configurations are observed as having flow stability and high flow rate benefits, the next task is to examine both critical and off-design, non-critical pipeline slopes that may occur if destination sites mandate higher or lower pipeline slopes to reach. As different pipeline slopes yield different flow rates, the task ahead is to determine what slope choices associated with different destination uses produce flow rates to match the 40,000 m³/day input aqueduct flow rate. Three possible pipeline configurations (A, B, C) determined by their slopes originating from the *castellum* basin wall are examined using FLOW-3D Computational Fluid Dynamics [20] CFD models (Figure 9). The use criteria involving different pipeline slopes are determined by computing the output flow rate (Table 1) from all 10 pipelines configured at different A, B or C slopes to determine whether the total output flow rate is lower, matches, or exceeds the input aqueduct flow rate.

Table 1. Flow rate results for Cases A, B and C.

Type	Flow Velocity (m/s)	Flow Volume (m³/s)	13 Ports Open (m³/day)	10 Ports Open (m³/day)
A	0.20	0.02	~22,400	~17,200
B	0.51	0.04	~51,500	~39,600
C	1.22	0.1	~52,000	~42,000

Figure 9 CFD model results show plane views of three-dimensional centerline interior pipeline flows for different pipeline slope conditions. The *centenum-vicenum* pipeline with a diameter of ~30 cm is used for the CFD model. The LHS model region represents the *castellum* entry port with a sluice gate position set to have critical entry flow to pipelines; the model RHS shows a submerged reservoir catchment at the end of a pipeline with a bottom drainage leading water by a further pipeline (or channel) to a destination site or intermediate reservoir. Although only a short pipeline length is illustrated in the models, the results are typical of flow patterns within longer pipelines as once a uniform flow profile is established, it continues over a long distance. Figures are characterized by an average full-flow 1.63 ft/s input velocity to a single basin wall pipeline; individual inlet velocities to pipelines are slightly different due to their relative locations with respect to the supply inlet.

The first figure (Case A, Figure 9) shows flow velocity conditions for a near level pipeline leading from the *castellum*. This configuration would provide water to hillside housing located at approximately the same height as the *castellum*) and water supply to upper-level reservoirs designed to store water at night to later discharge water through additional pipelines to sites with large, immediate water demands (such as baths) exceeding the continuous aqueduct supply rate over a given time period. Flow velocity is low in the near level pipeline due to full-flow wall friction effects and, for the low velocity subcritical (Fr < 1) flow, upstream influence exists so that distant exit reservoir flow stability conditions play a role in determining the delivery flow rate. This usage would require upper-level reservoirs to store water so that when fully charged, valves on pipelines to destination baths would open and have drainage rates higher than the aqueduct supply rate. Once water was delivered, then valves were closed and reservoirs refilled. This cyclical use could be made consistent with bath water change timing provided near-horizontal piping has the capability to transfer water at the aqueduct supply rate—a question addressed in the next section.

The second figure (Case B, Figure 9) represents pipeline slopes exceeding the optimum pipeline critical slope. Pipelines at these slopes are consistent with the height difference and the distance between the *castellum* and some of the nearby flat areas of present day Nîmes that once held the streets of Nemausus. Initial entry port flow is full subcritical flow and, at a downstream location, partial supercritical flow develops in the pipeline until a hydraulic jump (HJ) occurs due to large internal pipeline friction effects at high water velocity converting supercritical partial (Fr > 1) flow to subcritical (Fr < 1) full flow. As Figure 9, Case B indicates. A smooth internal pipeline wall would delay the appearance of a hydraulic jump but for the present example case, very rough pipeline internal walls and connection joints are assumed (for smooth wall roughness, θ_c values would decrease). For significant internal wall roughness, an internal hydraulic jump is created isolating a partial vacuum region between full entry and post-hydraulic jump flow regions as previously noted. If openings were placed along piping top regions over the partial vacuum region, then a stable flow rate would be enhanced. Again, as no extant pipelines exist, the presence of pipeline top openings is conjectural but well within Roman technology as observed on Ephesus pipelines and the Laeodocian site [3,4,15]. Pipeline designs with Case B flow characteristics without a terminal stilling basin would be devoted to lower priority sites that do not require a stable water delivery rate such as gardens, reservoirs, latrine flushing channels and intermittent household use. Other uses may include pipeline water transfer to fountain houses that have multiple chambers supporting different hydrostatic head values [4] to transfer water at different flow rates to different destinations; such terminal fountain houses were likely part of a city flow network. Figures 10 and 11 show typical Roman flow destinations still existing in present day Nîmes served by one or more of the *castellum*'s 10 pipelines. The Maison Carré Roman Temple to Apollo (Figure 11) constructed in the period 10–16 BC required a later water supply addition for ritual and ceremonial purposes; the second century AD Roman amphitheater (Figure 12) also required ample water supply for large public gatherings for events and spectacles and likely required the output of several of the *castellum* pipelines for this purpose. This could be accomplished by water storage in lower reservoirs elevated above the amphitheater level with suitable valve systems to control the flow rate to drinking fountains and water basins within and adjacent to the amphitheater.

For display fountains, *nymphaea*, high-status administrative buildings, and elite residential areas, critical flow designs (Figure 9, Case C) are preferred as all destinations would benefit from a stable, high delivery flow rate without the use of intermediate distribution reservoirs and stilling basins (Figure 8) and thus have an immediate economic benefit to reduce construction costs and system complexity. Case C represents the optimum critical slope condition for which the volumetric flow rate is the maximum possible and, as an air space exists over a long stretch of the pipeline length, pressurized pipeline leakage is largely eliminated thus producing lower maintenance requirements. This pipeline choice can be used for city sites reachable in the slope range ~4.6° < θ_c < ~7.1° depending on the number of basin ports open. This consideration helps city planners place structures requiring large flow rates and helps determine the placement of main reservoirs (Figure 8) from which additional pipeline branches emanate to destination sites. Given that deliberate use of the Case C pipeline design was within Roman hydraulic engineer's knowledge base, it may be surmised that the lower priority pipeline slopes of the Case B type required a stilling basin attachment before distribution to other destinations (Figure 8) and would be of secondary use while the higher priority pipelines requiring a high, steady flow rate directly to a destination site of the Case C type were preferable. As the slope difference between Cases B and C is small and direct use of a Case C design with a precise slope has the constraint of direct access to a destination site by a pipeline of that slope, most probably Roman engineers constructed pipelines as close as possible to a critical slope design to obtain the many benefits listed.

Based upon CFD results, an estimate of the volumetric flow rate is next made for each of the Case A, B, and C pipeline designs for 10 basin ports open and results compared to the aqueduct input flow rate of 40,000 m^3/day. If the input aqueduct flow rate exceeds any of the Case A, B and C 10 pipeline outlet flow rates, then such pipeline configurations are not feasible as *castellum* basin overflow would result. If the input aqueduct flow rate is equal to the total out flow from 10 pipelines for pipeline configurations given in Cases A, B or C, this gives indication of a probable pipeline slope usage. Here the pipeline slopes range from ~1.0 degree (near horizontal) for Case A, ~7.1° slope for Case B, to a critical angle slope for Case C of ~4.6°. Table 1 summarizes the CFD computed output flow rates for A, B and C pipeline configurations.

For Case A, full flow exists (Figure 9) in near-horizontal pipelines with a submerged exit into a reservoir. For Case B, full flow into the pipeline entrance transitions to supercritical, partial flow on a steep pipeline slope; a hydraulic jump is formed within the pipeline induced by the deceleration of flow by pipeline wall frictional effects together with submerged exit flow into a reservoir. For Case C, critical flow exists in a pipeline at a ~4.6° slope yielding the maximum pipeline flow rate. The high water velocity is consistent with low partial flow height thus lessening the water contact area with the rough interior surface of the pipeline—this largely minimizes the creation of a hydraulic jump from water-wall frictional effects. The pipeline exit flow is assumed to be free fall into a receiving reservoir.

From Table 1, Case A low-slope pipelines appear to be of minor (or no) use as the total of 10 pipelines open (three bottom ports closed) permit a much lower output flow rate (17,200 m^3/day) through all pipelines than the input aqueduct input flow rate of ~40,000 m^3/day. Even with all 13 ports open, the output flow rate of 22,400 m^3/day is well below the input 40,000 m^3/day aqueduct flow rate. The interpretation is that the input aqueduct flow rate far exceeds the capability of Case A near-horizontal pipelines to transport such high flow rates. The use of many near-horizontal pipelines filling high-level reservoirs then appears not to be the principal design intent of the *castellum*.

For three bottom ports closed, the calculated Case B flow rate is ~39,600 m^3/day which is close to the estimated aqueduct flow rate of ~40,000 m^3/day. This close flow rate matching produces a steady water height in the *castellum* that guarantees steady flow throughout the water distribution system; here partially open floor ports are useful to exactly match flow rates (Figure 10). This close match signals the Roman engineer's design intent of the *castellum* to provide water to city distribution locations by pipelines of slopes in the ~4.6° < θ_c < ~7.1° range. Although a hydraulic jump may occur due to wall roughness effects, it may largely discounted if Roman engineers utilized smooth inner wall pipelines to promote flow stability. The near flow rate match indicates the hydraulic technology to make a *castellum* design to closely match the input aqueduct flow rate in advance of the building and flow rate testing of installed pipelines. For a 0.25 km long pipeline sloped at ~4.6°, the altitude drop from the *castellum* to the city area is ~18 m which is a reasonable value given a personal downhill walking tour from the *castellum* to the city center.

Case C critical flow pipelines appear to have the delivery capacity close to the ~40,000 m^3/day aqueduct flow rate and preferably would be in use as there is only a minor slope difference between Cases B and C. Since Case C slopes would reduce the occurrence of an internal pipeline hydraulic jump, this design would be preferable, but not always achievable, due to surveying accuracy constraints or destination site requirements that dictate pipeline lengths and slopes. The critical slope on the order of ~4.6° may have played a role in locating city structures that demanded rapid, stable transfer of water, such as *nymphaea* and elite housing with internal water display structures. Given the ~4.6° pipeline slope and considering a height difference from the *castellum* to a potential city level reservoir, the pipeline lengths would be on the order on ~0.25 km; this may influence the placement of intermediate reservoirs.

From Table 1, the likely pipeline candidates in use were Cases B and C examples used in conjunction with three partially open (or closed) floor ports. The Case B flow delivery capacity approximates the aqueduct water supply water of ~40,000 m^3/day. Case B and C designs are practical for 10 basin wall entrances operating continuously as pipeline slopes on the order of ~4.6° to ~7.1° guarantee stable (or close to stable) water transfer from the high elevation *castellum* to lower city level sites. The remaining three floor ports, if open, would rapidly drain the *castellum* as flow from all 13 ports exceeds the input aqueduct flow rate as Table 1 indicates. It is important to note that for Case B and Case C the pipeline transfer flow rate approximates the input aqueduct flow rate and that a steady, smooth water height is maintained in the *castellum* basin close to the basin top rim which was the aesthetic design intent of Roman engineers. For situations for which several (but not all) of the pipelines are at the critical slope, then a mixed array of pipeline slopes with more than, less than, and equal to the critical slope exist. Here the use of the three floor ports for flow rate adjustment would be critical to match the aqueduct input flow rate to the sum of output flow rates from the *castellum* to maintain constant *castellum* water height.

7. Off-Design Hydraulic Function

To this point, flows exceeding the input design flow rate are adjusted back to the design flow rate by adjustable floor port openings. For cases for which the flow rate is somewhat less than the design maximum, flow into the *castellum* basin will back up in height until the water height reaches the basin wall lower pipeline port openings. This discharge causes an upstream transient water height wave to propagate upstream in the low slope aqueduct channel upstream of the entry port to the *castellum* basin. As the transient water height begins to discharge into the *castellum* basin wall ports, oscillatory wave motion on the basin water surface will follow and propagate upstream into the supply aqueduct channel. This effect induces a periodic discharge into the basin wall pipelines, destroying the desired aesthetic effect of a smooth *castellum* basin water surface as well as inducing periodic flow into the pipelines that, given their long length, may transition partial to full flow due to pipeline interior wall roughness. This flow condition then has transient flow instabilities that affect the stability and aesthetics of the water in the receiving *castellum* basin and the downstream reservoir. Provided baffles are installed in the reservoir, them flow oscillations can be damped thus providing a stable flow rate to further downstream destinations. On this basis, the importance of maintaining the design flow rate is of prime importance as lower flow rates flow rates fail to accomplish the main purpose to provide adequate water supply to the city.

For cases for which aqueduct flow into the *castellum* basin is significantly less than the design flow rate, again upstream flow backup occurs until basin water height reaches the lower parts of the basin wall port openings. In this case, a stable trickle flow occurs into the pipelines with no instabilities occurring. Thus for either of the off-design situations, the *castellum* still functions to provide limited water supply to final destinations but at an aesthetic disadvantage.

8. Conclusions

The challenge to Roman engineers was to eliminate sources of flow instability by a *castellum* and pipeline design that transferred input aqueduct water at the highest stable flow rate possible through the 10 basin wall pipelines. The *castellum* design demonstrates Roman hydraulic knowledge at work in many ways—particularly in the use of a shallow, wide-diameter basin with the retaining basin wall slightly higher than the top of the pipeline entrance ports. This design initiates basin input flow from the aqueduct water entrance port into pipeline entrances at critical, partial flow conditions; its continuance as critical or near-critical flow is guaranteed by appropriately sloped pipelines to important city destinations. Given Roman experience with flow instabilities associated with pipeline internal wall roughness, it is likely that selection of smooth interior walls was the design preference. As the 10 pipeline cumulative flow rate approximates the maximum aqueduct input flow rate for Cases B and C, it is clear that the design intent of the *castellum* recognized advantages in selecting pipeline slopes to largely limit or eliminate hydraulic jump occurrence—this occurs when Case B near-critical flow conditions apply. As the basin wall height is close to the top of the pipelines; partial entry flow into the 10 wall pipelines was the design intent of Roman engineers; this made possible by the positioned height opening of the sluice gate. Given the pipeline critical or near-critical slope range of ~4.6° to ~7.1°, CFD results indicate that the 40,000 m³/day aqueduct flow rate is transferred at the near maximum pipeline flow rate by Case B and C designs. Note that this θ_c range is only 2.5° difference to ensure critical or near-critical pipeline flow. This matching is necessary to maintain a constant water height in the *castellum* and minimize the number of pipelines that can successfully transmit the aqueduct input flow rate. If critical and near-critical designs were not considered, then a *castellum* design with more than 10 basin wall pipelines would be necessary thus increasing the size and cost of the present design.

The conclusion of the present analysis is that Roman hydraulic engineers designed the *castellum* to match the input aqueduct flow rate to the 10 basin wall pipeline transfer flow rate by employing a critical and/or near-critical flow conditions. This option presumes a series of lower-level reservoirs at pipeline termination locations that support pipeline branches to different sites (Figure 8). As different pipeline flow rates occur within different pipelines with slopes at critical, near-critical and higher and lower slopes to supply spatially dispersed sites with different water demands, this necessitates partial opening of the floor ports to match the design water output to the given aqueduct input flow rate. For situations where the flow rate may exceed 40,000 m³/day, again the floor ports' cover openings can be adjusted to eliminate excess flow over the design maximum of 40,000 m³/day that the critical and near-critical pipelines can convey.

Provided the 10 basin wall pipeline slopes can be maintained in the ~4.6° to ~7.1° slope range, this design option provides the most efficient and stable way to match the aqueduct input flow rate and eliminate maintenance problems associated with flow instabilities inducing pipeline joint leakage. The totality of the Pont du Gard aqueduct and *castellum* design demonstrates a coordinated engineering design of all subsidiary hydraulic components that supply the *castellum*.

In summary, the total Pont du Gard aqueduct and *castellum* design includes (1) flow rate regulation through a far upstream intersecting side channel to drain away flows exceeding the design intent aqueduct input of a ~40,000 m³/day flow rate. (2) The partially open bottom three *castellum* floor drains that can be used to remove excess aqueduct input water to achieve the 40,000 m³/day design input flow rate to the *castellum*. (3) Use of the low aqueduct slope preceding the *castellum* tunnel chosen so that pipelines emanating from the basin wall can be appropriately sloped to provide the pipeline maximum flow rate consistent with the supply aqueduct flow rate to the city below—this slope range is on the order of ~4.6° to ~7.1°, and this condition largely explains why the terminal channel slope to the *castellum* entrance port tunnel is very low to facilitate critical pipeline slopes to destination city sites. (4) Aqueduct channel width, depth, side wall height and slope dimensions were designed to contain the design intent of a ~40,000 m³/day flow rate without spillage as well as rainfall-induced temporary water flow rate overages. (5) The low channel slope entry (0.002°) to the *castellum* is designed to slow water velocity and raise its height to the near top of the supply tunnel (Figures 3 and 4). (6) The sluice gate (Figure 10) height adjustment is used to produce the desired basin entry pipeline water critical height to permit critical (or near-critical) entry flow to pipelines. (7) The ~40,000 m³/day aqueduct input flow rate closely matches the 10 basin wall pipelines output flow rate (with floor ports closed) under Case B near- and Case C exact-critical flow conditions, Table 1). This flow rate match condition is necessary to maintain a smooth, constant operational water height in the *castellum* basin. (8) Critical flow conditions at 10 pipeline entrances are produced by making the sluice gate opening height equal to half the pipeline diameter (Figure 10), and this ensures critical entry flow to pipelines [16–19]. Critical and near-critical flow in pipelines is continued by pipeline slopes of ~4.6° to ~7.1°. Note that this pipeline slope range is approximately equal to the hill slope angle down to the main flat part of the city showing the design intent to place the tunnel at a specific height to achieve this result. (9) Production of critical or near-critical flow conditions in pipelines eliminates the influence of downstream flow resistance elements (bends, chokes, pipeline angle change, hydraulic jumps) that can propagate upstream to produce unstable, transient oscillations and unstable flow delivery from, the *castellum* basin. (10) The use of the critical and near-critical flow pipeline slope design produces the maximum, stable flow rate possible in pipelines and reduces pipeline joint leakage as an atmospheric airspace exists over the partial critical flow eliminating pressurized full-flow conditions that induce pipeline joint leakage; critical and near-critical flow are maintained in the pipeline to produce the benefits indicated above. Note that if the pipeline declination angle is lower than the θ_c range (~4.6° to ~7.1°) then downstream disturbances can propagate upstream to cause unstable basin oscillations that destabilize steady state behavior and cause spillage from the top of the *castellum* rim. For pipeline declination angles greater than the θ_c range, upstream disturbances cannot propagate upstream but the output flow from the 10 basin wall pipelines is less than 40,000 m^3/day. Based upon these many technical considerations inherent to the *castellum* design, the integrated, coordinated design of all components of the Pont du Gard aqueduct reflect Roman engineer's hydraulic engineering knowledge and thus serve to add to the compendium of Roman practices and water engineering inventions thus far described in the open literature.

Funding: This research received no external funding.

Conflicts of Interest: The author declares no conflict of interest.

References

1. Lewis, M. *Surveying Instruments of Greece and Rome*; Cambridge University Press: Cambridge, UK, 2001; pp. 181–188.
2. Hodge, T. *Roman Aqueducts and Water Supply*; Bristol Classical Press: London, UK, 2011.
3. Green, M. *Dictionary of Celtic Myth and Legend*; Thames and Hudson Publishers: London, UK, 1997.
4. Sage, M. *Roman Conquests: Gaul*; Pen and Sword Books Ltd.: London, UK, 2011.
5. Hauck, G.; Novak, R. Water Flow in the Castellum at Nîmes. *Am. J. Archaeol.* **1988**, *92*, 393–407. [CrossRef]
6. Lugt, J. *Vortex Flow in Nature and Technology*; John Wiley & Son: New York, NY, USA, 1983.
7. Nielsen, M. *Pressure Vibrations in Pipe Systems*; NYT Nordisk Forlag: Copenhagen, Denmark, 1952.
8. Bennett, C. *Frontinus: Stratagems and the Aqueducts of Rome*; Harvard University Press: Cambridge, MA, USA, 1961.
9. Herschell, C. *The Water Supply of the City of Rome*; New England Water Works: Boston, MA, USA, 1973.
10. Ortloff, C.R.; Crouch, D. Hydraulic Analysis of a Self-Cleaning Drainage Outlet at the Hellenistic City of Priene. *J. Archaeol. Sci.* **1998**, *25*, 1211–1220. [CrossRef]
11. Henderson, F.M. *Open Channel Flow*; The Macmillan Company: New York, NY, USA, 1966.
12. Morris, H.; Wiggert, J. *Open Channel Hydraulics*; The Ronald Press: New York, NY, USA, 1972.
13. Bakhmeteff, B. *Hydraulics of Open Channels*; McGraw-Hill Book Company: New York, NY, USA, 1932.
14. Ortloff, C.R.; Crouch, D. The Urban Water Supply and Distribution System of the Ionian City of Ephesos in the Roman Imperial Period. *J. Archaeol. Sci.* **2001**, *20*, 843–860.
15. Ortloff, C.R. *Water Engineering in the Ancient World: Archaeological and Climate Perspectives on Societies of Ancient South America, the Middle East and South-East Asia*; Oxford University Press: Oxford, UK, 2010.
16. Flow Science, Inc. *FLOW-3D/Version 7.0 Users' Guide*; Flow Science, Inc.: Santa Fe, NM, USA, 2019.
17. White, K. *Greek and Roman Technology*; Cornell University Press: New York, NY, USA, 1984.
18. Rasmussen, C. A Comparative Analysis of Roman Water Systems in Pompeii and Nîmes. Master's Thesis, University of Arizona, Tucson, AZ, USA, 2017.
19. Ven, T.C. *Open-Channel Hydraulics*; McGraw-Hill Book Company: New York, NY, USA, 1959.
20. Ortloff, C.R. *The Hydraulic State: Science and Society in the Ancient World*; Routledge Press: London, UK, 2020.

Article
Roman Water Transport: Pressure Lines

Paul M. Kessener

Faculty of ArtRadboud, Radboud University, 6525 XZ Nijmegen, The Netherlands; lenlcloud@icloud.com

Abstract: In Roman times long distance water transport was realized by means of aqueducts. Water was conveyed in mortared open channels with a downward slope from spring to destination. Also wooden channels and clay pipelines were applied. The Aqua Appia, the oldest aqueduct of Rome, was constructed in the third Century BCE. During the Pax Romana (second Century CE), a time of little political turmoil, prosperity greatly increased, almost every town acquiring one or more aqueducts to meet the rising demand from the growth of population, the increasing number of public and private bath buildings, and the higher luxury level in general. Until today over 1600 aqueducts have been described, Gallia (France) alone counting more than 300. Whenever a valley was judged to be too wide or too deep to be crossed by a bridge, pressure lines known as 'inverted siphons' or simply 'siphons' were employed. These closed conduits transported water across a valley according the principle of communicating vessels. About 80 classical siphons are presently known with one out of twenty aqueducts being equipped with a siphon. After an introductory note about aqueducts in general, this report treats the ancient pressure conduit systems with the technical problems encountered in design and function, the techniques that the ancient engineers applied to cope with these problems, and the texts of the Roman author Vitruvius on the subject. Reviewers noted that the report is rather long, and it is. Yet to understand the difficulties that the engineers of those days encountered in view of the materials available for their siphons (stone, ceramics, lead), many a hydraulic aspect will be discussed. Aspects that for the modern hydraulic engineer may be common knowledge and of minor importance when constructing pressure lines, in view of modern construction materials. It was different in Vitruvius's days.

Keywords: Roman aqueducts; inverted siphons; static pressure; pressure surges; lead pipes; stone conduits; air entrapment; Vitruvius

Citation: Kessener, P.M. Roman Water Transport: Pressure Lines. *Water* 2022, 14, 28. https://doi.org/10.3390/w14010028

Academic Editors: Helena M. Ramos and Charles R. Ortloff

Received: 13 October 2021
Accepted: 6 December 2021
Published: 23 December 2021

Publisher's Note: MDPI stays neutral with regard to jurisdictional claims in published maps and institutional affiliations.

Copyright: © 2021 by the author. Licensee MDPI, Basel, Switzerland. This article is an open access article distributed under the terms and conditions of the Creative Commons Attribution (CC BY) license (https:// creativecommons.org/licenses/by/ 4.0/).

1. Some Aspects of Roman Aqueducts

In Roman times aqueducts with mortared open channels, often roofed to limit evaporation loss and temperature rise, had a regular (but not necessarily uniform) downward slope from source to destination (Figures 1 and 2) [1]. The longest aqueduct known is that of Constantinople, 250+ km. The Carthago and the Cologne aqueducts each were 95 km long, the Tempul aqueduct of Cadiz (Roman Gades, Spain) 83 km, while four of Rome's eleven aqueducts surpassed 50 km (Table 1) [2].

The route of an aqueduct usually ran, for at least a part, through mountainous and uneven terrains. To guarantee that the water would flow from source to destination, bridges were constructed, tunnels were dug, and, at times, the channel was cut right out of the vertical rock face (Figure 3).

The terrain complexity frequently required great expertise in surveying techniques, planning and design [3]. The often-anonymous engineer, who had been given the task to bring good quality water to a town and to solve the problems that were encountered on the way, usually is lost to history together with the available engineering tools [4]. While bricks and natural stones were the main materials the mortared channels were constructed from, only few wooden-lined channels have been identified. Supported by a solid foundation and with thick walls the mortared channels were covered by a barrel vaulting, less often by flat slabs. Dimensions of the channel ranged widely (Table 2, Figures 4 and 5).

Figure 1. Map of known aqueducts in the Roman Empire.

Figure 2. Remains of two of Rome's aqueducts, the Aqua Claudia, crossing the Roman Campania, piggyback carrying the Aqua Anio Novus on its top (Fratelli Allesandri no. 476, 19th c. photo).

Table 1. Length of some classical aqueducts. Constantinople: Çeçen 1996; Crow 2012; Sürmelihindi et al. 2021. Carthago: Rakob 1983. Cologne: Haberey 1972; Grewe 1986. Rome: van Deman 1934; Ashby 1935; Blackman 1978; Hodge 1992. Cadiz: Casado 1985, 319; Pérez et al. 2014. Lyon: Burdy 2002. Pergamon: Garbrecht 1978, 1987, 2001. Nîmes: Fabre et al. 2000. Arles: Leveau 1996. Cherchel: Leveau and Paillet 1976. Alcanadre: Mezquiriz 1979. Noviomagus: Kessener 2017a.

Location	Country	Length (km)
Constantinople	Turkey	>250
Carthage	Tunisia	95
Cologne	Germany	95
Rome (Aqua Marcia)	Italy	91
Rome (Anio Novus)	Italy	87
Cadiz (Tempul)	Spain	83
Lyon (Gier)	France	75
Rome (Aqua Claudia)	Italy	69
Lyon (Brevenne)	France	66
Rome (Anio Vetus)	Italy	64
Romen (Aqua Traiana)	Italy	58
Pergamon (Kaikos)	Turkey	50
Nîmes	France	50
Arles	France	48
Cherchel	Algeria	45
Alcanadre	Spain	30
Noviomagus	Netherlands	5

Figure 3. Side aqueduct (Turkey). Channel, alongside the Manavgat River, cut out of the vertical rock face (author's photo).

Table 2. Dimensions of some aqueduct channels.

Aqueduct	City/Country	Inside Channel Dimension (Height × Width, cm)
Degirmendere	Ephesos/Turkey	245 × 80
Aqua Marcia	Rome	240 × 90
Carthago	Tunesia	190 × 85
Brevenne	Lyon/France	166 × 120
Cologne	Germany	142 × 70
Aspendos	Turkey, south coast	90 × 50
Mont d'Or	Lyon/France	74 × 44
Patara	Turkey, south coast	35 × 40

Figure 4. Channel of the Degirmendere aqueduct of Ephesos (author's photo).

Figure 5. Channel of the Carthago aqueduct (author's photo).

For maintenance and repair channels were made accessible from the top by means of an opening, either round or square, spaced at regular intervals, at times combined with a vertical shaft in case of an underground course (manhole, inspection shaft, 'regard' as the French say, 'Einstieg-Schacht' for the Germans). Such vertical shafts were also dug when constructing tunnels, by connecting the shafts underground (Figure 6a,b).

Figure 6. (**a**,**b**) Inspection shaft of an underground section of the Degirmendere aqueduct of Ephesos (author's photo).

The planning of an aqueduct route was an involved task. Vitruvius (First Century BCE) describes an instrument, called 'chorobates', a 20-foot-long narrow table with legs at square angles and vision-sights at either end (Figure 7), which he esteems the most accurate for leveling [5]. Plumb bobs would guarantee the horizontal orientation of the table, and in case of 'windy weather' water could be poured in a 5 foot long, 1 inch wide, and 1.5 inch deep hollow in the table top. When the water level is the same at both ends of the groove, the table top is oriented horizontal, and 'one knows how large the slope (of the channel) is', Vitruvius adds.

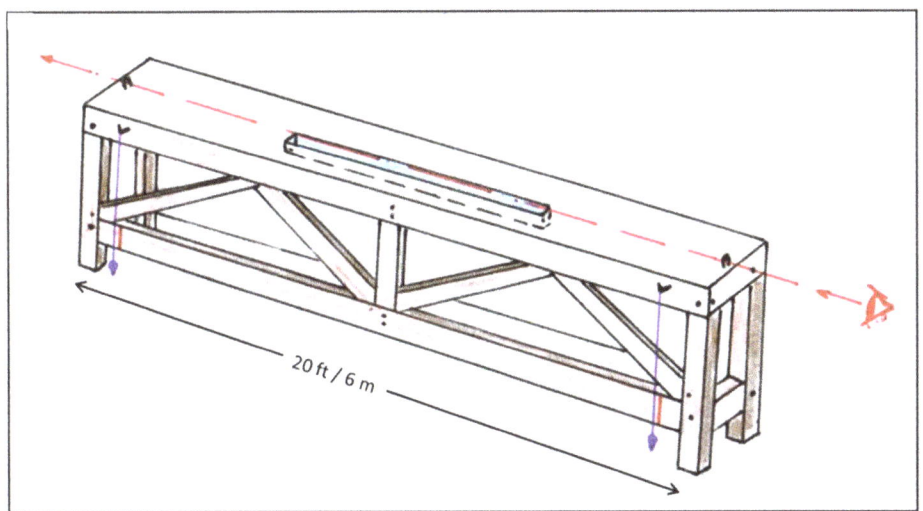

Figure 7. A *chorobates*.

In the same section Vitruvius refers to Archimedes 'who is known to state in his writings' that 'water surfaces coincide with a sphere concentric with the center of the sphere of the earth'. But that does not affect the leveling, Vitruvius adds, as long as the water level at both ends of the groove is at equal distance from the table top [6]. The consequence, however, is, that when one takes the horizontal level at location A, the sighting to location B will result—because of the curvature of the earth—in a position that is above the corresponding horizontal level at B (Figure 8). The greater the distance is, the worse this error becomes [7]. And for water to flow from A to B the channel should arrive below the corresponding horizontal level at B. For small distances the error is insignificant, increasing substantially for greater remoteness (Table 3). When surveying from a water source to the city, that is in downstream direction, repeated errors result to in failure for water to arrive where planned; when leveling upstream from city to source, the errors, if the source is reached, guarantee that water will arrive at destination.

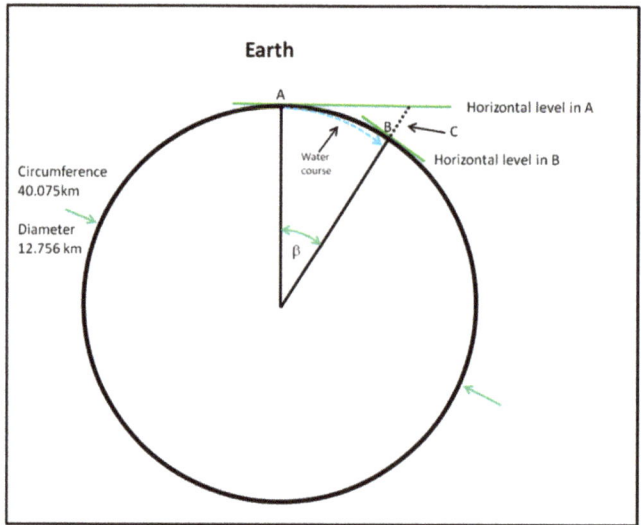

Figure 8. Taking the horizontal in A will result in an error C at B (C = R • (1/cosβ − 1), R = radius of the earth). The slope of a channel should be such that the channel arrives below the horizontal (at the same radius R of the earth) at B (drawing by author).

Table 3. Leveling error C due to the curvature of the earth. The error increases rapidly with the distance between A and B. Back-sighting from B to A will result in a similar error at A. Correcting for this error at B gives the horizontal at that point. One may then lower the level at B for the required slope of the channel.

Distance	Error
100 m	1 mm
500 m	2 cm
1 km	8 cm
5 km	2 m
10 km	7.8 m
100 km	784 m
250 km	4.9 km

Grewe 2014 is aware of the leveling problem and suggests that the course of an aqueduct should be leveled by having the chorobates repetitively installed by turning it around for 180 degrees and set it up at the end of its former position (Figure 9). Repeating the measurement will even out errors of the instrument itself as well as to allow for the curvature of the earth [8]. This method would theoretically do the job, but is quite if not entirely impractical in mountainous and wooded terrains, not to speak of the time consuming effort to repeatedly install the 6 m long chorobates. Some authors think that the chorobates is not at all suitable for leveling and suggest that the Roman engineers used the course of roads, rivers, and streams to plan their aqueduct [9].

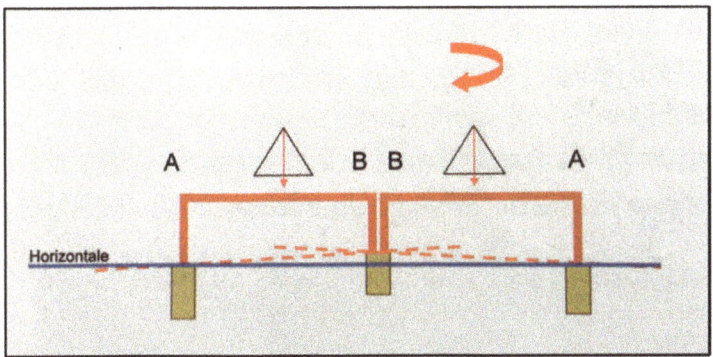

Figure 9. Proposal by Grewe for applying the chorobates eliminating errors of the instrument and compensating for the curvature of the earth: 'Durch Drehung des Chorobates nach jedem Messgang werden Gerätefehler eliminiert' ('By turning the chorobates after every measurement the errors of the instrument will be eliminated') (Grewe 2014, 38).

Yet a solution for eliminating the error is to have a chorobates put up at both locations A and B (or set up an indication of its level at position A), then perform a back-sighting from B to A, and correct at location B for the error C that may be determined from the back-sighting. Then the slope may be set by lowering this new level at B for the required slope-level. Also the unfinished channel itself may serve as a large leveling instrument to determine the slope, as was common practice for instance in mid 20th century Turkey (Figure 10) [10].

Figure 10. Constructing a channel with the required slope. By having temporary dams installed at regular distances (arrows) and introducing water in the channel as a water level, the required slope may be set straightforwardly (Büyükyıldırım 2017, 72).

In case low lying terrain had to be crossed, the channel was built on sequential arches so that the required slope could be maintained, the water arriving at the town at a height as planned. The aqueducts of Rome are known for the endless rows of arches in the Campania plains of which sections still stand (Figure 2), where at times channels were positioned on top of existing ones saving considerable costs and efforts. The 95 km Carthago aqueduct ran for 17 km on arches up to 30 m high to cross the Miliane plain (Figure 11) [11].

Figure 11. Remains of the Carthago aqueduct bridge crossing the 17 km Miliana plain south of Tunis. Further down, the piers, now destroyed, were up to 30 m high (author's photo).

The slope (gradient) of the channel varied for each aqueduct and along its trajectory. Often the initial section was steep, while on approaching its destination the slope became less. The 50 km aqueduct of Nîmes with its famous Pont du Gard (Figure 12) started at present day Uzès with a gradient of 67 cm per kilometer from the spring that is situated only 17 m above the aqueduct's end. After the 15 km stretch to the Pont du Gard it continued at only 7 cm/km for another 10 km, in extremely hilly terrain, incorporating another four bridges, a great achievement of precision even by today's standards [12].

Figure 12. The Pont du Gard in heavy weather (2002 flood). The bridge withstood nature's force undamaged; this was not so for the 1260 CE bridge crossing the river Gardon 15 km upstream (https://www.academie-pontdugard.com/le-gardon-lenfant-terrible/ (accessed on 15 May 2020)).

Not far downstream from one of its two springs the aqueduct of Aspendos (south coast of Turkey) had a slope of an estimated 160–170 m/km [13]. The Carthage aqueduct has a slope of 95 m/km for the first 6 km from the spring at Zaghouan. After French repairs in the 19th c. the Carthage aqueduct runs again today for a stretch of 70 km, providing Tunis with water and serving local populace on its way (Figure 13). The ruined Miliana bridge was substituted by a steel siphon. In 1995 its discharge was 150 L/s, almost 13,000 cubic meters/day.

Figure 13. Carthage aqueduct, supplying local villagers, late 1990's (author's photo).

To limit leakage, the inner walls of the channels were covered with 'opus signinum', a special type of mortar resistive to temperature changes and cracking [14]. Because the Romans often tapped karstic springs, over time calcareous incrustations ('sinter') accumulated on the walls and floor of the channels at a rate in the order of several mm's per year. This could lead to a substantial reduction of channel dimensions and discharge. The 120 cm wide channel of the Nîmes aqueduct has deposits up to 50 cm wide on both inner walls, indicating continuous operation for many centuries (Figure 14). In later times large sinter blocks taken from the Nimes channel were used for construction purposes. In contrast Lyon's Gier aqueduct shows no sinter at all as it was supplied from a non-karstic (granite) area.

Figure 14. Channel of the Nîmes aqueduct, not far downstream of the Pont du Gard. Substantial calcareous incrustations (*sinter*) on walls and floor of the 120 cm-wide channel. Dotted line: inner surface of channel wall and floor (author's photo).

The sinter of the Cologne aqueduct, having grown in thickness of just 1 mm/year, is of a high quality. In medieval times the up to 30 cm wide the incrustations in the channel were broken out to be reworked and polished, showing a travertine like structure from the seasonal deposits. It was regarded as a valuable material and served for altar slabs, ornaments, funerary ornaments, and pillars in churches in Cologne and surroundings. Known as 'aqueduct marble' the material was exported to Denmark, England, and Holland (Figure 15) [15]. Unworked sinter slabs from the Aspendos aqueduct were applied as grave stones in Selçuk cemeteries [16].

Figure 15. Columns of aqueduct marble at St. Servaas church, Maastricht, the Netherlands (Grewe 2014, 368).

By analyzing isotopes of oxygen and carbon the deposits may serve as a high resolution paleo-environmental record [17]. Sinter imprints collected from the Barbegal mill complex in the south of France served to reconstruct the functioning of the 16 water wheels [18]. The 246 km fourth and fifth century CE aqueduct of Constantinople, longest of all Roman aqueducts and functioning for at least 700 years, is said to have been regularly cleaned of its massive calcareous deposits [19]. The eleven aqueducts of Rome had an estimated joint discharge of over one million cubic meters per day, a figure similar to the water provision of Paris in the 1970s. The 95 km aqueduct of Cologne conveyed 21.000 cubic meters per 24 h from the Eifel. The four aqueducts of Lyon delivered 75.000 cubic meters/24 hr [20].

2. Pressure Conduits: Inverted Siphons

The Roman engineer had only one force available to transport water over long distances: gravity. He had to make sure that the water could flow downstream from start to finish. Whenever a valley was too wide to be circumvented or too deep to be crossed by a bridge an 'inverted siphon' was constructed. With a siphon the water was made to flow to the other side of the valley by means of a closed conduit according to the principle of communicating vessels, a technique already applied in Hellenistic times (Figures 16 and 17) [21,22]. Water from the aqueduct entered a 'header tank', from where it went into a pipe that ran down the slope of the valley, crossed the lowest part, then rose again on the other side. There the water ended in a 'receiving tank' at a level somewhat below the header tank. From the receiving tank the water continued its course to destination in an open channel again.

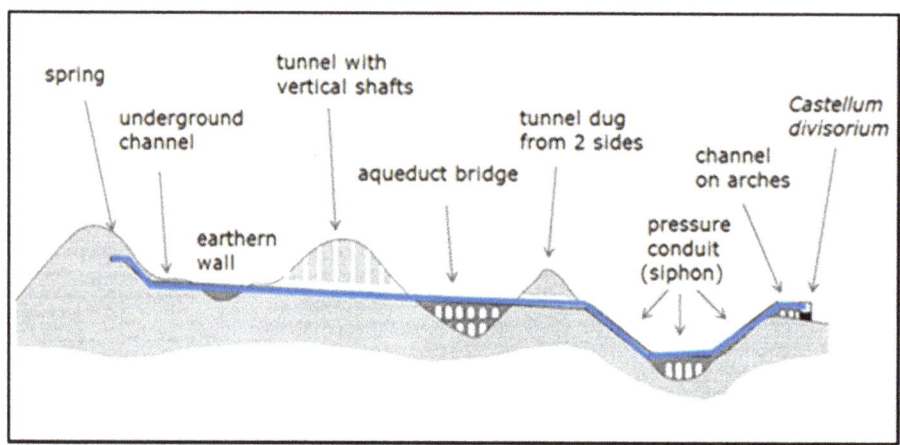

Figure 16. Elements of an aqueduct, including a pressure conduit, a 'siphon' (adapted from Fahlbusch 1992, 86).

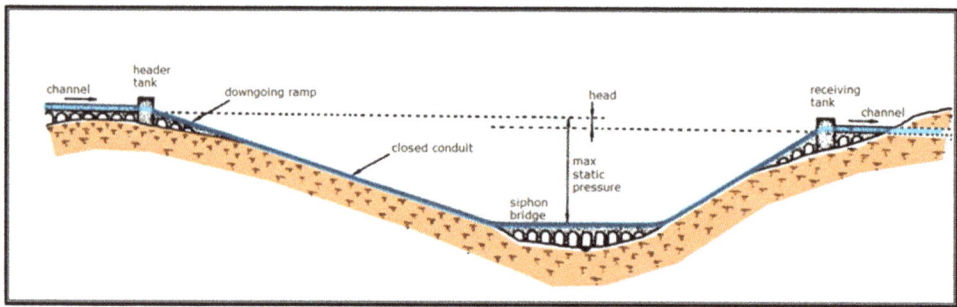

Figure 17. Elements of a siphon (adapted from Kessener 2004, 19).

In the lowest part of the valley, the conduit was often installed on a 'siphon bridge' to have the river or stream there pass without damaging the line. The pressure in the conduit down in the valley could be considerable (Table 4) [23].

Table 4. Length, maximum depth, and hydraulic gradient of some classical siphons.

City	Aqueduct	Siphon	Material of Conduit	Length (m)	Max Depth (m)	Hydraulic Gradient (m/km)
Gades (Spain)	Tempul	de la Playa	stone	19,500	20	not known
Smyrna (Turkey)	Kara-Bunar	Kara-Bunar	clay/stone	4400	158	1.1 (estimate)
Lyon (France)	Yzeron	Craponne-Lyon	lead	3600	91	9.2
Gades	Tempul	de los Arquilles	stone	3555	50	3.1
Lyon	Mont d'Or	d'Ecully	lead	3500	70	3.1
Lyon	Brevenn	Grange-Blanche	lead	3500	90	4–5.6 (estimate)
Pergamon (Turkey)	Madradag	Madradag	lead	3250	190	12.6
Alatri (Italy)	Alatri	Alatri	lead	3000	100	9 (estimate)

Table 4. Cont.

City	Aqueduct	Siphon	Material of Conduit	Length (m)	Max Depth (m)	Hydraulic Gradient (m/km)
Lyon	Gier	Beaunant	lead	2660	122	3.0
Lyon	Yzeron	Grezieux-Craponne	lead	2200	33	3.2
Aspendos (Turkey)	Aspendos	Aspendos	stone	1670	45	8.3
Termini Imerese (Sicily)	Cornelio	Barratina	lead	1300	40	3.8
Lyon	Gier	Soucieu (le Garon)	lead	1210	93.5	not known
Laodikeia ad Lykum (Turkey)	Laodikeia	Laodikeia	stone	800	50	26
Lyon	Gier	St. Genis	lead	700	79	8.3
Lyon	Gier	St. Irenée	lead	575	38	4
Oinoanda (Turkey)	Oinoanda	Oinoanda	stone	500–700	22	6–16 (estimate)
Lyon	Mont d'Or	Cotte-Chally	lead	420	30	19
Patara (Turkey)	Patara	Delik Kemer	stone	260	20	18.5

The highest pressure was reached in the 3150 m long Hellenistic Madradag-siphon at Pergamon, Turkey, 19 bar, corresponding to a max depth of 190 m (Figure 18). The siphon was fed by an over 40 km long ceramic unpressurized triple pipeline. Second with 15 bar is the Second Century BCE Kara-Bunar siphon of Smyrna of which virtual nothing has remained due to the enormous population increase of present day Izmir. In contrast the course of the Madradag siphon of Pergamon has not been disturbed and can be walked, in rather uneven and difficult terrain, where one passes bridges of later aqueducts and less deep Roman siphons [24]. From the header tank a free view of the acropolis exists (Figure 19). When leveling with a chorobates the error from the earth's curvature amounts to 1.66 m, which is small compared with the level difference of 41 m between header tank and receiving tank. Viewing with a chorobates from the envisaged receiving tank on top of the acropolis towards the northern mountains, one may locate a spot on the slope almost 40 m down from the extant header tank. It was then to decide the construction location of the header tank above this point, but how the engineers knew how high one had to go up the hill from that point to have sufficient hydraulic gradient for their siphon remains open to question.

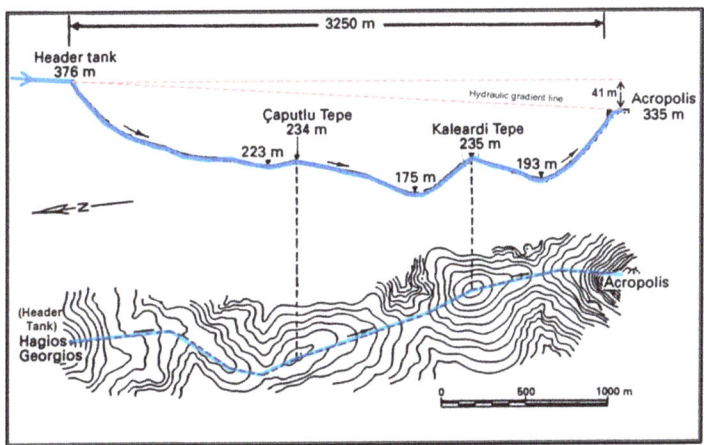

Figure 18. Profile and plan of the Madradag siphon of the Hellenistic Pergamon aqueduct (after Fahlbusch 1982).

Figure 19. View from the location of the header tank towards the Pergamon acropolis (centre), with intervening hills Çaputlu Tepe and Kaleardi Tepe just visible (photo courtesy M. Baykent).

The longest pressure line is that of Cadiz on the south coast of Spain, Roman Gades, ending at the distribution tank in the town (Sifón de la Playa): 19.5 km (Figure 20) [25]. Conduit stones of the Sifón de la Playa have been recovered from the shore line and put up for inspection in a small Cadiz park (Figure 21). Not much remains of the header tank of the siphon de la Playa (Figure 22). Half way down from the Tempul spring, there is a second siphon ('Sifón de los Arquillos'), to be discussed below.

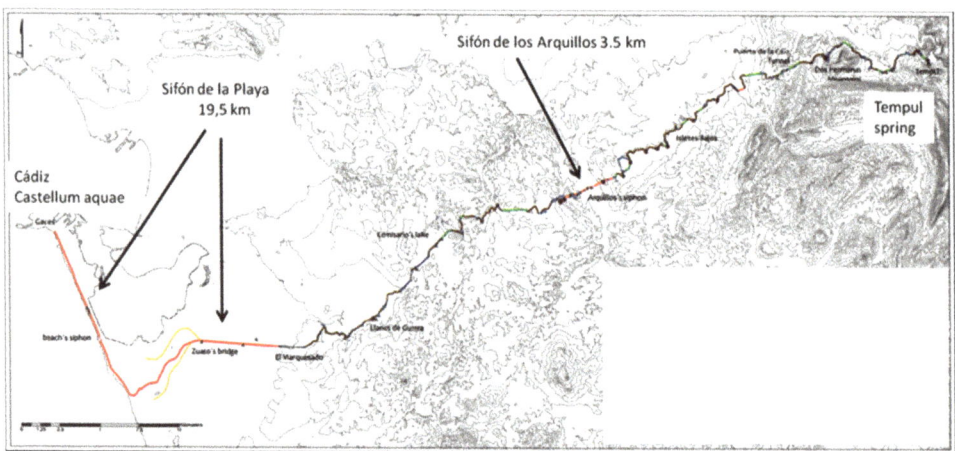

Figure 20. Aqueduct of Gadez, length 80 km, equipped with two siphons. The siphon 'de la Playa' ran for a substantial part on the Mediterranean shore line (Pérez, Bestué, Ruíz 2012).

Figure 21. Conduit stones of the 'de la Playa' siphon retrieved from a Cadiz beach (author's photo).

Figure 22. Remains of the presumed header tank of the 'de la Playa' siphon (author's photo).

The four aqueducts of Lyon (France) had a total of nine siphons (Figure 23). The Gier aqueduct, with 75 km the longest of Lyon's aqueducts, had four siphons, up to 120 m deep and 3.5 km in length [26]. For the required capacity, nine to eleven 20 cm diameter conduits made of lead were laid out in parallel. The amount of lead for these siphons which all together had a length of 16.6 km, is estimated to have been 10,000 to 15,000 tons [27]. The lead has disappeared to be reused for roofs of churches, dwellings and other purposes. A number of header tanks and receiving tanks have survived as well as siphon bridges and parts of sloping ramps (Figures 24 and 25). Some of the tanks have been partly restored.

Figure 23. Lyon's 4 aqueducts, du Mont d'Or, de l'Yzeron, de la Brevenne, and du Gier. Blue arrows point to the siphons of the 75 km Gier aqueduct (red line). From south to north the siphon of la Durèze, the Garon siphon, Yzeron siphon, and the Trion siphon (after Burdy 1996 and 2001).

Figure 24. Gier aqueduct upstream from header tank of Yzeron siphon (author's photo).

Figure 25. Remains of header tank of the Yzeron siphon with descending ramp (Gier aqueduct). The floor of the roofed header tank is at 10.5 m above ground level, vaulted roof partly destroyed. Nine parallel conduits of 20 cm diameter made from lead ran down the ramp towards the bottom of the valley 125 m below the header tank (author's photo).

The materials of the siphon-conduits were diverse: lead, stone, terracotta, or combinations of these have been used. They were assembled from prefab pipe elements having varying lengths: 40–70 cm for terracotta pipe elements, 50–100 cm for perforated stone blocks, up to 3 m for lead pipes. The pipe elements were joined by bringing the end of one pipe element into the somewhat larger end of the next (for terracotta conduits) or by means of socket and flange (for stone and terracotta conduits). The up to 1 m wide stones conduit elements of the siphons of Aspendos, Cadiz, Patara, Kibyra, and Laodikeia ad Lykum are of the latter type, as are the smaller 50 cm cubic conduit stones at Oinoanda (all in Turkey except Cadiz) (Figures 26 and 27). Lead pipes were either cast, or—more often—made of leaden sheets that were bent around a wooden pole and soldered at the seam (Figure 28) [28]. Cast lead pipes could be joined with a stone element in between as found in Ephesos, or by means of a lead sleeve slid over both ends [29].

Figure 26. Conduit stone of the Aspendos siphon, 90 × 90 × 50 cm, with a bore of 28 cm (author's photo).

Figure 27. Stone siphon of Oinoanda, Turkey, 50 × 50 × 50 cm cubicles, bore 15 cm (author's photo).

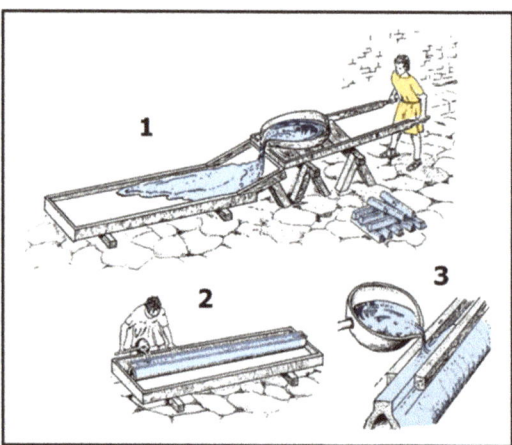

Figure 28. Production of lead pipes. Pouring molten lead to a sheet, bending it around a wooden pole, and soldering the seam (after Hodge 1992, 309).

The pipes, depending on the quality of the water, were subject to calcareous incrustations—just as the open channels. Over the years the inner cross section would be reduced hampering the flow (Figure 29). The estimated 3000 conduit stones of the 1650 m long Aspendos siphon, destroyed by an earthquake in the 4th century CE, served as construction material for a new bridge over the nearby Eurymedon river. They were reused again centuries later by the Seljuks for a bridge they built on the ruins of the Roman predecessor. A number of conduit stones, with incrustation, can be seen in the fabric of the remains of the Roman bridge, as well as of its Seljuk counterpart that still stands today (Figures 30 and 31) [30].

Figure 29. Conduit stone of the Laodikeia a/L siphon, with considerable incrustations (sinter) (author's photo).

Figure 30. Conduit stones of the Aspendos siphon reused for the mole of the Roman bridge crossing the Eurymedon River (author's photo).

Figure 31. Conduit stones of the Aspendos siphon in the fabric of the 13th century Seljuk bridge (author's photo).

The Karabunar siphon of Smyrna is said to have been made of stone elements alternated with terracotta pipes [31]. The joints of stone and ceramic conduits were sealed with a mixture of live chalk, oil, and herbs, which expands when moisturized [32].

Sections of lead siphon-conduits have been preserved. A 90 cm long fragment with diameter 31–34 cm, is all that remained from a find of about 10 tons of lead conduit that was retrieved in 1980 from the Rhône river near Vienne at extremely low water level (Figures 32 and 33). Regretfully the pipes have been melted down without prior investigation, only the 90 cm fragment remaining. The find shows that at Vienne the Rhône river was crossed with a siphon on the river bed. The pipe section that was put 'dans la four', in the oven, to be melted down is estimated to have been an astonishing 80 m long [33].

Figure 32. A 90 cm-long fragment of lead conduit retrieved from the Rhône River near Vienne, France, with one small and one large stamp. The large stamp reads IMP·M·AVR·ANTONINO·PIO·AUG·IIII·ET·BALBINO·II·COS (under the 4th consulate of Emperor Marcus Aurelius Antoninus Pius Augustus and the 2nd consulate of Balbinus). The second stamp reads L·VIR·SATTO·V (Lucius Virius Satto of Vienne made it). From the large stamp, it could be dated 213 CE (after Burdy and Cochet 1992).

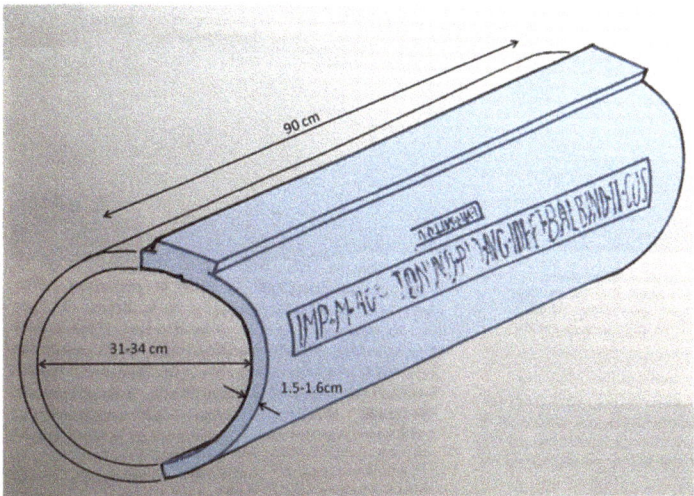

Figure 33. Vienne lead conduit fragment, dimensions. (adapted from Burdy & Cochet 1992).

Near Arles, France, 33 lead pipes were recovered from the Rhône river over the years 1570–1825. Now in the Arles museum they have been investigated by J. Hansen of Denmark (Figure 34) [34]. The pipes, with a length of 3 m and an inner diameter of 10–12 cm have a lead seam along their length and were part of a siphon that also crossed the Rhône, between Arles and Trinquetaille. The pipe elements were fixed to each other by inserting one end into the larger end of the next pipe, and driving a large nail through both ends. Subsequently the joint was sealed with a thick layer of soldered lead that also covered the nail that must have hampered the flow to some extent (Figure 35).

Figure 34. Lead pipes recovered from the Rhône river near Arles. They were part of a siphon crossing the river on the river bed between Arles and Trinquetaille (Haberey 1972, 157). A number of pipes are on exposition in the Arles Museum.

Figure 35. Connecting lead pipes with a nail driven through both ends and covering the joint with a soldered lead sheet (Haberey 1972, 140).

Hansen noted that the joints were no weak points in the line (*'waren nicht das schwache Glied der Kette'*), an indication of the superior soldering techniques of the Romans. Such conduits may be regarded as being made of a homogenous material. This is not the case for conduits with joints sealed with the expanding mix. Although this material is highly resistive to pressure (a high compressive strength), its tensile strength is much lower than that of stone, lead, or terracotta, the material the pipe elements are made of. These conduits are susceptible to bursting at the joints.

3. Siphon Conduits and Water Pressure

Ancient pressure conduits may thus be split into two categories. Cat.I conduits are made of lead, with soldered joints. These conduits can be regarded as 'homogenous', with a uniform material all along the conduit, that is, the tensile strength parallel to the conduit axis is equal to the tensile strength perpendicular to the conduit axis. Cat.II conduits have joints sealed with the weak expanding mix (resistive to pressure but having a low tensile strength, its breaks easily). Here, the tensile strength parallel to the conduit axis (at the joints) is lower than the tensile strength perpendicular to the conduit axis (of the stone material). These conduits must be considered as 'non-homogenous'.

Theoretically, a pipe can burst in two ways when the inside pressure becomes too high: along its length, or perpendicular to its length. The minimum pressure to burst along its length $P(l)$ equals (Figure 36):

$$P(l) = t(p) \cdot (d/2R)$$

where $t(p)$ = tensile strength of the material of the pipe wall perpendicular to the pipe axis (force/m^2);

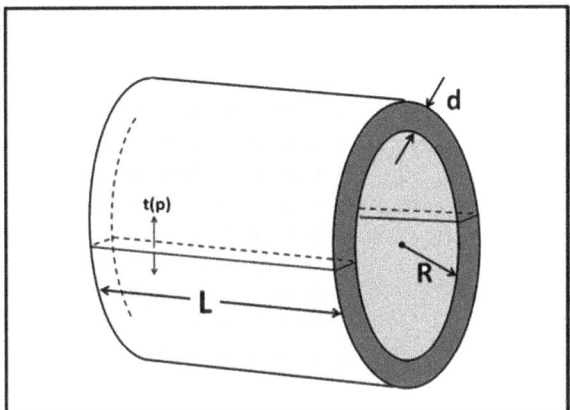

Figure 36. Lead pipe with inner diameter 2R and wall thickness d. Inside pressure P. For a length L of the pipe not to burst lengthwise, the strength of the wall section along this length (equal to its surface d·L times tensile strength t(p)) must be larger than the force that pulls on that wall section (equal to inside pressure P times cross-sectional surface along that section = P·2R·L). Thus, to make the pipe burst, P·2R·L > d·L·t(p), or P > t(p)·(d/2R).

d = pipe wall thickness;
R = inner radius of the pipe.

To have the pipe burst perpendicular to its axis, the pressure must at least be P(p), where

P(p) = 2 · t(l) · (d/R) · (1 + d/R) with t(l) = tensile strength along pipe axis (Figure 37).

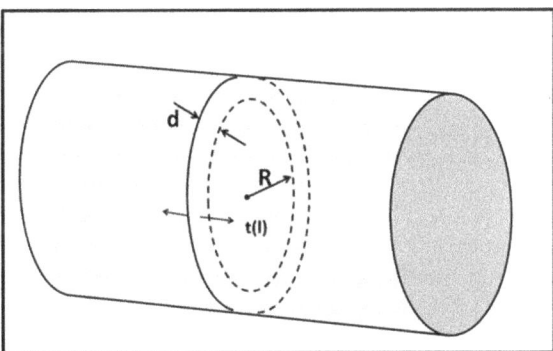

Figure 37. Lead pipe with inner diameter 2R and wall thickness d. Inside pressure P. Assume it is capped off at the ends. The inside pressure P generates a force pulling via the capped-off ends along the entire circumference of the pipe. This force is equal to the inner cross-sectional surface πR2 times pressure P. This force must be compensated by the tensile strength of the entire pipe wall along the circumference 2π(R + d)·d·t(l). To have the pipe burst perpendicular to its length, the force from the inside pressure must be larger than the force from the tensile strength, or πR2·P > 2π(R + d)·d·t(l).

For pipes made of homogenous material (Category I conduits) the tensile strength is equal in all directions: t(p) = t(l). From this follows P(p) > 2 · P(l). Homogenous conduits will always burst along their length when the inside pressure gets too high. For Category II conduits, with t(p) > t(l), this may not be the case as the low tensile strength along the conduit axis (the low tensile strength of the sealing mix) make the Cat.II pipes are prone to burst at the joints. Thus, the choice of the material for the pressure pipes requires that precautions must be taken to guarantee proper functioning of the siphon and to prevent damage. Factors influencing failure are static pressure, forces generated by the water flow, effects from presence of air in the conduit, and the possibility of the occurrence of pressure surges, which are described below.

4. Static Pressure

If a siphon is just filled with water—not flowing—only static pressure has to be reckoned with. The static pressure p at any point in the conduit is related to the vertical distance h between that point and the free surface of the water:

$$p = \rho \cdot g \cdot h$$

where

ρ = specific mass of water = 1000 kg/m^3;
g = gravitational acceleration = 9.81 m/s^2;
h = vertical distance below free surface of the water in m [35].

For a conduit full of water, the water pressure p exerts forces perpendicular to the pipe wall along its circumference, which are evened out as long as the tensile strength of the pipe material is sufficiently high. This is true for both categories of conduits provided that, for Cat.II conduits, the combined wall thickness of the male–female joints has a similar tensile strength as the pipe wall itself (and that displacement of a pipe element along the conduit axis is prevented by the next pipe element (Figure 38)).

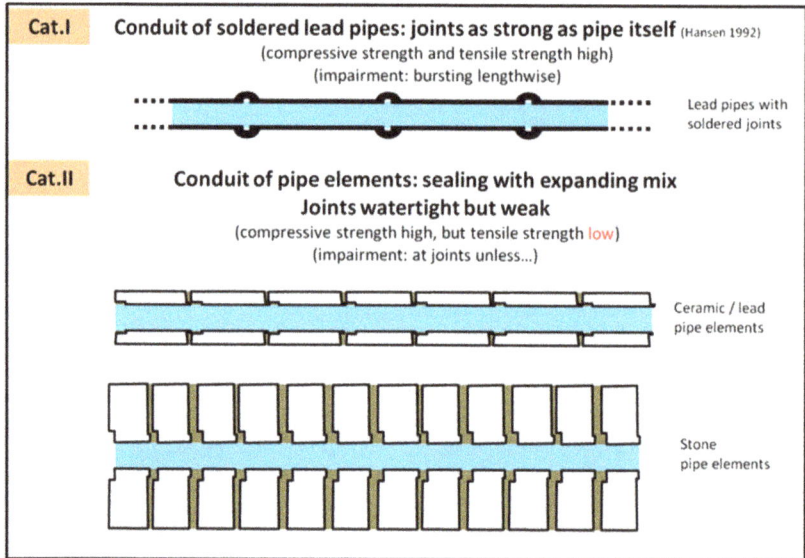

Figure 38. Two categories of conduits for siphons, Cat.I and Cat.II.

If indeed properly secured, Cat.II conduits are, in turn, susceptible to splitting along the length of the line, as an element from the Delik Kemer siphon of the Patara stone aqueduct shows (Figure 39; for this siphon on the south coast of Turkey, see below).

Figure 39. In situ conduit stone (90 × 90 × 50 cm) of the Patara aqueduct, top view, with crack parallel to conduit axis (author's photo).

5. Bends in the Line

Things change, however, where there is a bend in the line. On the pipe element at the bend, the static pressure exerts a net outward force along the bisector of the bend angle, with magnitude F, in newtons (Figure 40):

$$F = 2 \cdot p \cdot A \cdot \sin\left(\frac{\alpha}{2}\right)$$

where

p = static pressure (newton/m^2);
A = cross-section of conduit (m^2);
α = angle of bend.

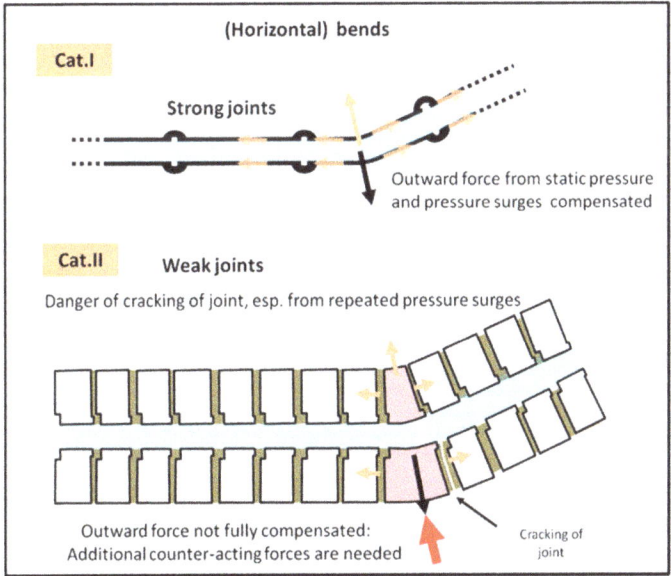

Figure 40. Forces on a bend; for Cat.II conduits, extra measures are required.

For an angle of 180 degrees, a 'U-turn', this force is at its maximum (sin(α/2) = 1), while for a straight conduit, with α = 0, this force is of course zero. For a bend of 30 degrees in a 28 cm-diameter conduit at a pressure of 40 m of water column, as in the case of the Aspendos siphon (for the Aspendos siphon, see also below), where the conduit turns from going down from the header tank to horizontal, this force F (in newtons) becomes considerable:

$$F = 2 \cdot 1000 \frac{kg}{m^3} \cdot 9.81 \frac{m}{s^2} \cdot 40m \cdot \frac{\pi}{4} \cdot (0.28m)^2 \cdot \sin\left(15°\right) = 12,507 \text{ N}$$

This is a force equivalent to a weight of 1275 kg (or 1275 kg-force (kgf)). For vertical bends changing from going down to horizontal or from horizontal to going up, such force may be readily countered by an adequate foundation of the conduit. For the 3250 m-long and 190 m-deep Madradag siphon at Pergamon, the forces were significant due to the high pressure (see Figure 18). The conduit consisted of pipe elements of cast lead, 2–3 m long, that were joined by means of lead sleeves slid over neighboring pipe ends. The 17.5 cm-inner diameter conduit was kept in place by having every individual pipe element fitted into a perforated trachite stone slab and burying the entire conduit (Figures 41 and 42).

Figure 41. Madradag siphon, Pergamon. View from header tank towards acropolis, anchor stones for the lead conduits, upper side broken out (after Gräber 1913).

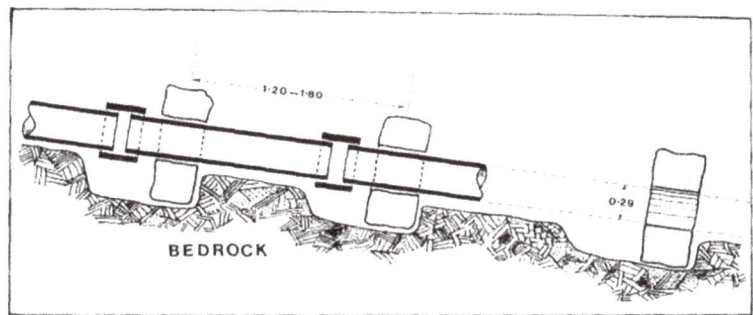

Figure 42. Anchor stones of the Madradag siphon, with depressions cut in bedrock to facilitate jointing of lead pipe elements (after Garbrecht 2001, 120).

At the two vertical 20-degree bends on top of two intervening hills, the Caputlu Tepe and the Kaleardi Tepe, 136 and 146 m below the header tank, the conduit changes from rising up to going down, meaning the force from static pressure is directed upward. Here, the fixation stones (anchor stones) are larger (respectively, 1.5 cubic meters and 3.4 cubic meters) to compensate for the upward force of about 11,000 and 12,000 newtons (over 1000 kgf) at these points [36]. The estimated weight of the anchor stones of 8500 kg and 3750 kg (Caputlu and Kaleardi) was sufficient, if not overdone, to compensate for the uplifting force from static pressure (Figures 43 and 44).

Figure 43. Madradag siphon. Fixation stone on top of Caputlu Tepe, broken into two fragments. Estimated weight 8400 kg (author's photo).

Figure 44. Anchor stone on top of Kaleardi Tepe, plus lesser one just below. In the back, the Pergamon acropolis (author's photo).

For pressure lines of Category II, the force from static pressure exerted on a pipe element that makes up a bend will be diverted to the neighboring pipe elements, but only as far as the sealing material keeps the pipe elements together. The tensile strength of the sealing material is comparatively weak, so that additional means were needed to prevent the pipe element(s) of the bend being pushed out of position. For vertical bends this could be achieved by having the conduit laid on a solid foundation (where a descending conduit changes to horizontal or v.v.) or by adding mass enlarging the weight, as was done at Pergamon on the top of the intervening hills. The Roman author Vitruvius indeed recommends for such pipelines to make the vertical bends of special red stone ('ex saxo rubro'), undoubtedly known for its strength [37]. However, for horizontal bends, even if minor, additional measures must be taken, such as increasing friction forces with the underground by sand ballast, by building a wall pushing back, or by fastening the conduit elements to each other with bands, as Vitruvius advises [38]. The last measure was carried out for the Delik Kemer siphon of the Patara aqueduct, which will be discussed here shortly.

6. The Delik Kemer Siphon of the Patara Aqueduct

The 500 m and 20 m-deep 'Delik Kemer' stone siphon of the Patara aqueduct on the south coast of Turkey was built on a 10 m-high and narrow cyclopean wall crossing a 30 m-deep depression. Over 90 m of the stone conduit is in situ on top of the wall today (Figure 45).

Figure 45. Patara siphon. In situ stone conduit on 10 m-high cyclopean wall. Flow direction from right to left (author's photo).

Although the wall has no bends, the in situ conduit on the narrow wall appears rather sinuous (Figure 46). In order to prevent sideways dislocation, the 80–105 cm-wide male–female jointed conduit elements were, at some time, fixed to one another by metal clamps, as is attested by fixation holes on the side and on top of the in situ conduit blocks (Figure 47a,b).

Figure 46. Delik Kemer siphon, view in flow direction. Sinuous course of the in situ pipeline (author's photo).

Figure 47. (a,b) Fixation holes in conduit stones, on the side (left) and on top of the stone elements (author's photo).

Two almost identical inscriptions, on either side of the Cyclopean wall above each of the two passages, recount of the destruction of the siphon by an earthquake in 68 CE, during the reign of Nero [39]. Patara city remained devoid of water for almost three years before the siphon was restored again (Figures 48 and 49).

Figure 48. Cyclopean wall, west view. Two passages, with inscription stone above passage on the right (flat stone). Similar inscription above left passage on the other side of the wall (author's photo).

Figure 49. Inscription stone on east side of cyclopean wall above passage (author's photo).

The inscription also recounts of the construction of a second siphon, consisting of three parallel ceramic pipelines that were installed after the earthquake as a safety measure.

Some intact 29 cm-outer diameter and 41 cm-long ceramic pipes with a wall thickness of 9.25 cm and a bore of 10.5 cm that presumably belonged to this 'security siphon' have been found in the early 1990s (Figure 50) [40]. The intact pipes are now lost, but fragments lie astray in the surrounding area.

Figure 50. Clay pipe element ('künk' in the Turkish language) of the triple siphon laid out after the earthquake as back up (Büyükyildirim 1994, 57–59).

Remarkably, an upstream section of the cyclopean wall, supporting the start of the in situ conduit elements, incorporates a large number of such fragments. Here, the wall was apparently restored using fragments of the triple ceramic siphon as construction material (Figures 51 and 52).

Figure 51. Breach in upstream section of cyclopean wall, with repair works. Arrow on the right indicates the start of the in situ conduit (author's photo).

Figure 52. Repair works of cyclopean wall, just ahead of the start of in situ conduit. Fragments of clay pipe elements applied as construction.

This may indicate that the siphon and part of the cyclopean wall was destroyed again, possibly by the 365 CE earthquake of Crete, and subsequently restored once more [41]. Conduit stones fallen of the wall were reinstalled but not all in the original order, the fixation holes no longer all corresponding, while a number of newly made conduit stones, some made from 'spolia' (reused construction material), replaced destructed ones. This may account for the rather sinuous course of the present line (and therefore susceptible to dislocation from water pressure and pressure surges) of which 182 elements over a length of over 90 m are on top of the wall today, a section that includes the siphon's lowest point.

The metal clamps were removed or stolen when the siphon was again in ruins, so supporting walls had to be constructed on either side of the line on the narrow top, some remains in situ today, to prevent sideway dislocation of the conduit stones from pressure and pressure surges (Figures 53 and 54) [42].

Figure 53. Remains of supporting wall on right-hand side of stone conduit, view in flow direction (author's photo).

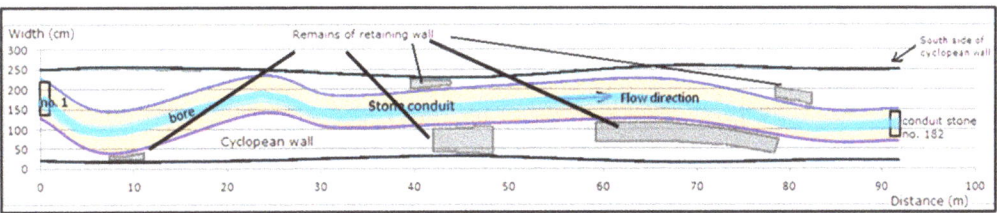

Figure 54. Schematic top view of the in situ stone conduit on Cyclopean wall, length (m) vs. width (cm), with approximate location of stone conduit and remains of retaining walls (Figure 53 refers to 2nd remains on right hand side). Flow direction from left to right. The width of the cyclopean wall varies from 230 to 260 cm. The width of the conduit stones varies from 80 to 105 cm. Average length of the conduit stones 0.5 m, with considerable variations (author's drawing).

Once in operation the siphon was of course susceptible to static pressure and pressure surges that could lead to the occurrence of conduit stones to crack, as indeed happened (Figure 39). Then the line had to be repaired, for which a exceptional technique was developed to insert new conduit stones without dismantling the line. The trick was to remove the damaged element plus the two neighboring conduit stones, and put two specially shaped elements back on either side of the opening. Then a third fitting element was slid from above into the remaining gap to have the line restored again. Such repaired spot is seen in the in situ conduit, while the special shaped stones (one broken in half) may be noted on the terrain close the Cyclopean wall (Figures 55–58).

Figure 55. Delik Kemer siphon, view from above. Insert stone in situ. Note rectangular cuttings of the two neighboring conduit stones (author's photo).

Figure 56. Specially shaped conduit stone for the insert stone (author's photo).

Figure 57. Insert stone, broken in half, reconstructed (author's photo).

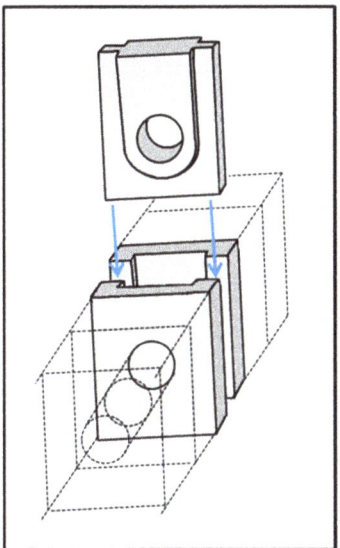

Figure 58. Reconstructed repair technique by lowering specially shaped insert stone from above (author's drawing).

A number of conduit stones show vertical holes from the top to the bore. The holes could be closed by mortared stone plugs and additionally secured with metal clamps, to prevent the plugs to blow out when operating the siphon (Figure 59). Similar holes have been observed for the siphons at Aspendos, Laodikeia ad Lykum (some stone plugs in situ), Roman Ankara (Ankara museum garden, with one stone plug in situ), Smyrna (Turkey), Hippos-Susita (Israel). The holes were probably made in relation to cleaning procedures and removing obstructions [43]. At the lowest point of the Patara siphon three holes in the side enabled emptying of the conduit (Figure 60) [44].

Figure 59. Conduit stone with funnel-shaped hole from the top. Fixation holes to secure stone plug with a metal clamp (author's photo).

Figure 60. Lowest point of the Patara siphon, west side. Side view of three conduit stones with holes to the bore for emptying the conduit. The fourth and lowest hole does not communicate with the bore (author's photo).

7. Effects from the Flow of Water

Forces exerted onto the conduit from the flow of water are generated by the friction between the water and pipe wall ('drag'), and, at bends, by the force that is needed to change the direction of flow ('inertial thrust'). Because the velocity of the water at the inner surface of the pipe wall is zero, the drag is mainly determined by viscosity of the water and roughness of the wall. Assuming that wall roughness and conduit diameter are similar all along the conduit, each pipe element will undergo a force in the direction of flow. To stay fixed the pipe elements must exert an equal force in opposite direction. This drag-force has a certain value per unit length of conduit. In an operating siphon the velocity of the water will have a fixed value, mainly determined by the head of the siphon, conduit diameter, roughness of the inner conduit surface. Because the water velocity is the same all along the line there is no increase in kinetic energy, so that the energy won by gravity between header tank and receiving tank must be equal the energy lost by drag along the conduit plus energy lost by turbulence [45]. Discarding the energy lost by turbulence, the energy won by gravity represents an upper limit of the energy lost by drag (Figure 61).

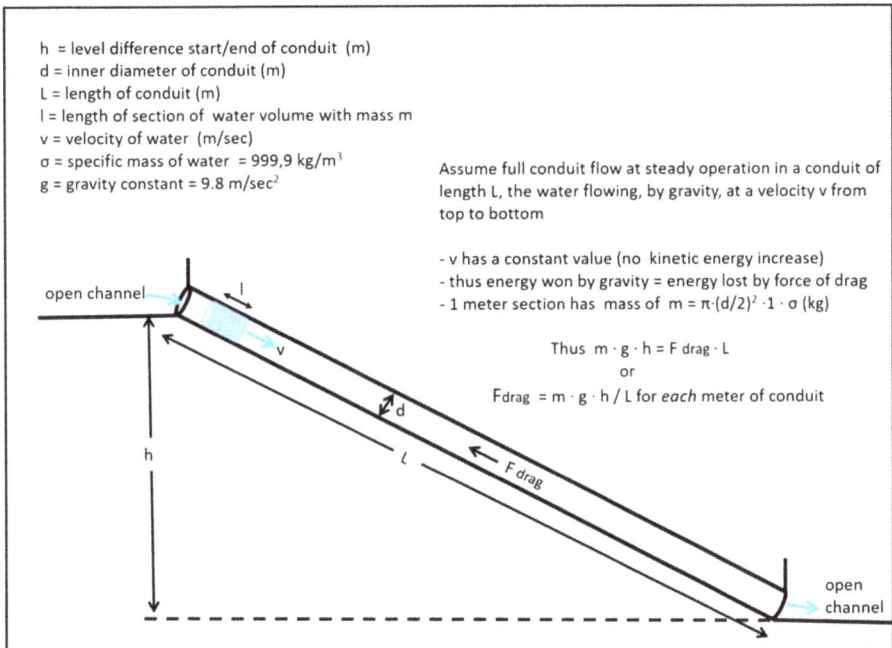

Figure 61. Forces from drag in a water conduit (author's drawing).

For the stone siphon of Aspendos, with a length of 1670 m, the header tank being 14.5 m above the receiving tank, this force is about 5.24 N per meter of conduit (diameter 28 cm). For the average length of a pipe element of 50 cm this represents a force of 2.62 N which is the equivalent of a weight of 270 g. This small force will be readily yielded by the friction forces between the heavy conduit stones and the underground.

At bends in the conduit the direction of flow changes whereby a force is exerted onto the conduit element that makes up the bend ('inertial thrust'). This force is related to the change of direction of the impulse of the water (impulse = mass times velocity) and tends to push the conduit element out of position. To be precise, this is the force that has to be exerted by the conduit element onto the flowing water to the effect that the direction of flow changes. It may be represented by a vector along the bisector of the angle of the bend, taking that the magnitude of the velocity of the water does not change (Figure 62).

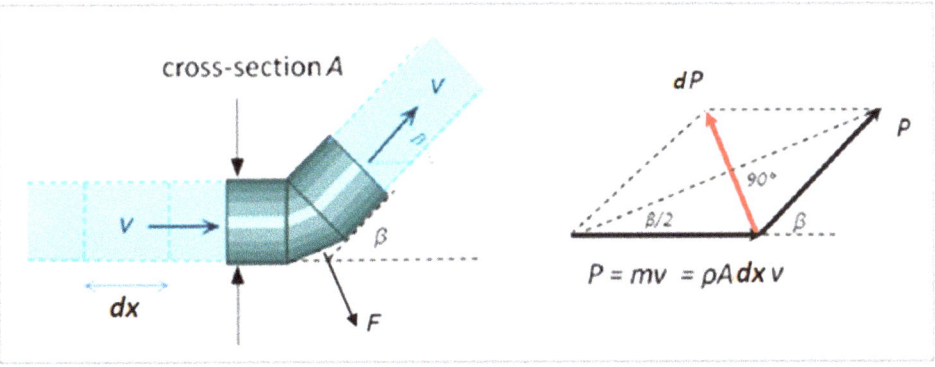

Figure 62. Forces on a bend from flow (author's drawing).

From Figure 62, it may be seen that the change in impulse dP of a volume of water A·dx (A is the cross-section of the conduit, and dx is the thickness of a corresponding slice of water) that goes around the bend equals

$$dP = 2 \cdot \rho \cdot A \cdot dx \cdot v \cdot \sin\left[\frac{\beta}{2}\right]$$

where

ρ = specific mass of water (kg/m^3);
$\rho \cdot A \cdot dx$ = the mass of the slice of water going around the bend;
v = mean flow velocity of the water (m/s);
β = angle of the bend.

The mean flow velocity v of the water in the conduit may be estimated with the formula of Darcy–Weisbach [46]:

$$v^2 = \frac{8 \cdot g \cdot \Delta H \cdot Rh}{\lambda \cdot L} = C \cdot \frac{\Delta H}{L}$$

where

g = gravitational acceleration = 9.81 m/s^2;
ΔH = difference in level between start and end of the conduit, the 'head' (m);
Rh = hydraulic radius of the conduit = D/4 for full flow;
D = inner diameter of the conduit (m);
λ = friction factor related to the roughness of the inner wall of the conduit (dimensionless, related to the irregularities of the inner surface of the pipe, to be determined from handbooks);
L = total length of the conduit (m);
C = $\left(\frac{8 \cdot g \cdot Rh}{\lambda}\right)$ which has a fixed value for a specific conduit.

The formula simply means that the longer a conduit is, the lower the velocity of the water will be, while the larger the difference in level between start and end of the conduit the faster the water will flow. The flow velocity in ancient siphons was not very high. For the 1.670 m long Aspendos siphon with its 28 cm diameter stone conduit, a wall roughness of at least some mm's (stone conduit, $\lambda \approx 0.043$) and a ΔH of 14.5 m, as well as for the 3.250 m long Madradag pressure line at Pergamon, with wall roughness less than 1 mm (lead conduit, $\lambda \approx 0.026$) and a ΔH of 41 m, the flow velocity for maximum discharge is about 1 m/sec (which is average walking speed).

This means that for a horizontal bend of 55 degrees, which is the case at Aspendos (see below for the course of the Aspendos siphon) the required force F is 57 Newtons, equivalent to a weight of about 6 kg [47]. The magnitude of this force is small compared to the forces from static pressure (as discussed above for a 30 degrees bend in the Aspendos line: 12,507 Newtons; for a bend of 55 degrees the force from static pressure would be even 22,262 Newtons, the 57 Newton being virtually nil in comparison). The friction forces between the heavy conduit stones and the foundation on which they are positioned may be assumed much larger than the force from inertial thrust. The consequence is that for ancient siphons the effects of flow can be neglected, and no measures were needed preventing the line to break apart at the bends from forces of flow. This may of course not be the case for high flow velocities that occur in modern systems.

Several authors [48] discuss oscillations of the water column occurring during rapid start-up ('sloshing') that may contribute to initial flow instabilities. The Aspendos siphon has two 'hydraulic' towers with open basins on top by which the siphon was split up into three consecutive siphons. One possible reason for the hydraulic towers would be the elimination of any entrained air pockets as well as helping to damp initial start-up flow instabilities. To possibly limit initial flow instabilities a procedure involving a very slow water filling rate was prescribed by Vitruvius in the first Century BCE. See below for the Aspendos siphon and its hydraulic towers [49].

8. The Presence of Air in the Line

There are several reasons why air in the conduit may interfere with the operation of a siphon. At the start-up of a siphon air may accumulate in air pockets at the downstream side of high points (Figure 63). These air pockets, depending on how far below the header tank they occur, reduce the pressure difference between header tank and the end of the siphon, the receiving tank. The siphon then delivers less water than envisaged, and it may even be so that the siphon does not start at all and the header tank just overflows. In deep siphons this effect is reduced as air pockets will be compressed, while some of the air may be absorbed into the water [50].

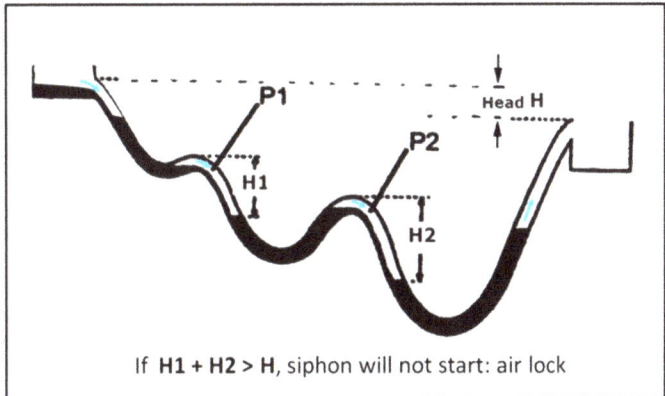

Figure 63. When filling a siphon with high points, air pockets that form may prevent start-up (H1 + H2 > H).

As seen above the Madradag siphon of Pergamon has two high points corresponding with the two intervening hills, the Çapultu Tepe end the Kaleardi Tepe. Because of the high static pressure at these points, some 140 m below the header tank and 100 m below the receiving tank, the air pockets were compressed to the extent that the discharge of the siphon was reduced to only 90% of its maximum value (Figure 64) [51].

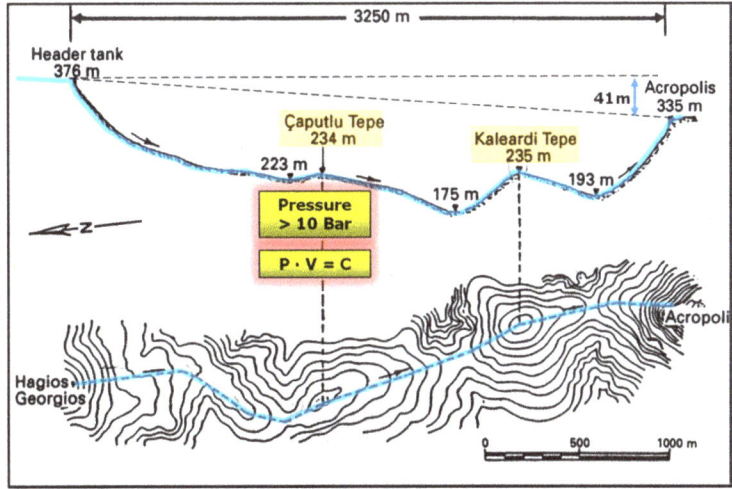

Figure 64. Strong compression of air pockets, volume reduced by factor of 10. Siphon starts at 90% capacity (after Garbrecht 2001, 116, adapted by author).

One may ask whether the engineers were aware of it: the siphon functioned as expected. But it is a problem that plays a role even today [52]. For a modern 60 km long and 2 m-diameter waste water conduit in the Netherlands functioning at a 30% capacity loss, it appeared, after 10 years of investigation, that air entrainment at the inlet was the cause of air pockets forming at local high points that reduced the debit [53].

However, there is an additional problem that has to be reckoned with. Once filled and functioning, air may be entrained into the conduit at the header tank and transported to the air pockets, enlarging their volume. The result is a further reduction of the head driving the siphon, which in the end may even lead to a total stop. The intake at the header tank of the Madradag siphon was positioned not far below the upper edge of the tank, some 10–15 cm. The tank is in derelict state today, but a 1906 photo by Gräber shows more details (Figures 65 and 66). The 1.57 m deep Hellenistic tank was originally divided in two parts by a separating wall C, a settling section B supplied by a triple clay conduit A, and a tank for the supply of the Madradag pressure line. Originally there were three openings in wall C to sufficiently supply this second section and with it the Madradag pressure line starting at E. The tank was adapted by the Romans to supply the lower city of Pergamon, adding an extra 3 holes in wall C (5 are visible in Gräber's photograph), and a wall D in order to separate the supply to the Pergamon acropolis by the Madradag line from their new waterway to the lower city (with two Roman ceramic pipe lines starting at F) [54].

Figure 65. Header tank of the Madradag siphon (adapted from Gräber 1906). A, arriving triple clay pipeline. B, settling compartment. C, separating wall with holes. D, dividing wall. E, start of Madradag siphon. F, orifices for a later Roman twin pipeline.

Figure 66. Header tank of the Madradag siphon, condition 2021 (courtesy Murat Baykent).

The upper side of the 18 cm-wide conduit of the Madradag siphon was located just 14–15 cm below the water level in the tank, as set by an overflow ridge. Entrainment of air into the line could not be avoided here. Moreover, entrainment of air in the header tank into the siphon conduit was common practice because the intake in the wall of the header tank was, as a rule, positioned at or not far below the water level in the tank [55]. When entering the conduit, the water may form a vortex, pulling air bubbles into the conduit (compare the emptying of a bathtub) (Figures 67 and 68). The deeper the intake is below the free water surface, the less readily vortices will form.

Figure 67. Vortex pulling air bubbles (https://www.pinterest.com/pin/48835977186173061/, accessed on 17 December 2020).

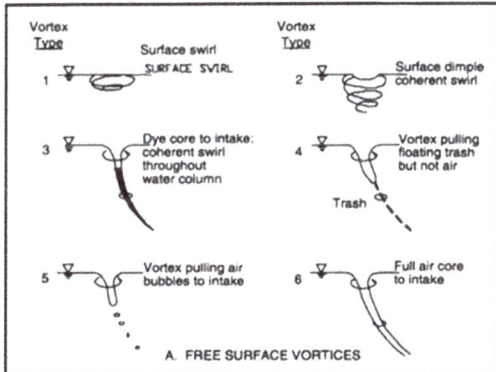

Figure 68. Types of free surface vortex forming (Amin 2018, Figure 1). Type 5 and 6 are the cause of the entrainment of air.

The upper side of the 18 cm wide conduit of the Madradag siphon was located just 14–15 cm below the water level in the tank as set by an overflow ridge. Entrainment of air into the line could not be avoided here. Moreover, entrainment of air in the header tank into siphon conduits was common practice, because the intake in the wall of the header tank was as a rule positioned at or not far below the water level in the tank [55]. When entering the conduit the water may form a vortex pulling air bubbles into the conduit (compare the emptying of a bathtub) (Figures 67 and 68). The deeper the intake is below the free water surface, the less readily vortices will form.

There is a minimum submergence required to avoid vortex formation with entrainment of air (the so-called 'submergence law') [56]:

$$S = D + c \cdot D^{1/2} \cdot v$$

where
S = minimum submergence to avoid vortexing (m);
D = diameter of inlet pipe (m);
v = velocity of water in conduit;
c = a constant.

For the Madradag siphon, and also for the siphons of Lyon (see below) and Aspendos for instance the submergence was much less than required (Figure 69). Once the siphons were running, air thus entered the conduit together with the water and moved down with the flow.

	submergence (cm)	
	needed	estimated
Pergamon	75	15
Lyon	80	30
Aspendos	100	20

Figure 69. Required and estimated submergence for some siphons.

There is archaeological proof that air was indeed entrained in siphons. A conduit stone of the in situ remains of the twin siphon at Laodikeia ad Lykum (Turkey) shows substantial incrustations. The sinter (calcareous incrustation) is thickest on the sides but less thick at the top (see Figure 70). That is because at the top side entrained air has prevented sinter deposits to a considerable extent. At the bottom there is a groove, where sinter building was reduced because of debris rolling with the water [57].

Figure 70. In situ conduit stone of the twin siphon at Laodikeia a/L. Bore 40 cm, *sinter* on the sides 2 × 11 cm wide.

At Aspendos a few con¬duit stones are still lying along the siphon's course. In 1890 Lanckoronski noted about one of them: 'Einen der Wasserleitungsquadern ... $\frac{3}{4}$ von Sinter gefüllt ... die gliebene Öffnung gleich einem Dreieck mit gerundeten Ecken und eingezogenen Seiten' [58], ('One of the conduit stones ... for $\frac{3}{4}$ filled with calcareous incrustation ... the remaining bore like a triangle with rounded edges and caved-in sides'). A conduit stone with this remarkable sinter was indeed found in the 1990's (Figure 71). One side of the triangular sinter shows accumulation of debris that was caused by some malfunctioning at the header tank, subsequently covered again with a thin layer of sinter. Obvi¬ously that must have been the bottom side of the conduit element. Thus the top side of the triangle was at the upper side, and again, entrained air, always at the top of the conduit, had prevented deposition of sinter to some degree.

Figure 71. Bore of stone conduit of Aspendos siphon, triangular form of *sinter* deposits.

For the Madradag siphon air will be entrained at the header tank and transported to the air pockets at the high points formed when filling the siphon, enlarging their volume. Yet, at the downstream side of the air pockets there is a transition of a partly filled conduit (the water passes underneath the air pocket) to full conduit flow. At this point, of considerable turbulence, air may again be entrained further down the conduit with the water flow, reducing on its turn the volume of the air pockets. Whichever process is more important determines what will happen [59]. The conduit of the Madradag siphon at the header tank had a slope angle much larger than downstream of the high points, 18 degrees vs 6 and 8 degrees [60]. Therefore air was less readily transported down the conduit at the header tank but rather more easily at the downstream side of the air pockets at the hill tops. This means that after start-up the air pockets became depleted and the siphon developed to full capacity on its own. Whether the designers were aware of this phenomenon is questionable, but the siphon operated as expected and brought water to the acropolis on top of the hill, no doubt to the amazement and wonder of her people.

A quite different situation existed for one of the siphons of the Yzeron aqueduct of Lyon (Figure 23). The Grezieux-Craponne-Lyon siphon of the this aqueduct (not to be confused with the Yzeron-siphon of the Lyon's Gier aqueduct) has a high point in the line, that could not be avoided because of the topography (Figure 72) [61]. But here the sloping of the terrain just downstream of the high point is much steeper than at the header tank. Air would thus readily be entrained at the header tank but only sparingly from the air pocket that had formed at the high point when filling the siphon. In case a closed conduit would run along the entire trajectory including this high point (alternative course in Figure 72), the siphon would start up at only 60% of its maximum discharge. And that, because of air increasingly accumulating at the high point, the siphon would after some time come to a complete and definite stand-still.

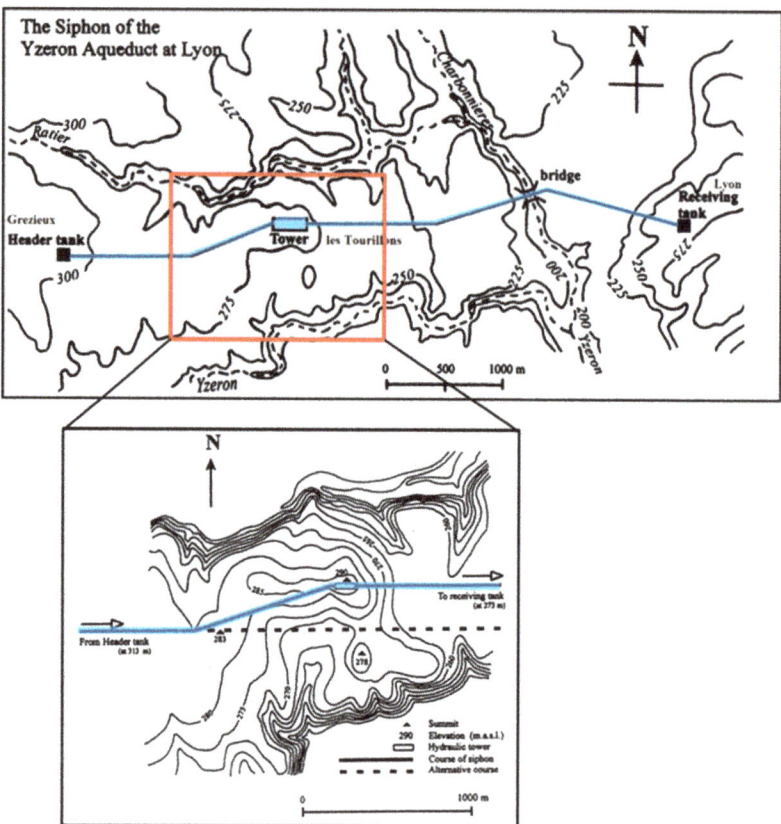

Figure 72. The Grezieux-Craponne-Lyon siphon of the Yzeron aqueduct of Lyon. Inset, with alternative course (author's drawing).

The Romans solved this problem building a 16 m-high tower on the nearby hill, with sloping ramps and with an open tank on top. The conduits (the siphon consisted of a number of parallel lead pipes) discharged into the open tank where formation of an air pocket at the high point was prevented, when filling the siphon, and where, when operating the siphon, air entrained at the header tank was released. By means of this 'hydraulic tower' the siphon was divided into two subsequent siphons, one with a length of 2200 m (Grezieux-Craponne) and the next 3600 m long (Craponne-Lyon). Regarded as a single siphon it is with 5800 m one of longest from Roman time (Figures 73 and 74).

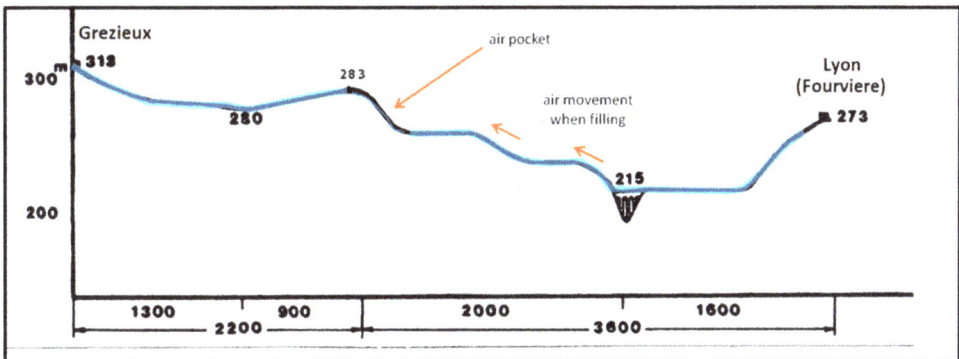

Figure 73. Profile of the Gezieux-Craponne-Lyon siphon along alternative course of Figure 72. High point at elevation 283, accumulation of air at high point when filling the siphon. Initial discharge 60%, evolving to a total standstill from air entrained at the header tank (adapted from Burdy 1991, Figure 59).

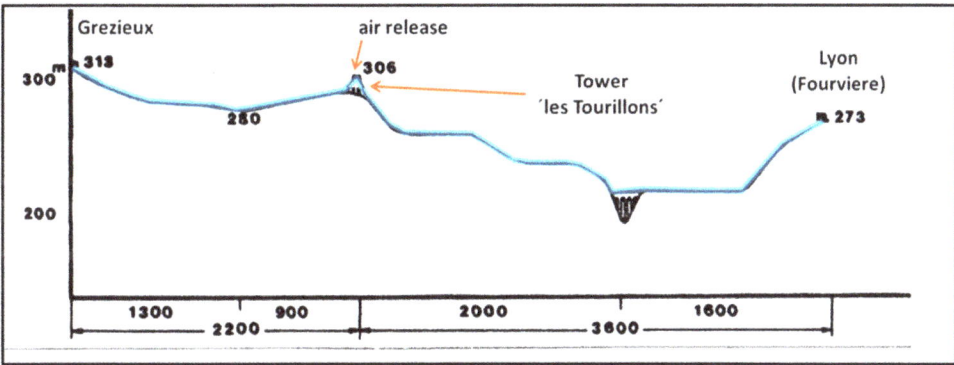

Figure 74. Profile of the Gezieux-Craponne-Lyon siphon. Hydraulic tower 'les Tourillons de Craponne' to release the air at the high point (adapted from Burdy 1991, Figure 59).

Of the hydraulic tower two impressive piers have been preserved today, locally known as 'les Tourillons' (Figures 75–77). Remnants of further piers are still visible at ground surface. According to Jean Burdy, who investigated Lyon's aqueducts, the top of the Craponne tower would not correspond exactly to the overall hydraulic gradient line between header tank and receiving tank at Lyon but some distance above it. A gradient of 7 m between Grezieux and Craponne for a distance of 2200 m (3.2 m/km), and 33 m between Craponne and Lyon for 3600 m (10.3 m/km). Thereby the level difference between the container on top of the tower and the receiving tank at Lyon, of the longer and more problematic section of the siphon, was increased, adding to its capacity [62].

Figure 75. Les Tourillons de Craponne, 1910.

Figure 76. Les Tourillons de Craponne, early 21st century (author's photo).

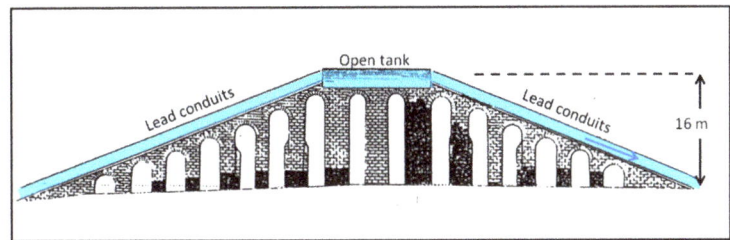

Figure 77. Reconstruction of les Tourillons de Craponne (Burdy 1991, 94 Figure 55, adapted by author). Burdy envisages from the width of the remains of the towers that 6–7 lead conduits, 25 cm in diameter, ran parallel up to the tank and down again (Burdy 1991, 102).

The conduits of the Lyon pressure lines were made of lead: Cat.I siphons. This means that horizontal bends in the line, present in the Grezieux-Craponne-Lyon siphon, do not constitute danger points in the line from pressure or pressure surges. This is not so much the case for the Cat.II siphons as discussed above. Additionally, some Cat.II siphons do have horizontal bends. Therefore, the question is, what are, apart from static pressure, further causes of pressure problems. As discussed, entrainment of air into siphons was unavoidable, sometimes leading to problems at start-up or even to a standstill, due to air pockets at high points. However, any siphon, when running, will have a mixture of air and water flowing in the line because of the entrainment of air at the header tank. What type of phenomena may occur is determined by the conduct of this mixture of air/air bubbles and water in a running conduit.

9. Air in Water Conduits

The behavior of air bubbles in conduits transporting water is related to the size of the air bubbles, the diameter of the conduit, the slope of the conduit, the velocity of the water in the conduit, the roughness of the inner wall of the conduit, and the viscosity of the water (Figure 78). These represent a large number of variables that, nonetheless, may be condensed to a single formula. For a conduit sloping down at an angle α, an air bubble will be transported with the flow if the flow velocity is larger than a so-called critical value Vcr. It can be deduced that [63]

$$\text{Vcr} = \sqrt{4 \cdot g \cdot \text{Db} \cdot \frac{\sin \alpha}{3 \cdot \text{Cb}}}$$

where
  g = gravitational acceleration = 9.81 m/s^2;
  Db = bubble diameter (m);
  α = slope angle;
  Cb = 'drag coefficient' of air bubble; for convenience, the value of Cb is often set to 1.

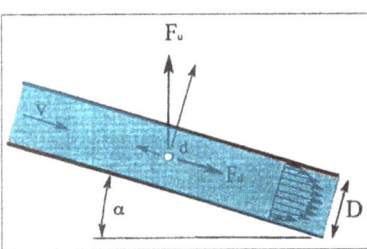

Figure 78. Air bubble in downward-sloping conduit with water flow. The air bubble will be stationary (but will tend to rise to the top of the conduit) if the drag force from the flowing water is compensated by a component of the rising force of the air bubble against the direction of flow (Fd = Fu · sin(α)).

From the formula, it can be seen that the larger the air bubble, the faster the water must flow to transport it. Additionally, the steeper the slope, the less readily air bubbles will go with the flow, which is all sensible enough. As air bubbles tend to accumulate at the upper side of the conduit, one must correct for the fact that the flow velocity near the conduit wall is lower, which has a greater effect for small bubbles than for large bubbles. The result is that the mean critical velocity Vm,cr above which small bubbles close to the conduit wall will go with the flow is higher than for larger air bubbles, and that Vm,cr is related to the diameter of the air bubbles, to the diameter of the conduit, and to the wall roughness [64]:

$$Vm,cr = ((\log(3.4 \cdot Dc/k)/(\log(15.1 \cdot Db/k)) \cdot (4 \cdot g \cdot Db \cdot \sin\alpha/3)^{1/2}$$

where
   Db = diameter of air bubble;
   Dc = diameter of conduit;
   k = wall roughness;
   g = gravitational acceleration = 9.81 m/s^2.

The formula is rather complex, but things become increasingly so when air bubbles coalesce to form large air pockets. It is known from experiments that large air pockets, also called 'slugs', rise faster in conduits sloping up than in vertical conduits (large air pockets rise vertically faster than small air bubbles anyway), with a maximum for a slope of about 40 degrees (Figures 79 and 80) [65].

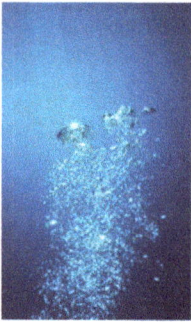

Figure 79. Rising air bubbles in water (HTML ZSi5R2Rg, May 2021).

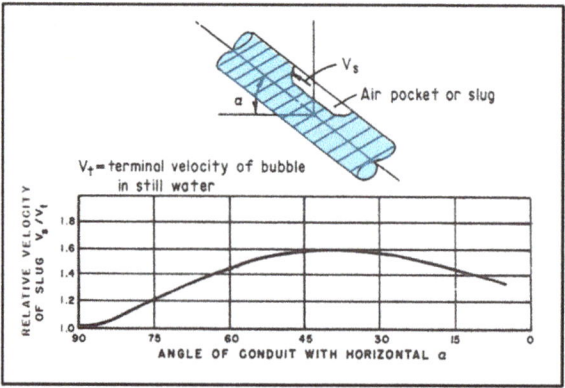

Figure 80. Velocity of large air pockets in conduits sloping up is larger as compared to vertical conduits. Maximum at a slope of about 40 degrees (after Falvey 1980, 52).

In conduits sloping down with small air bubbles moving with the flow, slugs may move upward against the flow, a process called 'blow back'. The slugs may at their front side collect small air bubbles that move with the flow, increasing slug size. At the downstream end of the slug, a change to full conduit flow occurs, a point of high turbulence (a 'hydraulic jump'), where small air bubbles may be entrained again with the flow thereby reducing slug volume [66]. The created air bubbles there, moving down with the flow, may coalesce to form slugs, on their turn rising against the flow (Figure 81). The behavior of air bubbles and slugs as function of the slope angle and flow velocity is given in Figure 82. Note that for certain slopes and flow rates air bubbles move with the flow, but large air pockets and slugs move against the flow.

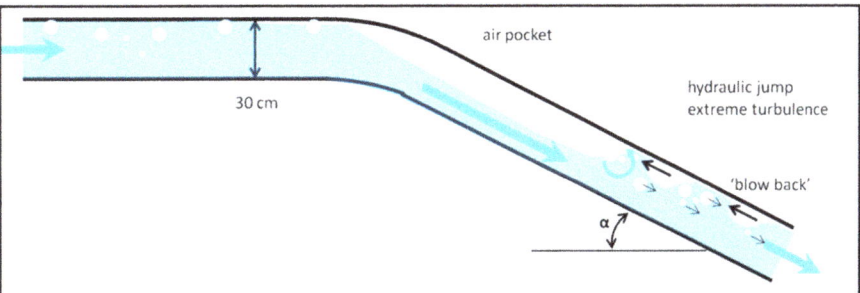

Figure 81. Air bubbles transported with the flow may form an air pocket at a high point or at a change to a steeper slope. The water then passes below the air pocket at a high velocity. At the end of the air pocket, a change to full conduit flow occurs, with extreme turbulence: a hydraulic jump. Air bubbles are here entrained again and go with the flow, at some point coalescing to slugs that may rise up against the flow.

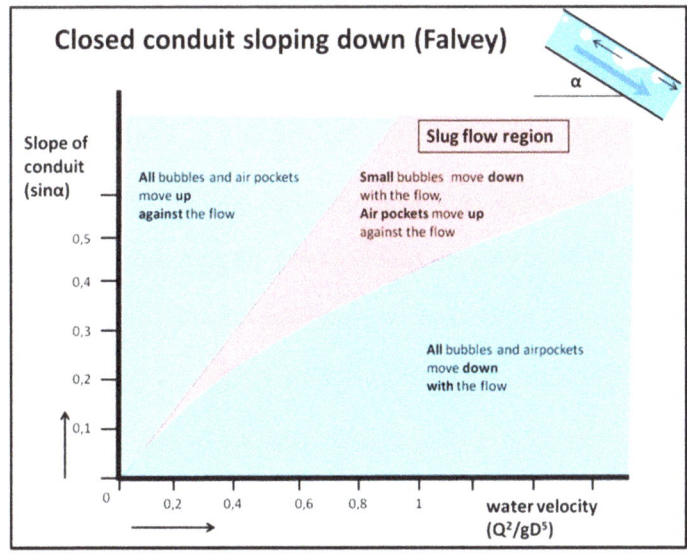

Figure 82. Conduct of air bubbles and air pockets (slugs) as function of downward slope angle The problematic range is the 'Slug flow region' where small air bubbles move with the flow, and slugs move against the flow (Falvey 1980, 29, adapted by author).

This phenomenon plays a role at an uncontrolled start-up of siphons, when there is much air in the line. While air pockets/slugs may rise against the flow in a conduit sloping down, they will rise faster than the flow in a conduit rising up. In such situation water is forced to flow back underneath the air pocket, giving rise to pressure surges in the entire line. And all siphons have a section rising up. It leads to back flow of water with extreme turbulence, irregular water discharge, and pressure surges (Figure 83).

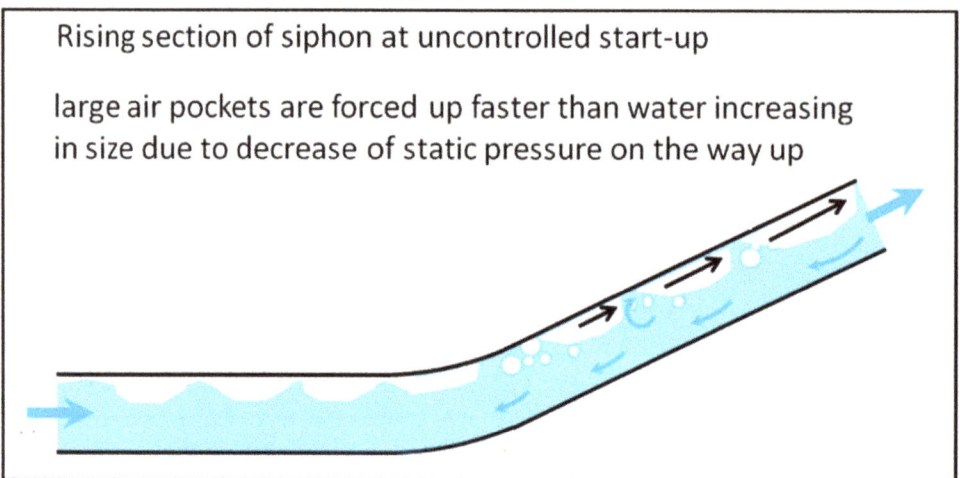

Figure 83. Air pockets in rising section giving rise to back flow of water, extreme turbulence, irregular water flow, irregular discharge, and pressure surges.

Pressure surges develop by this proces in the horizontal part of the siphon as well, with surprisingly rapid moving spray plugs alternated with stagnant flow, and forceful expulsions of air and water at the end of the conduit, endangering the integrity of the line, as has been shown by large scale experiments in the late 1990's at the Delft Hydraulic Laboratory of Deltares Institute at Delft in the Netherlands

In one of the Deltares experiments a 500 m siphon was constructed from 15 cm diameter steel pipes. For convenience the conduit was laid as two parallel lines with a U-turn at 250 m (Figure 84). At three locations Perspex windows were installed to be able to inspect the flow, located at the start and the end of the horizontal section, and 50 m upstream from the end. At the very end of the siphon the conduit consisted of a reinforced flexible hose of similar diameter that sloped up and was laid across the 2 m high edge of a steel cargo container which served as receiving tank. Difference in height between inlet and outlet ('header tank and receiving tank') was about 8 m (Figure 85). When water was flowing, air could be introduced into the line close to the inlet.

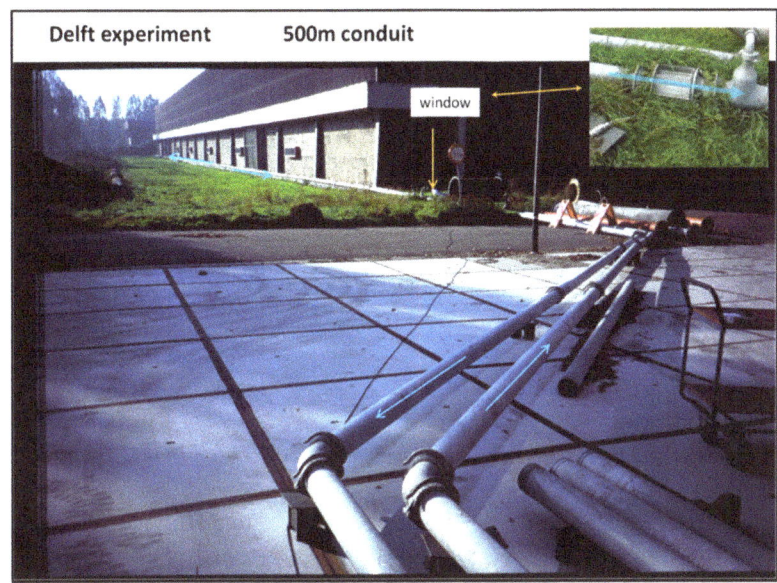

Figure 84. Delfares experiment at Delft, 500 m steel conduit with a 15 cm diameter, overview.

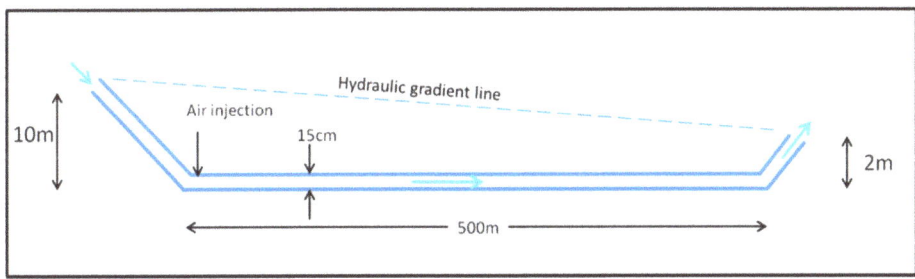

Figure 85. Delfares experiment, 500 m steel conduit with a 15 cm diameter, a siphon with a head of 8 m.

Small amounts of air introduced into the conduit formed air bubbles went with the flow and duly came out at the other end together with a steady water stream. But with larger inputs of air the flow became irregular, with periods of extremely turbulent air-water flow ('spray plugs') moving rapidly downstream in the horizontal part as observed through the windows, alternated with periods of almost standstill in a partly filled conduit with a distinct water-air interface.

At the receiving tank water was expelled abruptly and at high velocity in periods, or not at all, the heavy reinforced flexible hose across the 2 m high edge of steel 'receiving tank' experiencing major pressure transients, moving up and down with high force (Figure 86). Yet, when this hose was laid flat on the ground at the level of the horizontal conduit, the flow became regular, the water, with the air above it, flowing out in a continuous and regular fashion.

Figure 86. Deltares experiment, 'receiving tank'. Flexible reinforced hose laid over the edge of a steel cargo container. With much air in the line, the hose moved violently up and down from pressure surges, while the discharge of water was extremely irregular. Once the hose was laid out flat on the ground, the pressure surges ended and the discharge became steady.

It shows that it is the rising section of siphons, and all siphons have a rising section, in combination with the presence of air, which is the cause of problems. Especially the Category II siphons, with their conduit elements sealed with the weak expanding mix, are at risk due to such pressure surges, at rapid start-up, as well as during operation from abundant entrainment of air at the header tank, and that especially at horizontal bends [67].

10. Leaking Conduits

Air in pressurized conduits may be a further cause of pressure surges: water hammer from air escaping through leaking spots. Water hammer may be defined as a 'pressure surge due to a substantial and sudden change in the velocity of the water', for instance, by rapid closure of a valve. The water hammer effect is the driving principle of the 18th-century invention of the 'hydraulic ram'. The hydraulic ram applies a self-repeating automatic shutter that closes a valve to recurrently create a short pressure increase. By means of a non-return valve, the pressure increases may add up to the extent that part of a water stream that comes down from a height H may be lifted to a height ten to forty times H [68].

Ancient siphons were not fitted with valves or shutters. However, water hammer may also be the result of just the presence of air in the line, that is, from air escaping through leaking spots [69]. If, in a pressurized conduit, a compressed air bubble transported with the flow passes a leaking spot (at the top of the conduit), air will be released out of the conduit. Because the compressed air escapes much faster than an equal volume of the much heavier water, the water column behind the air bubble will be accelerated as long as air escapes. When the air bubble is depleted or has passed the leaking orifice, water will leak out again at a lower pace, and the water flow in the conduit will be decelerated, with a pressure surge/shock wave as a result (Figure 87). The magnitude of the pressure surge that is caused by this can be estimated with the so-called Joukowski law [70]:

$$dH = 0.5 \cdot c \cdot dV \cdot g^{-1} = C_1 \cdot dV$$

where
  dH = pressure increase in meters of water column;
  c = sound velocity in water $\approx$ 1000 m/s;
  dV = difference in flow velocity of the water upstream from the leaking spot just before and just after the air escapes out of the conduit (m/s);
  g = gravitational acceleration = 9.81 m/s^2;
  $C_1 = 0.5 \cdot c \cdot g^{-1}$.

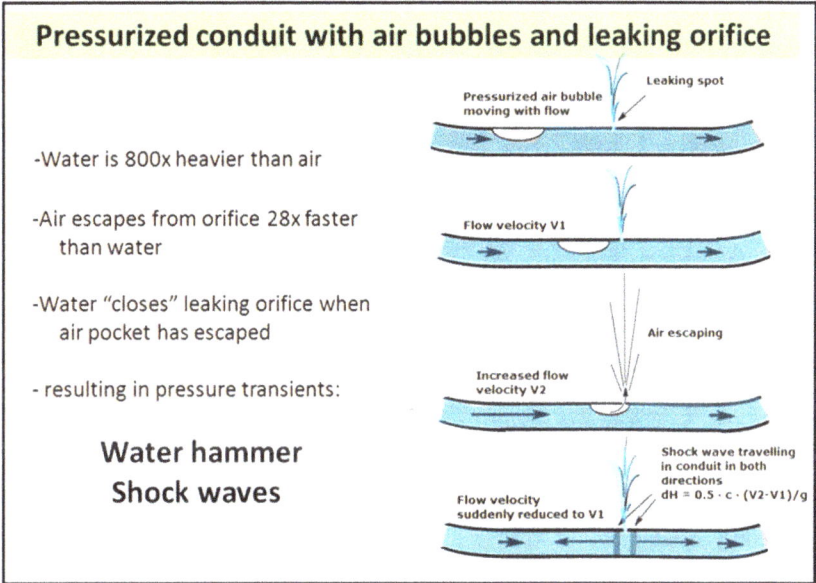

Figure 87. Pressure surges from air escaping through leaking spot.

The pressure surge is directly related to dV: the larger the difference in flow velocity, the more forceful the pressure increase will be. dV may be determined from continuity arguments, as the decrease in the volume of air in the conduit is related to the compressed air escaping through the leaking spot:

$$dV \cdot A_c = V_a \cdot A_h$$

where
  A_c = cross-section of the conduit in m^2;
  V_a = velocity of the compressed air that escapes through the leaking spot in m/s;
  A_h = cross-section of the leaking spot in m^2.
  V_a is set by characteristics of air flow through small orifices under high pressure [71]. For instance, for a leaking orifice 12 mm wide and conduit pressure of 40 m of water column (as for the Aspendos siphon, a pressure of about 400 kPa) the air flow through the orifice is about 0.05 cubic meters per second. Thus dV = $V_a \cdot A_h/A_c \approx$ 0.8 m/sec, hence dH = 0.5 · 1000 · 0.8/9.81 = 39 m of water column.
  This means that for the Aspendos siphon, 40 m deep, and apart from factors that may have a reducing effect such as the presence of more air bubbles in the conduit, water

hammer from escaping of air through a leaking spot of 12 mm diameter leads to a sudden pressure increase—a pressure surge—of almost 100% [72].

Did Category II siphons have leaks? Leaking spots could not be avoided in stone pipe lines—in contrast to lead conduits—because when installing the pipe elements, the joints were covered with the expanding mixture after which the elements were pushed against each other, the joints disappearing from view. A rigorous check to see whether the joints were watertight was not possible. Only after the entire siphon had been finished and put to the test the water tightness became clear. At Aspendos calcareous deposits (sinter) hanging down from one of the arches of the over 500 m long siphon bridge indicate that the siphon must have leaked considerably. At the twin siphon of Laodikea ad Lycum a mass of sinter can be found at several locations on the sides of in situ conduit stones adhering from their top, a sign of leaking spots at the upper side of the bore (Figures 88 and 89).

Figure 88. Calcareous incrustation/*sinter* hanging down from arch of the 500 m siphon bridge at Aspendos (author's photo).

Figure 89. Left-hand line of win stone siphon at Loadikeia a/L. Yellow line: sinter mass on the side from top to bottom, from a leak at the top of the bore (author's photo).

Inside the conduit the resulting pressure surges traveled both upstream and downstream, causing a sudden increase of the forces that tend to push pipe elements that make up a bend out of position, on top of the forces from static pressure. Reflection of pressure waves on the interface between air and water (from air bubbles and slugs) may cause pressure waves to be superimposed resulting to even more forceful pressure surges. Such pressure surges may occur repeatedly, in the end exceeding the forces that keep the conduit intact, especially at bends. A minor displacement of a Category II conduit element at a bend could result to cracking of the sealing material and the occurrence of an additional leak, whereby the inflow into the conduit at the header tank is enhanced, and with it the entrainment of air, so that pressure surges/water hammer effects would occur more frequently. In the end the leaks could get such that water entered the conduit faster than the incoming aqueduct supplied, so that large slugs would periodically form and move with the water flow, to the total wreck of the siphon.

The Category II siphons with their weak cemented joints between pipe elements are at risk, especially at horizontal bends, from static pressure, from pressure surges at (uncontrolled) start-up, from air entrainment at the header tank during operation (pressure surges and air pockets at high points reducing discharge if air is not released), and from pressure surges by air escaping through leaking spots. This is not so for the Category I siphons, with conduits made of lead pipes soldered together, the joints being as strong as the pipes themselves. Leaking joints of lead conduits could easily be repaired from the outside, while at high points air was released if necessary (les Tourillons). In modern pressure lines automatically operating air release valves may thus result to—at times detrimental—water hammer effects, which effects may be reduced by proper dimensions of the orifice through which air escapes.

11. Cat.II Siphons with Horizontal Bends

As seen above the Category II siphon of Patara, without horizontal bends, had its own problems, some of which occurred during operation from pressure problems, as the ingenious method to exchange broken conduit elements illustrates (Figure 58). The question is, are Category II siphons known having horizontal bends? Yes. Two examples: Aspendos on the south coast of Turkey and Gades (Cadiz), on the south coast of Spain.

The stone siphon the siphon at Aspendos (50 km east of Antalya) has two horizontal bends, of 16 degrees and of 55 degrees [73]. At each bend a 40 m high tower was constructed, with sloping ramps and an open tank on top (Figures 90–92). But unlike the situation at les Tourillons de Craponne there are no natural high points along the course of this 1670 m siphon. Yet the stone conduit was just as well led over sloping ramps to the open tank on top of the two subsequent towers.

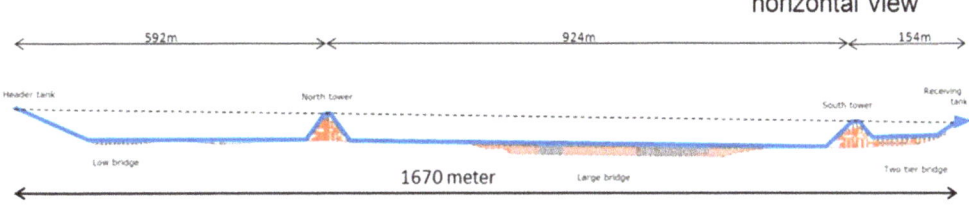

Figure 90. Profile of the Aspendos siphon (author's drawing).

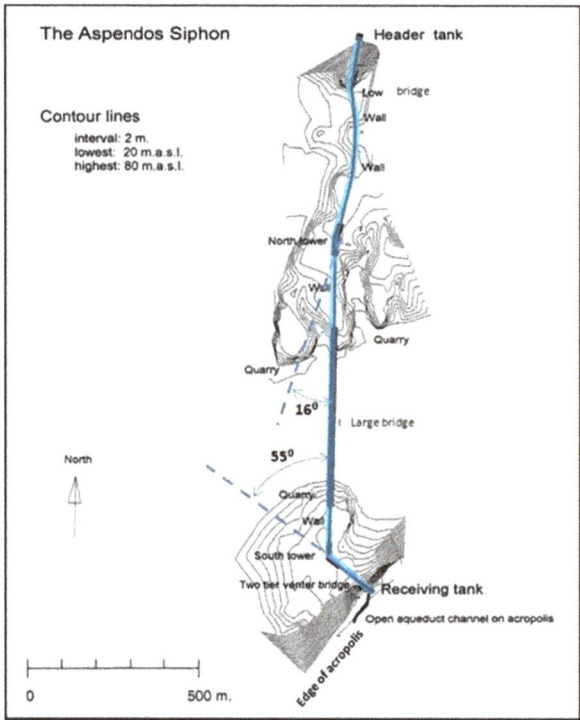

Figure 91. Aspendos siphon, horizontal bends of 16 and 55 degrees (author's drawing).

Figure 92. View from acropolis edge to the north. Remains of south tower in front; 15 m high large bridge and north tower plus header tank on the hill in the back (author's photo).

The towers, north of the acropolis, are among the highest buildings of Roman times. They constitute horizontal bends in the siphon's course, 16 degrees for the 'north tower', and 55 degrees for the 'south tower' (Figure 91). The reconstructed level of the receiving tank stands at 14.5 m below header tank. In case a closed conduit would have run over the towers, at their present height of 28 m, the siphon would not start up because of air pockets forming at these high points when filling the siphon. Originally the towers must therefore have been higher, and have been equipped with open tanks to release the air, and, consequently, the open tanks were positioned along the hydraulic gradient line between header tank and receiving tank [74].

The tanks on top had to be accessible on behalf of maintenance and repair, for which the staircase had been constructed in the central part of the towers. Today the stairs are accessible to a height 15 m giving an impressive view over the surroundings (Figure 93). However, there are no natural high points in the siphon's course, so there is no basis to build such enormous towers with open tanks on top in order to release air to guarantee flow as was the case for the Grezieux-Craponne-Lyon siphon. Yet, the towers of Aspendos were built for a reason that also relates to air.

Figure 93. Aspendos siphon, north tower, with entrance to stairs in central part. Right: person standing above arch (arrow) (author's photo).

By the towers the siphon was split up into three consecutive siphons, effectively taking the bends out of this Category II siphon. This prevented damage at the bends from static pressure, and, more important, from pressure surges during rapid start-up and the unavoidable entrainment of air at the header tank during operation, and from the possible water hammer effects from air escaping through leaks. The choice of the Roman engineer to design the siphon with horizontal bends led to the construction of the hydraulic towers. The towers had to be fitted with open tanks on top that had to reach up to the hydraulic gradient line and reduce static pressure to zero at the bends. From topographical arguments it can be deduced that the siphon, with two horizontal bends and with the two enormous towers, was cheaper to build than a siphon that went in a straight line to the acropolis, without bends and without towers, but requiring a much longer 15 m high siphon bridge [75].

The 80 km aqueduct of Gades (present day Cadiz, on the south coast of Spain) was equipped with two siphons, the 3.5 km long and 50 m deep 'de los Arquilles' siphon, and the 19.5 km low pressure 'sifón de la Playa' that had its end point in Gades itself (Figure 20) [76]. The de los Arquilles siphon, made of an estimated 11,500 conduit stones having a 30 cm perforation, had two intermediate 'hydraulic towers' where the line made horizontal bends of 11 and 13 degrees (Figures 94 and 95). In contrast to the Aspendos siphon the Gades towers are positioned on top of hills.

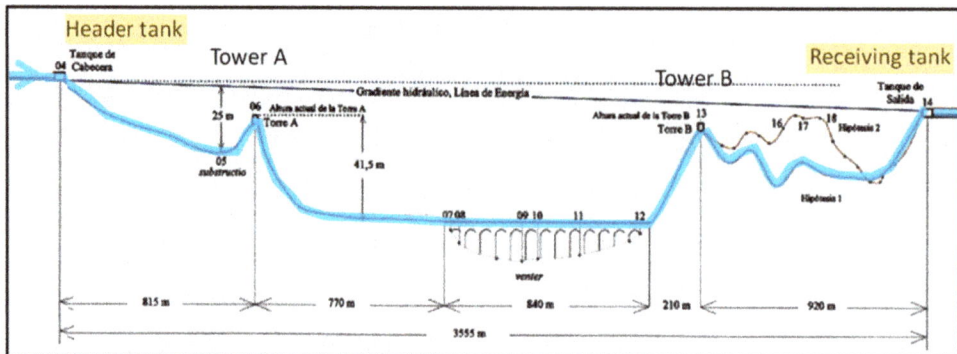

Figure 94. Profile of the *de los Arquillos* siphon of Gades. Water flow from left to right. Towers A and B on top of intermediate hills at about 10 m below the hydraulic gradient line (adapted from Pérez 2012, Figure 6).

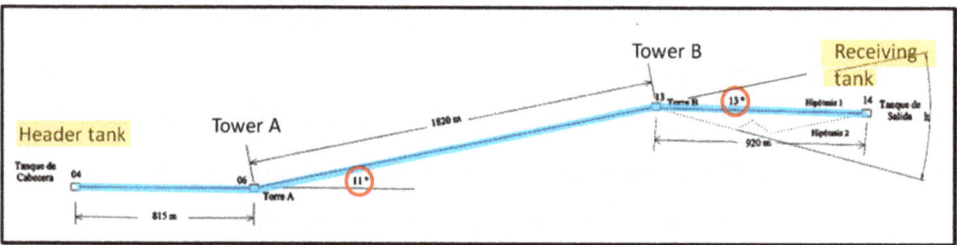

Figure 95. Plan of the *de los Arquillos* siphon of Gades. Bends at tower A (11 degrees) and tower B (13 degrees) (adapted from Pérez 2012, Figure 6).

There is some uncertainty about where the siphon started and where it ended, but according to the present state of knowledge the towers were probably 14 and 10 m high, and must have been equipped with open tanks on top to have the siphon start up at all, similar to the situation at the Grezieux-Craponne-Lyon siphon as discussed above, and at the same time prevent impairment of the siphon due to pressure surges at uncontrolled start-up and during operation from air entrained at the header tank. The present remains of the tower are some 3–4 m high (Figures 96 and 97). Between the two towers the conduit was carried for 840 m on a 15 m high bridge across the deepest part of the valley linking the hills with the towers. The route between the Tower B and the envisaged receiving tank appears problematic because of high points in the terrain, but that part has not been fully investigated yet and may be subject to future research.

So there are two situations where one of more 'hydraulic towers' to release air were built in siphons incorporated in aqueducts. First to release air because of air pockets forming at high points, during start up and due to entrainment of air at the header tank, that may prevent the siphon to function or to develop to a total stand-still (siphons of Grezieux-Craponne-Lyon and Gades). And second to prevent wrecking of Category II siphons at sharp bends in the line (Aspendos and Gades) because of pressure surges at (rapid) start up, and during operation because of the presence of air and entrainment of air in the line. Although the Grezieux-Craponne-Lyon does have sharp bends in the line, these bends are not to be regarded as danger points because the conduits were made of lead soldered together (Category I). These phenomena were evidently known to the Romans and their predecessors, and have been described by Vitruvius in the First Century BCE.

Figure 96. Remains of tower A of the *de los Arquillos* siphon of Gades (author's photo).

Figure 97. Remains of tower B of the *de los Arquillos* siphon of Gades (author's photo).

12. Vitruvius

The Roman author Vitruvius (first Century BCE) treats in book VIII of his Ten Books on Architecture ('De Architectura Libri Decem') the means for water conveyance of his days [77]. Translations of Vitruvius' manuscript are available in several languages [78]. Vitruvius also describes the technique of siphons and the problems that may occur, and mentions proposals of how to solve these problems. He discerns between conduits made of lead with soldered joints (Category I) and those made of ceramic pipes and stone elements (Category II). He explicitly advises to fill siphons carefully and slowly, as otherwise a 'very strong air(pressure)' ('vehemens spiritus') may arise that endangers the line (as we have seen above). Before starting Category II siphons Vitruvius recommends to introduce ashes into the conduit to seal possible leaking spots (and thus prevent water hammer from air escaping). This sealing procedure, based on the principle of dry organic material expanding when moisturized getting stuck in the leaking orifice closing it off, has survived into our time as a recipe for mending leaking car radiators. For the ancient siphons it was not the loss of water that was the problem, but the water hammer effects and pressure surges from air escaping that could endanger the line. Not without reason does Vitruvius advise to strengthen the bends of Category II siphon with bands or sand ballast (not so for Category I siphons), and—for vertical bends—to have the bend made of large stone blocks of special quality, 'ex saxo rubro', undoubtedly known for its weight and strength [79].

Furthermore, Vitruvius recommends to install 'colliviaria' in the line, means to release air from the conduit: 'colliviaria facienda sunt, per quae vis spiritus relaxetur' ('colliviaria are to be installed, by which the force of air will be reduced') (Figure 98) [80]. The expression colliviaria (usually emended to colliquiaria) does not occur elsewhere in all of Latin literature (such expression is known as a 'hapax legomenon'), so that its meaning is not evident which had led to many speculations [81].

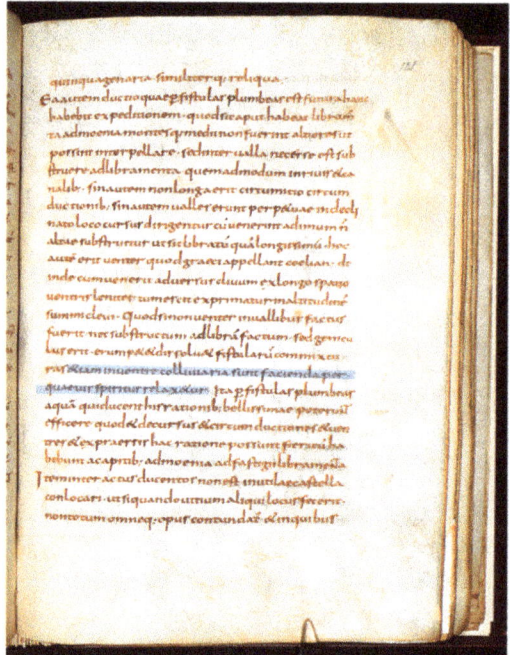

Figure 98. Excerpt of MS Harley 2767, Vitruvius' *De Architectura Libri Decem*, book VIII, 6, 5–6, section with *colliviaria* indicated.

From the hydraulic arguments above it will be clear that Vitruvius refers to provisions to release air from the conduit at a high point to guarantee a continuous water flow. Examples: les Tourillons for the Grezieux-Craponne-Lyon siphon of the Yzeron aqueduct and the two towers for the de los Arquilles siphon of the Gades/Cadiz aqueduct at the south coast of Spain, and also the towers for the Aspendos siphon.

On the subject of siphons Vitruvius has been criticized for not understanding what he was writing about. A.T. Hodge 1992: 'his siphon account ... does not show any real understanding on the part of the author'. M. Lewis 1999: Vitruvius' book VIII is '... bitty and discursive, and the sections ... on aqueducts, hardly convey the impression of a writer who is master of his subject' [81]. As shown above, Vitruvius' treatise on aqueducts and siphons very well meets the physics of gravity driven pressure conduit systems. It reads like a general manual of how to build water conduits, including pressure lines, with the construction materials available at the time. Beyond Vitruvius' descriptions lies the water technology base of Roman water engineers that has escaped written text—here certain manuscripts in the Water Special Issue 'Water Engineering in Ancient Societies' present indications of a deeper knowledge by use of modern hydraulic engineering principles, to uncover what lies behind Roman water system designs albeit in Roman pre-scientific terminology yet to be discovered.

13. Discussion

The problems that occurred in the ancient pressurized conduits (siphons) were caused by static water pressure, and, more important, by the effects from the presence of air in the conduit, both at start-up as well as during operation. The kind of problems that occur are related to the properties of the conduit, consisting either of homogenous material (soldered lead conduits, Category I), or of non-homogenous materials (conduits put together from prefab pipe elements made of stone or ceramics, Category II). The archaeological findings show that the ancient engineers were well aware of the problems and knew how to cope with them. For the Category I siphon of the Yzeron aqueduct at Lyon a hydraulic tower was built at a high point in the siphon's course, to release air so that a stand-still of the siphon was prevented. In the Category II siphon at Aspendos and the de los Arquillos siphon at Gades two hydraulic towers were incorporated at horizontal bends, to prevent damage from static pressure and pressure surges from uncontrolled start-up and from water hammer, as well as to prevent the de los Arquillos siphon from failing to start up successfully. For the ancient siphons only gravity was available as driving force, whereby the head—the difference in level between header tank and receiving tank—was not very large. Problems caused by air became quickly manifest. In modern systems, with high-pressure pumps and superior conduit characteristics such problems may get unnoticed for long periods of time and confront the engineer with unexplained capacity reductions. But the principles that lie at the base of modern pressure lines do not differ from those of the old days: the laws of nature have not changed in 2000 years.

Funding: This research was funded by Dutch Research Council NWO and the Austrian Archaeologocal Institute ÖAI.

Data Availability Statement: Related information are mentioned in Appendix A.

Conflicts of Interest: The authors declare no conflict of interest.

Appendix A

1. www.romaq.org (7 April 2021); Hodge 1992, 1.
2. Aqueducts exclusively built for water mills are known. Near Barbegal in France, not far from Arles, remains of a complex with 16 water-driven grain mills have been identified. The vertical wheel mills were located on a hill side in two parallel rows of eight mills one below the other. The mills were supplied by a separate aqueduct (Sellin 1983, Hodge 1992, Leveau 1996, Sürmelehindi et al. 2019. For a history of water mills, see, e.g., Reynolds 1983, also Ritti-Grewe-Kessener 2007).
3. For Roman surveying, see Lewis 2001; also Grewe 1998, 2017.
4. This was not always the case. In the first century CE, a 17 km aqueduct was planned for ancient Saldae, present-day Bejaja in Algeria. In order to pass a hill, a 428 m-long tunnel was planned, to be dug from two sides to meet in the middle. At one instance, the two stretches that were excavated had a joint length that exceeded the distance to be covered. Then, an engineer from the Roman army, Nonius Datus, was called in for help, and he solved the problem. This was commemorated in an inscription, still to be seen in Bejaja, mentioning the three virtues that an able engineer should possess:

Spes, Virtus, Patientia. Copies of the inscription are in the Museo della Cività in Rome, and in the Museum für antike Schiffahrt in Mainz (de Waele 1996; Grewe 2002).
5. Marcus Vitruvius Pollio, first century BCE, is the author of *'De Archtectura Libri Decem'* (*'Ten Books on Architecture'*). His book VIII treats issues on water and water transport.
6. Vitruvius VIII, 5, 1–3.
7. For a distance between A and B, corresponding to an angle β in Figure 7, the error C becomes = R (1/cosβ − 1), where R = radius of the earth. β may be determined by dividing the distance between A and B by the circumference of the earth, and multiplying it by 360 (degrees).
8. Grewe 2014, 38.
9. Lubbers 2018.
10. Büyükyıldırım 2017, 72.
11. Rakob 1983.
12. Hodge 1992, 187–190.
13. Kessener 2000, 109.
14. Malinowski 1979; also 1996.
15. Grewe 2014, 298/382 for an overview.
16. Kessener 2016, 263.
17. Passchier et al. 2016.
18. Sürmelihindi et al. 2019.
19. Sürmelihindi et al. 2021.
20. See, e.g., Hodge 1992, 347–348.
21. http://www.romanaqueducts.info/siphons/siphons.htm (17 February 2021).
22. For a discussion see Smith 1979, 2007a/b; Hodge 1983, 1992; Kessener 2004, 2016.
23. For the Gades and its two siphons, see Perez et al. 2012; Smyrna: Weber 1899; for the Lyon aqueducts and their nine siphons: Burdy 2002; Pergamon: Fahlbusch 1982, Garbrecht 2001, Manvroudis 2015; Alatri: Lewis 1999; Aspendos: Kessener 2000, 2011; Termini Imerese: Belvedere 1986; Laodikeia ad Lykum: Weber 1989; Oinoanda: Stenton and Coulton 1986; Patara: Passchier et al. 2016.
24. For the water provision of Pergamon both in Hellenistic and Roman times, see the work of Günther Garbrecht (Garbrecht 2001).
25. Pérez and Bestué 2008, 2010.
26. Lyon (Lugdunum) was founded in 43 BCE. The aqueducts are dated 20 BCE (Mont d'Or), 10 BCE (Yzeron), CE (Brevenne), and 120 CE (du Gier). Burdy 1991, 1996, 2002.
27. Hodge 1992, 156.
28. For Roman soldering techniques, see, e.g., Hodge 1992, 307–309.
29. As assumed for the lead pipe elements of Pergamon's Madradag siphon, Garbrecht 2001, 128.
30. Grewe/Kessener/Piras, 1999. The Seljuk bridge was restored in the 1990s, losing some of its charm.
31. For the Hellenistic Karabunar siphon of Smyrna, see Weber1899, also Lewis 1999, 158–162.
32. Malinowski 1979, 1996.
33. Burdy and Cochet 1992.
34. Hansen 1992.
35. The unit of pressure is Pa(scal), where 1 Pa = 1 newton/m^2; 100,000 Pa = 1 bar ≈ 1 Atm ≈ 14.5 pound/sq inch.
36. Fahlbusch 1982, 73; Garbrecht 2001, 121.
37. Vitr. VIII, 6, 8. It is not known what type of stone Vitruvius points to. Calabat 1973, 179, suggests a porphyry-type stone; Lewis suggests trachyte or andesite for its strength (Lewis 1999, 169, n.90).
38. Vitr. VIII, 6, 9.
39. Şahin 2007: the inscription reads: 'the Emperor Caesar Flavius Vespasianus Augustus restored the analemna [the cyclopean wall] of the aqueduct, which was destroyed by

an earthquake, from its base with the from stone blocks constructed conduit on top; along the analemna of the pressure line he had installed three four-inch wide ceramic conduits, in order that, because of the two conduits, in case repairs are needed, the water flow and its use remain continuously possible. Furthermore he restored other parts of the water conduit and brought water [to the town]—after an interruption of thirty months—by his legate and propraetor Sextus Marcius Priscus; the costs were accounted from saved taxes of the town, and the union made *** *denarii* [unknown amount] available, without that a written request was drawn up. The construction was already started by Vilius Flaccus, the legate and proprietor of Claudius Caesar Augustus, it was finished and water flowed to the town during the administration of Epirus Marcellus, legate and proprietor of Claudius Caesar Augustus (after German translation by Şahin 2007). See also Passchier et al. 2016.

40. Büyükyildirim 1994, 57–59.
41. The July 21, 365 CE earthquake of Crete of magnitude 8.6, lifted the west of the island 10 m upwards. It was responsible for extensive destructions throughout the Eastern Mediterranean as far as Cyprus, Palestine and Egypt, generating a giant tsunami (Pararis-Carayannis 2010). The Roman historian Ammianus Marcellinus (330–400 CE) described the effects on Alexandria: 'Slightly after daybreak, and heralded by a thick succession of fiercely shaken thunderbolts, the solidity of the whole earth was made to shake and shudder, and the sea was driven away, its waves were rolled back, and it disappeared, so that the abyss of the depths was uncovered and many-shaped varieties of sea-creatures were seen stuck in the slime; the great wastes of those valleys and mountains, which the very creation had dismissed beneath the vast whirlpools, at that moment, as it was given to be believed, looked up at the sun's rays. Many ships, then, were stranded as if on dry land, and people wandered at will about the paltry remains of the waters to collect fish and the like in their hands; then the roaring sea as if insulted by its repulse rises back in turn, and through the teeming shoals dashed itself violently on islands and extensive tracts of the mainland, and flattened innumerable buildings in towns or wherever they were found. Thus in the raging conflict of the elements, the face of the earth was changed to reveal wondrous sights. For the mass of waters returning when least expected killed many thousands by drowning, and with the tides whipped up to a height as they rushed back, some ships, after the anger of the watery element had grown old, were seen to have sunk, and the bodies of people killed in shipwrecks lay there, faces up or down. Other huge ships, thrust out by the mad blasts, perched on the roofs of houses, as happened at Alexandria, and others were hurled nearly two miles from the shore, like the Laconian vessel near the town of Methone which I saw when I passed by, yawning apart from long decay.' (https://en.wikipedia.org/wiki/365_Crete_earthquake (19 May 2021)).
42. For siphons made of lead pipes soldered together, Cat.I conduits, such precautions were not needed as, at bends, the forces from static pressure are transferred away from the bend via the pipe wall (see Figure 40). The conduit, as a whole, may need to be fixed to prevent sliding out of place, but the conduit would only burst, along its length, where pressure is highest: in the lowest part of the siphon.
43. Number of explanations for the presence of these holes have been proposed, such as safety valves, devices to alleviate pressure surges, and giant whistles when filling the siphon (the tone would indicate the degree of filling). Most probably, the holes were cut in a diagnostic procedure to pinpoint obstructions, after which the clog-up could be removed by poking with a rod and the holes closed off with a stone plug (Kessener 2017, 364–370).
44. Patara and Laodikeia a/L: Kessener 2017, 364–371; Smyrna: Lewis 1999, Figure 3, on a hole tapered to a 1 cm opening, presumably related to an air release valve; Hippos Susita: Tsuk et al. 2002, 208.
45. Narrow spots in the conduit may cause additional drag forces at such points.

46. After http://www.pipeflow.co.uk/public/articles/Darcy_Weisbach_Formula (19 May 2021).
47. The force may be calculated as follows: F = dP/dt = $(2 \cdot \rho \cdot A \cdot dx \cdot v \cdot \sin(\beta/2))/dt$ = $2 \cdot \rho \cdot A \cdot dx/dt \cdot v \cdot \sin(\beta/2) = 2 \cdot \rho \cdot A \cdot v^2 \cdot \sin(\beta/2)$, where $\rho = 1000$ kg/m^3, diameter of conduit = 28 cm, water velocity = 1 m/sec, and β = 55 degrees; it follows that F = $2 \cdot 1000 \cdot \pi/4 \cdot (0.28)^2 \cdot 1^2 \cdot \sin(55°/2)$ = 57 newtons, or a force of about 6 kgf.
48. Ortloff and Kassinos 2003.
49. Ganderberger 1957, 99: 'Alte Praktiker füllen ihre Rohrleitungen sehr langsam. Sie benötigen oft 2 bis 3 Tage für eine Rohrleitung van 6 bis 10 km Länge' ('Experienced practiciens fill their piped conduits very slowly. They often use 2 to 3 days for a pipe line 6 to 10 km long'). For the 3250 m Madradag siphon of Pergamon the filling procedure would thus take at least a 24 h day. Comment by Charles Ortloff [48]: This low input flow rate would not have been prescribed by Roman engineers unless flow disturbances were observed at high filling rates during start-up operations of the Aspendos aqueduct system. The low and slow input flow rate mandate likely derives from prior start-up observations of instabilities generated at higher slow rates. For example, for a too-high input flow rate into the first siphon, water flow would accelerate to high velocity flowing down the siphon slope, causing a partial flow with air space above at a high supercritical Froude number value. Upon the water flow reaching the horizontal section at the bottom of the siphon, a hydraulic jump is formed with the subcritical post hydraulic jump flow entering the upward part of the siphon. As flow continues, the increased static pressure and frictional flow resistance in the upward section of the siphon causes the hydraulic jump to rise in the declination part of the siphon until it reaches the inlet to reinstate full flow conditions and a possible inlet overflow. Note here that the open basin at the top of the first tower/end of the first siphon limits static back pressure and regulates the hydraulic jump position. A siphon entry overflow will then signal that the input flow rate is too high as a back flow now may exist at the siphon inlet. Now as water height increases in the upward part of the siphon, it reaches the open basis at the top—this limits the backflow static pressure and stabilizes the hydraulic jump position close to the inlet and like permits continuation of full flow into the next sequential siphon with similar flow conditions as noted in the first upstream siphon. Some backflow into the open top basin occurs that may cause static pressure instabilities to occur that can effect subcritical flows within the first siphon. Now backflow problems can be largely eliminated by reducing the input flow rate to a stable value, where the hydraulic jump effect is now minor and no backflow exists at the siphon inlet—this lower flow rate promotes stable flow in the entire system after initial start-up instabilities occur.
50. See, e.g., Corcos, 1989.
51. Volume V of the air pocket is inversely related to air pressure P: $P \cdot V$ is a constant.
52. Knauss 1983.
53. Kamma and van Zijl 2002.
54. For details, see Garbrecht 2001, 149–151; 203–206; Tafel 20-1 and 26-2.
55. Kessener, 2016.
56. https://paulbrimhall.com/newsletter-archives/submergence-law-why-it-matters/; also http://www.pumpfundamentals.com/help11.html (24 July 2019).
57. See also Kessener 2001.
58. Lanckoronski 1890, 93–94.
59. One may of course ask why the course of the Madradag siphon runs over the tops of intermediate hills and not around them in order to avoid these high points in the line. For one thing, by choosing the hill tops, the static pressure was reduced as much as possible. However, more probably, the trajectory was chosen to prevent damage by environmental events. Along the entire course, the terrain slopes down on either side of the conduit: it ran on the local watershed. Damage from torrential rains and storms was prevented.

60. Garbrecht 2001, 121–123.
61. Burdy 1991, 100–103.
62. Burdy 1991, 101.
63. Falvey 1980, 48. A similar formula for Vcr was used by Kamma and van Zijl 2002, 56: Vcr = $1.23 \cdot (g \cdot Db \cdot \sin\alpha)^{1/2}$, where Cb = 0.88 in Falvey's formula.
64. Aksoy 1997.
65. Falvey 1980, 50–52; 61–65.
66. Baines and Wilkinson 1986.
67. For further literature on pressure surges from air in conduits, see e.g., Burrows et al.1995; Zhou et al. 2002.
68. Stern 1983, 177–179.
69. Schnapauff 1966.
70. Falvey 1980, 57–77.
71. Falvey 1980, Figure 45.
72. In modern pressure lines automatically operating air release valves may thus result to—at times detrimental—water hammer effects, which effects may be reduced by proper dimensions of the orifice through which air escapes.
73. Kessener, 2000.
74. Assuming that the conduit between the header tank and receiving tank had a fixed inner diameter.
75. Which would have been an option (Kessener 2000, 258–259).
76. Pérez 2012; Pérez et al. 2014.
77. Not much is known about Marcus Vitruvius Pollio. He was born ca. 80 BCE, and his manuscript was presumably written ca. 20 BCE. The oldest manuscript is a Carolingian copy from about 800 CE, the Harleianus, kept at the British Library (MS Harley 2767) (www.bl.uk/catalogues/illuminatedmanuscripts/record.asp?MSID=8557&CollID=8&NStart=2767, accessed on 24 May 2018).
78. Among others, Rowland 2001 (English), Peters 1997 (Dutch), Fensterbusch 1976 (German), Callebat 1973 (French), Choisy 1971 (French), Granger 1962 (English), Morgan 1960 (English), Krohn 1912 (Latin text).
79. See note 37. Additionally, Kessener 2001, 150 n.77.
80. Vitr, VIII, 6, 6.
81. Fahlbusch and Peleg 1992; Lewis 1999; Kessener 2001; 2002; Ohlig, 2006.
82. Kessener 2003, 2016.
83. Hodge 1992, 124; Lewis 1999, 145 and 171.

References

1. Amin, A.; Kim, C.G.; Kee, Y.H. Numerical analysis of free surface vortices behavior in a pump sump. In *IOP Conference Series: Earth and Environmental Science*; IOP Publishing: Busan, Korea, 2017.
2. Aksoy, S. Gedanken über Luftblasen in Druckleitungen. *Mitt. des Leichtweiss-Institut für Wasserbau*, 1997; Unpublished work.
3. Baines, W.D.; Wilkinson, D.L. The motion of large air bubbles in ducts of moderate slope. *J. Hydraul. Res.* **1986**, *25*, 157–169. [CrossRef]
4. Belvedere, O. *L'Acquedotto Cornelio di Termini Imerese*; L'Erma di Bretschneide: Rome, Italy, 1986.
5. Burdy, J. *L'Aqueduc Romain de l'Yzeron*; Conseil Général du Rhône: Lyon, France, 1991.
6. Burdy, J. *L'Aqueduc Romain du Gier*; Conseil Général du Rhône: Lyon, France, 1991.
7. Burdy, J. *Les Aqueducs Romains de Lyon*; Presses Universitaires de Lyon: Lyon, France, 2002.
8. Burdy, J.; Cochet, A. Une date consulaire (213 après J.-C.) sur un tuyau de plomb viennois. *Gallia* **1992**, *49*, 89–97. [CrossRef]
9. Burrows, R.; Qiu, D.Q. Effect of air pockets on pipeline surge pressure. *Proc. Inst. Civ. Eng.-Water Marit. Energy* **1995**, *112*, 349–361.
10. Büyükyıldırım, G. *Tarihi Su Yapıları*; Antalya Bölgesi: Ankara, Turkey, 1994.
11. Büyükyıldırım, G. *20.Yüzyılda Su İşleri ve Antalya*; Devlet Su İşleri: Antalya, Turkey, 2017.
12. Callebat, L. *Vitruve de l'Architecture, Livre VIII*; Les Belles Lettres: Paris, France, 1973.
13. Çeçen, K. *The Longest Roman Water Supply Line*; Türkiye Sınai Kalkınma Bankası: Istanbul, Turkey, 1996.
14. Choisy, A. *Vitruve*; F. de Nobele: Paris, France, 1971.
15. Corcos, G. *Air in Water Pipes*; Agua Para La Vida: Berkeley, CA, USA, 1989.

16. Fahlbusch, H. Vergleich antiker griechischer und römischer Wasserversorgungsanlagen. *Mitt. des Leichtweiss-Institut für Wasserbau* **1982**, *73*. Available online: https://www.oeaw.ac.at/resources/Record/990000779540504498#usercomments (accessed on 12 October 2021).
17. Fahlbusch, H. Die Wasserversorgung in der Antike. *Schr. Frontinus-Gesell.* **1992**, *17*, 85–106.
18. Fahlbusch, H.; Peleg, J. Die Colliviaria Vitruvs, Möglichkeiten der Interpretation. *Mitt. des Leichtweiss-Institut für Wasserbau* **1992**, *117*, 105–140.
19. Falvey, H.T. Air-Water flow in Hydraulic structures. *Eng. Monogr.* **1980**, *81*, 26429.
20. Fensterbusch, C. *Vitruv zehn Bücher Über Architektur*; Wissensch. Buchgesell.: Darmstadt, Germany, 1976.
21. Ganderberger, W. *Über Die Wirtschaftliche und Betriegssichere Gestaltung von Fernwasserleitungen*; Oldenburg Verlag: München, Germany, 1957.
22. Garbrecht, G. Die Wasserversorgung des antiken Pergamon, die Druckleitung. *Mitt. des Leichtweiss-Institut für Wasserbau* **1978**, *60*. [CrossRef]
23. Garbrecht, G. *Die Wasserversorgung des antiken Pergamon, In Die Wasserversorgung Antiker Städte*; Philip von Zabern: Mainz am Rhein, Germany, 1987; Volume II, pp. 13–47.
24. Garbrecht, G. Altertümer von Pergamon, Band 1-Teil 4. In *Die Wasserversorgung Von Pergamon*; DAI: Berlin, Germany; New York, NY, USA,, 2001.
25. Gräber, F. *Vorläufigen Bericht über Untersuchung der Pergamenischen Wasserleitungen*; Reimer Verlag: Berlin, Germany, 1906.
26. Gräber, F. *Altertümer Von Pergamon-Stadt und Landschaft-Die Wasserleitungen*; Gesellschaft der Freunde Universität Heidelberg e.v.: Berlin, Germany, 1913; (accessed on 12 October 2021). [CrossRef]
27. Granger, F. *De Architectura*; Heinemann: London, UK, 1962.
28. Grewe, K. Der antike Vermessungsingenieur Nonius Datus und sein Platz in de Geschichte der Technik. *Schr. Front.-Ges.* **2002**, *25*, 39–52.
29. Grewe, K. *Aquädukte, Wasser für Rom's Städte*; Regionalia Verlag: Rheinbach, Gernamy, 2014.
30. Grewe, K. Neues aus der Aquäduktforschung. In *Wasserwesen zur Zeit des Frontinus*; Wiplinger, G., Letzner, W., Eds.; Babesch Supplements: Leuven, Belgium; Peeters Publishers: Leuven, Belgium, 2017; Volume 32, pp. 15–30.
31. Grewe, K.; Kessener, H.P.M.; Piras, S.A.G. Im Zickzack-Kurs über den Fluß, die römisch/seldschukische Eurymedon-Brücke von Aspendos (Türkei). *Antike Welt* **1999**, *30*, 1–12.
32. Haberey, W. *Die Römischen Wasserleitungen Nach Köln*; Rheinland Verlag: Bonn, Gernamy, 1972.
33. Hansen, J. Die antike Wasserleitung unter der Rhône bei Arles. *Mitt. des Leichtweiss-Institut für Wasserbau* **1992**, *117*, 470–531.
34. Hodge, A.T. *Siphons in Roman Aqueducts*; Publications of the British School at Rome (PBSR): Rome, Italy, 1985; p. 51.
35. Hodge, A.T. *Roman Aqueducts and Watersupply*; Duckworth: London, UK, 1992.
36. Kamma, P.S.; van Zijl, F.P. De weerstand in persleidingen voor afvalwater tijdens de gebruiksfase. *Rioleringswetenschap* **2002**, *5*, 45–64.
37. Kessener, H.P.M.; Piras, S.A.G. The pressure line of the Aspendos Aqueduct. *Adalya* **2002**, *2*, 159–187.
38. Kessener, H.P.M. The Aqueduct at Aspendos and its inverted siphon. *J. Rom. Archaeol.* **2000**, *13*, 105–132. [CrossRef]
39. Kessener, H.P.M. Vitruvius and the Conveyance of Water. *BABesch* **2001**, *76*, 139–158.
40. Kessener, H.P.M. Vitruvius rehabilitandus est. In *Cura Aquarum in Israel*; Ohlig, C., Peleg, Y., Tsuk, T., Eds.; Books on Demand: Siegburg, Germany, 2002; pp. 187–200.
41. Kessener, H.P.M. Roman Watertransport: Problems in Operating Pressurized Pipeline Systems. In *Wasserhistorische Forschungen—Schwerpunkt Antike*; Schriften der Deutschen Wasserhistorischen Gesellschaft: Solingen, Germany, 2003; Volume 2, pp. 147–160. Available online: https://www.abebooks.com/9783833003400/Wasserhistorische-Forschungen-Schwerpunkt-Antike-Crais-3833003405/plp (accessed on 12 October 2021).
42. Kessener, H.P.M. Moderne persleidingen en Romeinse hydraulische technieken. *Rioleringswetenschap–Tech.* **2004**, *15*, 10–44.
43. Kessener, H.P.M. The Aspendos Siphon and Roman Hydraulics. In *De Aquaeductu atque Aqua Urbium Lyciae Pamphiliae Pisidiae*; Wiplinger, G., Ed.; BABesch Supplements: Leuven, Belgium; Peeters Publishers: Leuven, Belgium, 2016; Volume 27, pp. 261–274.
44. Kessener, H.P.M. *Roman Water Ditribution and Inverted Siphons*; Ipskamp: Nijmegen, The Netherlands, 2017.
45. Knauss, J. Wirbelbildung an Einlaufwerken-Luft-und Dralleintrag. In *Schriftenreihe des Deutschen Verbandes für Wasserwirtschaft und Kulturbau e.V.*; DVWK: Bonn, Germany, 1983; Volume 63.
46. Krohn, F. (Ed.) *Vitruvii de Architectura Libri Decem*; Nabu Press: Leipzig, Germany, 1912.
47. Lanckoronski, K. Grafen Städte Pamphiliens und Pisidiens; Band 1; Vienna, Austria, 1890. Available online: https://books.google.com.sg/books?id=8yghEAAAQBAJ&pg=PT188&lpg=PT188&dq=Grafen+St%C3%A4dte+Pamphiliens+und+Pisidiens&source=bl&ots=nC-OokaEOM&sig=ACfU3U3dQrUwlioFZMvbFunSbjWMOeMm3g&hl=en&sa=X&ved=2ahUKEwjAzMnP9Pj0AhVBTGwGHUqaDu8Q6AF6BAgDEAM#v=onepage&q=Grafen%20St%C3%A4dte%20Pamphiliens%20und%20Pisidiens&f=false (accessed on 12 October 2021).
48. Leveau, P. The Barbegal water mill in its environment: Archaeology and the economic and social history of antiquity. *J. Rom. Archaeol.* **1996**, *9*, 137–153. [CrossRef]
49. Lewis, M.J.T. *Vitruvius and Greek Aqueducts*; Publications of the British School at Rome (PBSR): Rome, Italy, 1999; Volume 67, pp. 145–172.
50. Lewis, M.J.T. *Surveying Instruments of Greece and Rome*; Cambridge University Press: Cambridge, UK, 2001.

51. Lubbers, J.P. *Aquaducten in Ancient Rome*; Plannen en Bouwen van Aqueducten in Ancient Rome zonder Hoogtemeetinstrumenten: Hoorn, The Netherlands, 2018.
52. Malinowski, R. Concretes and Mortars in Ancient Aqueducts. *Concr. Int.* **1979**, *1*, 66–75.
53. Malinowski, R. Dichtungsmörtel und Betone in der Antike. In *Cura Aquarum in Campania*; de Haan, N., Jansen, G., Eds.; Peeters: Leiden, The Netherlands, 1996; pp. 191–198.
54. Manvroudis, D. Precedent Ancient Hydraulic Works & Reconstruction of Operations of the Hellenistic Madradag Aqueduct & Inverted Siphon of Pergamon. Thesis, Belgrade, Serbia, 2015. Available online: https://www.academia.edu/36816887/Aspects_of_Roman_Water_Provision_and_related_Hydraulics_in_Aquam_Ducere_II_Proceedings_of_the_second_international_Summer_School_Water_in_the_City_Hydraulic_Systems_in_the_Roman_Age_held_in_Feltre_August_24_28_2015_Seren_del_Grappa_2018_31_52 (accessed on 12 October 2021).
55. Morgan, H. *Vitruvius, the Ten Books on Architecture*; Dover Publcations: New York, NY, USA, 1960.
56. Ortloff, C.H.; Kassinos, A. Computational Fluid Dynamics Investigation of the Hydraulic Behaviour of the Roman Inverted Siphon System at Aspendos, Turkey. *J. Archaeol. Sci.* **2003**, *30*, 417–428. [CrossRef]
57. Ohlig, C.J.P. Vitruvs *colliviaria* und die vis *spiritus*–(keine) Luft in Wasserleitungen? In *Cura Aquarum in Ephesos*; Wiplinger, G., Ed.; Peeters: Leuven, Belgium, 2006; pp. 319–326.
58. Pararis-Carayannis, G. The Earthquake and Tsunami of July 21, 365 AD in the Eastern Mediterrenean Sea. *J. Tsunami Soc. Int.* **2011**, *30*, 253–292.
59. Passchier, C.W.; Sürmelihindi, G.; Spötl, C. A High-resolution palaeoenvironmental record from carbonate deposits in the Roman Aqueduct of Patara, SW Turkey, from the time of Nero. *Sci. Rep.* **2016**, *6*, 28704. [CrossRef]
60. Peters, T. *Vitruvius*; Polak & van Gennep: Amsterdam, The Netherlands, 1997.
61. Pérez Marrero, J.; Bestué Cardiel, I. Avance del Estudio Hidraulico del Acueducto Romano de Gades. In Proceedings of the Actas IV Congreso de Obras Públicas en la Ciudad romana (Lugo-Guitiriz 2008), Lugo-Guitiriz, Spain, 6–8 November 2008; 2008; pp. 1–24.
62. Perez Marrero, J.; Bestué Cardiel, I. Nuevas Aportaciones al estudio Hidraulico del Acueducto Roman de Tempul. In *Aquam Perducendam Curavit, Captación*; Barrios, L.G.L., Palacios, J.L.C., Pujol, L.P., Eds.; Imprenta La Isla: Cádiz, Spain, 2010; pp. 184–196.
63. Pérez Marrero, J. Nuevo Análisis del Sifón Invertido de los Arquillos, Acueducto Romano de Gades. *CPAG* **2012**, *21*, 91–126.
64. Pérez Marrero, J.; Bestué Cardiel, I.; Lucas Ruíz, R. 2014 Route of the Roman Aqueduct of Cádiz. Poster Universidád de Cádiz: Cadiz, Spain, 2012. Available online: https://www.academia.edu/22378624/route_of_roman-aquaduct_of_C%C3%A1diz (accessed on 15 May 2021).
65. Rakob, F. Das Quellenheiligtum in Zaghouan und die römischen Wasseleitung nach Karthago. In *Mitteilingen des deutschen achäologischen Instituts*; DAI: Berlin, Germany, 1983; Volume 81, pp. 41–88. Available online: https://books.google.com/books/about/Das_Quellenheiligtum_in_Zaghouan_und_die.html?id=BtMTuQAACAAJ (accessed on 12 October 2021).
66. Ritti, T.; Grewe, K.; Kessener, H.P.M. A relief of a water-powered stone saw mill at Hierapolis. *J. Rom. Archaeol.* **2007**, *20*, 101–125.
67. Rowland, I.D. Noble Howe. In *Vitruvius: Ten Books on Architecture*; Cambridge University Press: Cambridge, UK, 2001.
68. SŞahin, S. Die Bauinschrift auf dem Druckrohraquädukt von Delikkemer bei Patara. In *Griechische Epigrafik in Lykien*; Schuler, C., Ed.; Börsedruck: Vienna, Austria, 2007; pp. 99–110.
69. Schnapauff, J. Luftblasen in Rohrleitungen—Vereinfachend betrachtet. *Neue Deliwa Zeitschrift* **1966**, *10*, 371–373.
70. Smith, N.A.F. Attitudes to Roman Engineering and the question of the Inverted Siphon. In *History of Technology*; Routledge: London, UK, 1979; Volume I, pp. 45–71.
71. Smith, N.A.F. The Hydraulics of Ancient Pipes and Pipelines. *Trans. Newcomen Soc.* **2007**, *77*, 1–49. [CrossRef]
72. Smith, N.A.F. The Roman Aqueduct at Aspendos. *Trans. Newcomen Soc.* **2007**, *77*, 217–244. [CrossRef]
73. Stenton, E.C.; Coulton, J.J. Oinoanda: The Watersupply and Aqueduct. *Anatol. Stud.* **1986**, *36*, 15–59. [CrossRef]
74. Stern, P. *Field Engineering*; Intermediate Technology Publications: London, UK, 1983.
75. Sürmelihindi, G.; Passchier, C.W.; Leveau, P.; Spötl, C.; Bourgeois, M.; Bernard, V. Bargegal: Carbonate imprints give a voice to the first industrial complex of Europe. *J. Archaeol. Sci. Rep.* **2019**, *24*, 1041–1058.
76. Sürmelihindi, G.; Passchier, C.; Crow, J.; Spötl, C. Carbonates from the ancient world's longest aqueduct: A testament of Byzantine water management. *Geoarchaeology* **2021**, *36*, 643–659. Available online: Wileyonlinelibrary.com/journal/gea (accessed on 8 September 2021). [CrossRef]
77. Tsuk, T.; Peleg, Y.; Fahlbusch, H.; Meshel, Z. A new survey of the aqueducts of Hippos-Susita. In *The Aqueducts of Israel*; Thomson Shore: Portsmouth, UK, 2002; pp. 207–209.
78. de Waele, J.A.K.E. Een Romeins ingenieursproject—De tunnel van Nonius Datus. *Hermeneus* **1996**, *68*, 173–181.
79. Weber, G. Die Hochdruck-Wasserleitung von Loadikeia ad Lycum. *Jahrb. Dtsch. Inst. (JdI)* **1898**, *13*, 1–13.
80. Weber, G. Die Wasserleitungen von Smyrna. *Jahrb. Dtsch. Inst. (JdI)* **1899**, *14*, 4–24.
81. Zhou, F.; Hicks, F.E.; Steffler, P.M. Transient Flow in a Rapidly Filling Horizontal Pipe Containing Trapped Air. *J. Hydraul. Eng.* **2002**, *128*, 625–634. [CrossRef]

MDPI
St. Alban-Anlage 66
4052 Basel
Switzerland
Tel. +41 61 683 77 34
Fax +41 61 302 89 18
www.mdpi.com

Water Editorial Office
E-mail: water@mdpi.com
www.mdpi.com/journal/water

www.ingramcontent.com/pod-product-compliance
Lightning Source LLC
LaVergne TN
LVHW070511100526
838202LV00014B/1831